Structural Equation Modeling

Applications in ecological and evolutionary biology

Structural equation modeling (SEM) is a technique that is used to estimate, analyze, and test models that specify relationships among variables. The ability to conduct such analyses is essential for many problems in ecology and evolutionary biology. This book begins by explaining the theory behind the statistical methodology, including chapters on conceptual issues, the implementation of an SEM study, and the history of the development of SEM. The second section provides examples of analyses on biological data including multi-group models, means models, P-technique and time-series. The final section of the book deals with computer applications and contrasts three popular SEM software packages. Aimed specifically at biological researchers and graduate students, this book will serve as a valuable resource for both learning and teaching the SEM methodology. Moreover, data sets and programs that are presented in the book can also be downloaded from a web site to assist the learning process.

BRUCE H. PUGESEK is a research statistician in the US Geological Survey – Biological Resources Division. He is the author of numerous scientific papers in the fields of ecology, behavior, evolution, and applied statistics.

ADRIAN TOMER is an Associate Professor at the Department of Psychology at Shippensburg University, Pennsylvania, where he teaches the psychology of aging and developmental psychology. He has a particular interest in the application of structural equation modeling to the behavioral and biological sciences.

ALEXANDER VON EYE is a Professor in the Department of Psychology at Michigan State University, where much of his research is dedicated to the development and application of statistical methods. He is the author of the book *An Introduction to Configural Frequency Analysis* (1990, ISBN 0 521 38090 1).

Structural Equation Modeling

Applications in ecological and evolutionary biology

Edited by

BRUCE H. PUGESEK

US Geological Survey – Biological Resources Division,
Northern Rocky Mountain Science Center, Bozeman

ADRIAN TOMER

Department of Psychology, Shippensburg University, Pennsylvania

and

ALEXANDER VON EYE

Department of Psychology, Michigan State University

CAMBRIDGE
UNIVERSITY PRESS

CAMBRIDGE UNIVERSITY PRESS
Cambridge, New York, Melbourne, Madrid, Cape Town, Singapore, São Paulo, Delhi

Cambridge University Press
The Edinburgh Building, Cambridge CB2 8RU, UK

Published in the United States of America by Cambridge University Press, New York

www.cambridge.org
Information on this title: www.cambridge.org/9780521104029

First published 2003
This digitally printed version 2009

A catalogue record for this publication is available from the British Library

ISBN 978-0-521-78133-6 hardback
ISBN 978-0-521-10402-9 paperback

To our wives with love

Yolanda, Daniela, and Donata

Contents

Contributors

Alexander von Eye
Department of Psychology
119 Snyder Hall
Michigan State University
East Lansing, MI 48824-1117
USA

Bret E. Fuller
Department of Public Health and Preventive Medicine
Oregon Health and Sciences University
3181 SW Sam Jackson Park Road
Mail Code CB-669
Portland, OR 97201-3098
USA

James B. Grace
US Geological Survey – Biological Resources Division
National Wetlands Research Center
700 Cajundome Blvd
Lafayette, LA 70506
USA

Patricia H. Hawley
Department of Psychology
Yale University
Box 208205
New Haven, CT 06520-8205
USA

Scott L. Hershberger
Department of Psychology
California State University – Long Beach
1250 Bellflower Blvd
Long Beach, CA 90840
USA

Peter S. Hovmand
Institute for Children, Youth, and Families
Kellogg Center, Suite 27
Michigan State University
East Lansing, MI 48824
USA

Bobby D. Keeland
US Geological Survey – Biological Resources Division
National Wetlands Research Center
700 Cajundome Blvd
Lafayette, LA 70506
USA

Todd D. Little
Department of Psychology
Yale University
Box 208205
New Haven, CT 06520-8205
USA

George A. Marcoulides
California State University, Fullerton
Department of Management Science
Langsdorf Hall no. 540
Fullerton, CA 92634
USA

Peter C. M. Molenaar
University of Amsterdam
Faculteit der Maatschappij – en Gedragswetenschappen, Afd. Psychologie
Roetersstraat 15
1018 WB
Amsterdam
The Netherlands

Makeba M. Parramore
Department of Human Development
Cornell University
Ithaca, NY 14853-2602
USA

Bruce H. Pugesek
U.S. Geological Survey – Biological Resources Division
Northern Rocky Mountain Science Center
1648 S. 7th Street, MSU
Bozeman, MT 59717-2780
USA

Michael J. Rovine
Director, Center for Development and Health Research Methodology
S-159 Henderson Human Development Bldg
Pennsylvania State University
University Park, PA 16802
USA

Bill Shipley
Département de Biologie
Université de Sherbrooke
Sherbrooke (Quebec)
Canada J1K 2R1

Adrian Tomer
Department of Psychology
Shippensburg University
Shippensburg, PA 17257
USA

Philip K. Wood
Department of Psychology
210 McAlester Hall
University of Missouri
Columbia, MO 62511
USA

Preface

This book describes a family of statistical methods known as structural equation modeling (SEM). SEM is used in a variety of techniques known as "covariance structure analysis", "latent variable modeling", "path modeling", "path modeling with LISREL", and sometimes it is mistaken for path analysis. This book will help biologists to understand the distinction between SEM and path analysis. The book consists of contributed chapters from biologists as well as leading methodologists in other research fields. We have organized the chapters and their content with the intent of providing a volume that readers may use to learn the methodology and apply it themselves to their research problems. We give the basic formulation of the method as well as technical details on data analysis, interpretation, and reporting. In addition, we provide numerous examples of research designs and applications that are germane to the research needs and interests of organismal biologists. We also provide, as a learning aide, the simulation programs, analysis programs, and data matrices, presented in the book at a website (http://www.usgs.gov/) so that readers may download and run them.

The book is divided into three sections. The first section, "Theory", describes the SEM model and practical matters of its application. Chapter 1 lays out the mathematics of SEM in a comprehensible fashion. Using an example from behavioral genetics, the authors express their model in what is called LISREL notation, a symbolic language that is commonly used to express SEM models. Chapter 2 describes SEM in a nonmathematical fashion. It will provide the reader with insight into how SEM differs from other methods and the benefits that may be obtained by using it. In Chapter 3, the author uses Huston's classic conceptual model of Shiras moose population dynamics to demonstrate that a complex model can be estimated and inferentially tested with SEM. The chapter provides examples of nonzero fixed parameters, measurement and structural models, and illustrates the distinction between exploratory and confirmatory models, the use of computer-generated information for model modification, and the concept and use of nested models. Chapter 4 provides a historical account of the development of SEM beginning with its origins in correlation and path analysis, and ending with the formulation of the LISREL model and its more recent expansion. Chapter 5 describes the numerous epistemological considerations that accompany an SEM study and provides guidelines for the implementation and reporting of SEM results. Details on the development of measurement instruments, sample size, model identification, fit indices, and other considerations necessary to the successful implementation of SEM are provided and well referenced. For those readers who wish only to understand SEM so that they can read and appreciate research that utilizes the method, we recommend that they read Chapters 1 through 3 of this section. For those who wish to implement an SEM study we recommend also Chapter 5. Everyone will benefit from reading the historical account of SEM, Chapter 4, especially those who seek a review of key papers past and present on the theoretical aspects of confirmatory factor analysis, maximum likelihood estimation, and other key components of the SEM methodology.

Section 2, "Applications", provides a sampling of the numerous ways that SEM can be employed. In Chapter 6, a confirmatory factor analysis of elephant behavior is presented. The authors provide examples of P-technique where data from a single individual are analyzed on a number of variables across a number of discrete points in time. Chapter 7 contrasts ordination techniques commonly employed in plant biology with an SEM approach. In Chapter 8, the author explores the notion of equivalent models in which more than one model may explain a data set. In Chapter 9, the author contrasts dynamic modeling, a method that is frequently employed in the study of complex ecological systems, with SEM. The strengths, limitations, and weaknesses of both methods are discussed. In Chapter 10, the author describes means modeling with SEM. Three examples of ANOVA applications, including, time-series analysis are presented. Chapter 11 addresses multigroup models, a method that allows comparisons of complex systems of variables from two or more groups. This approach has significant value for use in studies of multiple populations, habitat restoration, and situations where experimental versus control settings are desired at levels of organization such as the system or landscape level. Chapter 12 describes means modeling with latent variables. An example is provided for the study of natural selection in which environmental variables may impact phenotypic responses to a selection event. Chapter 13 provides an example of longitudinal analysis with SEM. The authors analyze tree growth data with

SEM and contrast results with an analysis of the data using latent growth curve methods.

Section 3, "Computing", contains Chapter 14, which discusses the relative merits of three popular software packages that perform SEM analysis. The authors compare performances on an analysis of R. A. Fisher's Iris data as well as compare features available in the software packages.

Section 1 Theory

1 Structural equation modeling: an introduction

Scott L. Hershberger, George A. Marcoulides,
and Makeba M. Parramore

Abstract

This chapter provides an introduction to structural equation modeling (SEM), a statistical technique that allows scientists and researchers to quantify and test scientific theories. As an example, a model from behavioral genetics is examined, in which genetic and environmental influences on a trait are determined. The many procedures and considerations involved in SEM are outlined and described, including defining and specifying a model diagrammatically and algebraically, determining the identification status of the model, estimating the model parameters, assessing the fit of the model to the data, and respecifying the model to achieve a better fit to the data. Since behavioral genetic models typically require family members of differing genetic relatedness, multisample SEM is introduced. All of the steps involved in evaluating the behavioral genetic model are accomplished with the assistance of LISREL, a popular software program used in SEM.

Introduction

Structural equation modeling (SEM) techniques are considered today to be a major component of applied multivariate statistical analyses and are used by biologists, economists, educational researchers, marketing researchers, medical researchers, and a variety of other social and behavioral scientists. Although the statistical theory that underlies the techniques appeared decades ago, a considerable number of years passed before SEM received the widespread attention it holds today. One reason for the recent attention is the availability of specialized SEM programs (e.g., AMOS, EQS, LISREL, Mplus, Mx, RAMONA, SEPATH). Another reason has been the publication of several introductory and advanced texts on SEM (e.g., Hayduk, 1987, 1996: Bollen, 1989; Byrne, 1989, 1994, 2000; Bollen & Long, 1993; Hoyle, 1995; Marcoulides & Schumacker, 1996; Schumacker & Lomax, 1996; Schumacker & Marcoulides, 1998; Raykov & Marcoulides, 2000), and a journal, devoted exclusively to SEM, entitled *Structural Equation Modeling: A Multidisciplinary Journal.*

In its broadest sense, SEM models represent translations of a series of hypothesized cause–effect relationships between variables into a composite hypothesis concerning patterns of statistical dependencies (Shipley, 2000). The relationships are described by parameters that indicate the magnitude of the effect (direct or indirect) that independent variables (either observed or latent) have on dependent variables (either observed or latent). By enabling the translation of hypothesized relationships into testable mathematical models, SEM offers researchers a comprehensive method for the quantification and testing of theoretical models. Once a theory has been proposed, it can then be tested against empirical data. The process of testing a proposed theoretical model is commonly referred to as the "confirmatory" aspect of SEM (Raykov & Marcoulides, 2000). Another aspect of SEM is the so-called "exploratory" mode. This aspect allows for theory development and often involves repeated applications of the same data in order to explore potential relationships between variables of interest (either observed or latent).

Latent variables are hypothetical or theoretical variables (constructs) that cannot be observed directly. Latent variables are of major importance to most disciplines but generally lack an explicit or precise way of measuring their existence or influence. For example, many behavioral and social scientists study the constructs of aggression and dominance. Because these constructs cannot be measured explicitly, they are are inferred through observing or measuring specific features that operationally define them (e.g., tests, scales, self-reports, inventories, or questionnaires). SEM can also be used to test the plausibility of hypothetical assertions about potential interrelationships between constructs and their observed measures or indicators. Latent variables are hypothesized to be responsible for the outcome of observed measures (e.g., aggression is the underlying factor influencing one's score on a questionnaire that attempts to assess offensive driving behavior). In other words, the score on the explicit questionnaire would be an indicator of the construct or latent variable – aggression. Researchers often use a number of indicators or observed variables to examine the influences of a theoretical factor or latent variable. It is generally recommended that researchers use *multiple indicators* (preferably more than two) for each latent variable considered in order to obtain a more complete and reliable "picture" than that provided by a single indicator (Raykov & Marcoulides, 2000). Because both observed and latent variables can be independent or dependent in a proposed model, a more detailed description of this issue will be provided later in this chapter.

Definition and specification of a structural equation model

The definition of a SEM model begins with a simple statement of the verbal theory that makes explicit the hypothesized relationships among a set of studied variables (Marcoulides, 1989). Typically, researchers communicate a SEM model by drawing a picture of it (Marcoulides & Hershberger, 1997). These pictures, or so-called *path diagrams*, are simple mathematical representations (but in graphical form) of the proposed theorical model. Figure 1.1 presents the most commonly used graphical notation for the representation of SEM models. As will become clear later, path diagrams not only aid in the conceptualization and communication of theoretical models, but also substantially contribute to the creation of the appropriate input file that is necessary to test and fit the model to collected data using particular software packages (Raykov & Marcoulides, 2000).

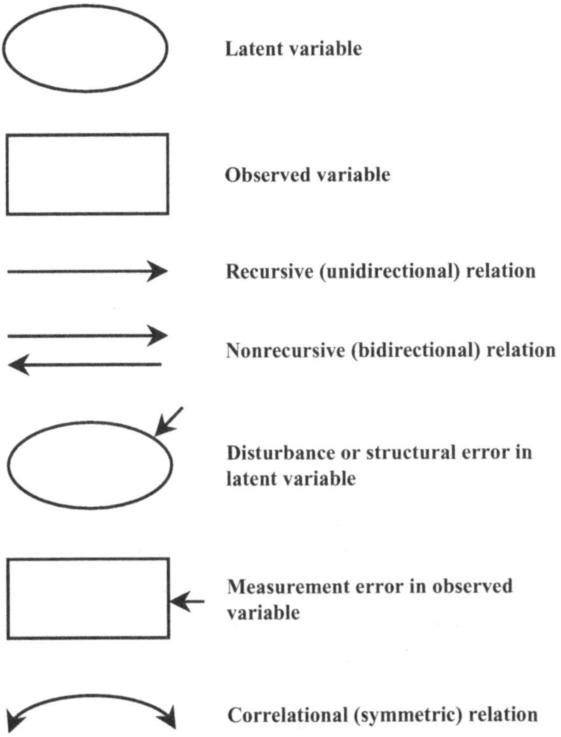

Figure 1.1. Commonly used graphical notation for the representation of SEM models.

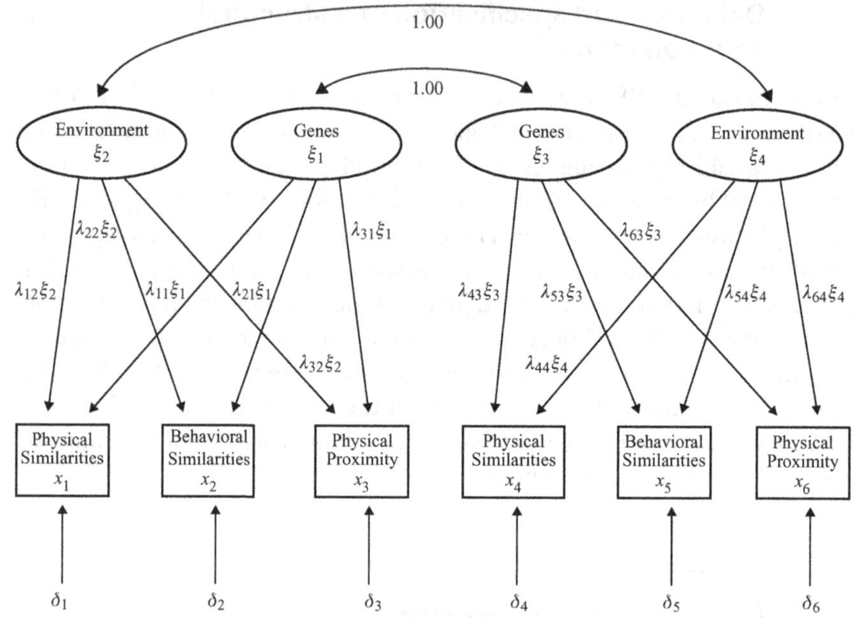

Figure 1.2. A model of sibling relatedness, in which the squares denote observed variables; the circles denote latent variables; the $\lambda_{ij}\xi_j$ are paths connecting latent with observed variables; and the δ_i are errors in the observed variables.

Figure 1.2 presents a simple example of a proposed theoretical model about sibling relatedness from the field of behavioral genetics. For years researchers have tried to understand the the "nature–nurture" phenomena by studying monozygotic twins, dizygotic twins, and nontwin siblings. To assess the amount of "relatedness" between siblings, researchers often use different types of questionnaire, standardized scales and tests, and independent observations. Two possible sources of relatedness between siblings are each sibling's genotype and environment. One may therefore define two different latent variables (i.e., genotype and environment) for each sibling, and denote each latent variable in the model by using the Greek letter ξ (ksi). Three possible observable variables (measures) of genotype and environment might be physical similarities, behavioral similarities, and physical proximity (Segal *et al.*, 1997). As it turns out, the scores or results observed for individuals on these variables will make up the correlation or covariance matrix that is analyzed to test a proposed model. The x values, which represent the observed variables or so-called *indicators*, are representative of the latent variables and make up the LAMBDA ($\mathbf{\Lambda}_x$) matrix. The error terms (error of measurement in each indicator) are denoted by the Greek letter δ (delta) and are assumed to be associated with each indicator.

As indicated previously, the hypothesized relationships among the various observed and/or latent variables in a model are typically the primary focus of most SEM investigations. These relationships are represented graphically by one-way and two-way arrows in the path diagram. These arrows or *paths* are often interpreted as symbolizing a functional relationship. In other words, the variable at the end of the arrow is assumed to be affected by the variable at the beginning of the path. Two-way arrows are representative of a covariance or association between the connected variables. These paths are not directional in nature, but are interpreted as correlational. Note that in Figure 1.2 the two-way arrow between the latent genotype variables has been set to 1, based upon known genetic relatedness between monozygotic twins, and that the two-way arrow between the latent environment variables has been set to 1 as well. This setting of the two-way arrow between the environments of the monozygotic twins forces the environment latent variables to be interpreted as the twins' shared environment, or those environmental influences completely common to the twins.

The path coefficients from the proposed model are subsequently derived from the following *model definition equations*:

$$x_1 = \lambda_{11}\xi_1 + \lambda_{12}\xi_2 + \delta_1$$

$$x_2 = \lambda_{21}\xi_1 + \lambda_{22}\xi_2 + \delta_2$$

$$x_3 = \lambda_{31}\xi_1 + \lambda_{32}\xi_2 + \delta_3$$

$$x_4 = \lambda_{43}\xi_3 + \lambda_{44}\xi_4 + \delta_4$$

$$x_5 = \lambda_{53}\xi_3 + \lambda_{54}\xi_4 + \delta_5$$

$$x_6 = \lambda_{63}\xi_3 + \lambda_{64}\xi_4 + \delta_6$$

where x_1 is the observed physical similarities for twin 1; x_2 is the observed behavioral similarities for twin 1; x_3 is the observed physical proximity for twin 1; x_4 is the observed physical similarities for twin 2; x_5 is the observed behavioral similarities for twin 2; x_6 is the observed physical proximity for twin 2; $\lambda_{11}\xi_1$ to $\lambda_{64}\xi_4$ are the factor loadings that will be estimated based on the observed data; ξ_1 and ξ_3 are the genetic latent variables for twins 1 and 2, respectively; ξ_2 and ξ_4 are the environmental latent variables for twins 1 and 2, respectively; and δ_1 through δ_6 are the measurement errors attributed to a particular variable.[1]

[1] In the context of behavioral genetic modeling, the errors-in-variables (δ) not only represent measurement error but environmental influences *unique* to each twin.

These coefficients or parameters can be *free*, i.e., to be estimated from the collected data; *fixed*, i.e., set to some selected constant value; or *constrained*, i.e., set equal to one or more other parameters. In this model, both the correlations between the monozygotic twins' genotypes and their environments have been fixed to 1 on the basis of quantitative genetic theory (Plomin *et al.*, 1997). Further, note that the variances of ξ_1 to ξ_4 have been fixed to 1 as well. This is done to establish a metric for the latent variables. Since latent variables cannot be measured directly, it is difficult to work numerically with them without first assigning them some scale of measurement. A natural choice is to standardize these variances to a value of 1. In addition, comparable paths between the two twins should be constrained to be equal, since there is no reason to believe genetic or environmental effects will be stronger for one twin or the other:

$$\lambda_{11}\xi_1 = \lambda_{43}\xi_3$$
$$\lambda_{21}\xi_1 = \lambda_{53}\xi_3$$
$$\lambda_{31}\xi_1 = \lambda_{63}\xi_3$$

for the genetic paths, and

$$\lambda_{12}\xi_2 = \lambda_{44}\xi_4$$
$$\lambda_{22}\xi_2 = \lambda_{54}\xi_4$$
$$\lambda_{32}\xi_2 = \lambda_{64}\xi_4$$

for the environmental paths.

Comparable measurement errors should similarly be constrained as equal between the two twins; i.e.,

$$\delta_1 = \delta_4$$
$$\delta_2 = \delta_5$$
$$\delta_3 = \delta_6.$$

Model identification

With the definition and specification of the model complete, the next important consideration is the identification of the model. It is important to note that once the model and the parameters to be estimated are specified, the parameters are combined to form a model-implied variance–covariance matrix that will be tested against the observed variance–covariance matrix (i.e., the variance–covariance matrix obtained from the empirical data). In a general way, the amount of unique information in the observed variance–covariance matrix is what will determine whether the model will be identified, and this verification procedure must be performed before any model can be appropriately tested. As it turns out, there are three levels of identification in SEM. The first and most problematic is that of an *under-identified* model. An *under-identified* model exists if one or more parameters cannot be estimated from the observed variance–covariance matrix. This type of model should be looked at with skepticism because the parameter estimates are most likely quite unstable. A *just-identified* model is a model that utilizes all of the uniquely estimable parameters. This type of model will always result in a "perfect fit" to the empirical data. Since there is no way one can really test or confirm the plausibility of a *just-identified* model (also referred to as a *saturated* model), this type of model is also problematic. As it turns out, the most desirable type of identification is the *over-identified* model. This type of model occurs when the number of available variance–covariances (units of information) is greater in number than the number of parameters to be estimated (Marcoulides & Hershberger, 1997). In other words, there is more than one way to estimate the specified parameters. The difference between the number of nonredundant elements of the variance–covariance matrix and the number of model parameters to be estimated is known as the *degrees of freedom* (df) of the model. For example, if the number of nonredundant elements of a variance–covariance matrix was 20 and 10 parameters were required to estimate the model, the degrees of freedom would be 20 minus 10, i.e., 10.

Model identification is an extremely complicated topic and requires several procedures to verify the status of a proposed model (for further discussion, see Marcoulides & Hershberger, 1997, or Raykov & Marcoulides, 2000). The t-rule, $\frac{1}{2}\,p(p+1)$ as cited by Marcoulides & Hersberger (1997), is one of the most frequently used necessary identification rules. Basically, the t-rule for identification is that the number of nonredundant elements in the variance–covariance (or correlation) matrix of the observed variables (p) must be greater than or equal to the number of unknown parameters in the proposed model (Marcoulides & Hersberger, 1997, p. 225). For example, Figure 1.2 has six observed variables or ($p = 6$), so there are $6(7)/2 = 21$ nonredundant elements in the variance–covariance matrix. If we attempt to estimate each path from an observed variable (x_1 to x_6) to each latent variable (ξ) and each error term associated with each observed variable (δ), we are estimating a total of 12 parameters. However, the six paths for twin 1 have been constrained to equal the six paths of twin 2 (e.g., $\lambda_{12}\xi_2 = \lambda_{44}\xi_4$), and the three indicator errors of twin 1 have been constrained to equal the three indicator errors of twin 2 (e.g., $\delta_1 = \delta_4$), resulting in a reduction of six parameters to be estimated, or, altogether, only six parameters are to be estimated. Therefore, we have an *over-identified* model with 15 degrees of freedom (i.e., 21 unique elements of the variance–covariance matrix minus six parameters to be estimated = 15 df). Of course, it is important to note that having positive degrees of freedom in a proposed model is only a necessary condition for identification; it is not a sufficient condition. There can be cases in which the degrees of freedom for a proposed model are positive and yet some parameters remain under-identified (Raykov & Marcoulides, 2000).

Suppose we wanted to expand our proposed model and incorporate another latent variable. In behavior genetic modeling, each of the twins' observed variables is corrected for age, since twins within a pair are necessarily of the same age – age creating a spurious source of twin similarity. If age is incorporated as a latent variable, the new model appears as Figure 1.3. Note that the genetic and environmental latent variables are now symbolized by

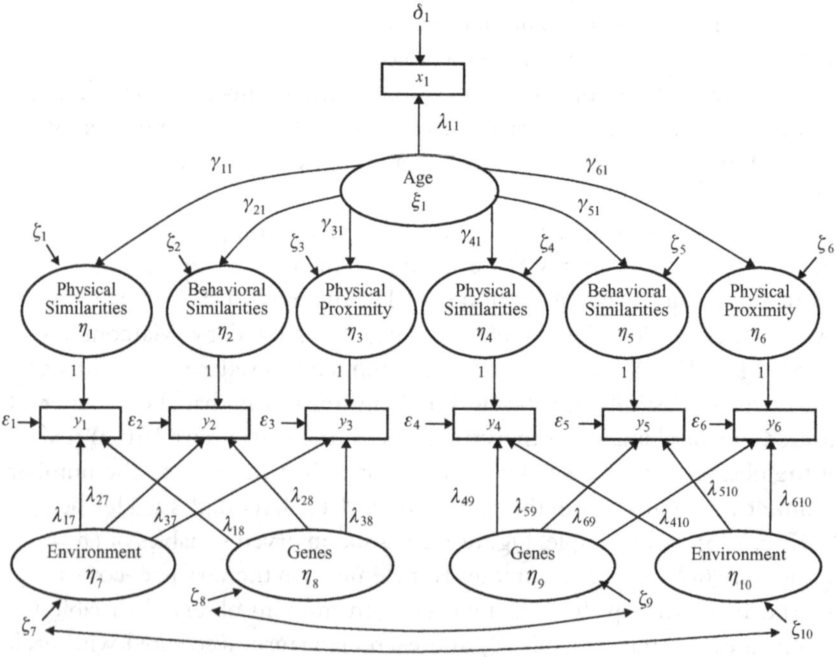

Figure 1.3. The model of sibling relatedness extended, with age as a covariate. Symbols are explained in the text.

the Greek letter η (eta). The indicators of these two latent variables, previously designated as x variables, are now designated as y variables. Further, these y values of the latent variables make up the so-called LAMBDA-y ($\mathbf{\Lambda}_y$) matrix. The error terms for the y variables are now denoted by the Greek letter ε (epsilon) and are associated (and correspondingly numbered) with each indicator. Age is now the single x variable in the model, with its latent variable denoted by ξ (ksi) and its error as δ (delta). The path connecting the observed variable age with the latent variable age is now the only entry in the LAMBDA-x ($\mathbf{\Lambda}_x$) matrix.

Let us now consider the questions "Why have we changed the symbolism of the x variables to y, and why has age been incorporated into the model as an x variable?" The answer to these two questions lies in the distinction between *dependent* and *independent variables. Dependent* (or "endogenous") *variables* are those variables that receive at least one path (one-way arrow) from another variable in the model. *Independent* (or "exogenous") *variables* are those variables from which paths only emanate but to which none is directed. Independent variables can be correlated among each other (i.e., connected in the path diagram by two-way arrows). It is important to note that a dependent variable may act as an independent variable with respect to one variable, but this does not change its dependent variable status. As long as there is at least one path ending into the variable, it is considered to be a dependent variable, no matter how many other dependent variables in the model are explained by that variable (Raykov & Marcoulides, 2000).

It was necessary to incorporate age as an independent variable based on our desire to have age act as a covariate of the original six observed indicators. Note the path that connects the latent age variable to each of the six original indicators. That parameter is expressed with the one-way arrow from one latent factor to the other and is represented by the Greek letter γ (gamma)[2]. There are other features to note about the new model in Figure 1.3. The loading of the age indicator variable on the latent age factor in the $\mathbf{\Lambda}_x$ matrix has been fixed to 1, with the indicator age error term (δ) set

[2] The reader will also note another change made to the original six indicators. Now each indicator has been the sole indicator of a latent η variable. For theoretical reasons, this change was unnecessary, but for practical reasons, this change was required. The LISREL program used to solve the model parameters only defines a parameter (γ) connecting latent independent with latent independent variables and not a parameter that connects latent independent variables with observed dependent variables. Again, this restriction requires a symbolic reformulation of the model but not one that is either theoretical or substantive.

to be zero and its variance freed. Due to identification difficulties, whenever a single indicator exists for a latent variable, a choice must be made between solving for either the value of the loading or the variance of the variable. Measurement errors are not generally identifiable with a single indicator. Another important feature to note is the two-way arrows connecting the dependent latent variables (η). This would seem to be in contradiction to the statement made above that only independent, and not dependent, latent variables may be correlated in a path model. Examining the two-way arrows in Figure 1.3, it is apparent that they do *not* directly connect the latent dependent variables, but rather connect another parameter denoted by the Greek letter ζ (zeta). Each dependent latent variable (η) has one ψ which represents the *residual error* in the variable. In other words, ψ represents all of the influences on the latent dependent variables not explicitly accounted for in the model. Some authors refer to these residual errors as *structural errors*. The two-way arrows between the ζ values of the twins' latent dependent variables is algebraically equivalent to the original formulation in Figure 1.2 of having the two-way arrows directly connect the twins' latent independent variables. Further, now the variance of the ζ values has been fixed to 1 in order to establish a metric for the latent dependent variables.

Our original model in Figure 1.3 was over-identified with 15 df. The addition of an observed age variable requires that we recalculate the degrees of freedom of the model. As before, comparable paths connecting the indicators to the latent variables should also be equated between the two twins (e.g., $\lambda_1\eta_1 = \lambda_4\eta_2$), as well as comparable indicator errors (e.g., $\varepsilon_1 = \varepsilon_4$) and comparable gammas (e.g., $\gamma_1 = \gamma_4$). In addition, we will be estimating the variance of age, and not its error or loading on ξ. In total, the model requires that 10 parameters be estimated, with the proper constraints imposed on the model (i.e., $3\,\mathbf{\Lambda}_y$, $3\,\varepsilon$, and $3\,\gamma$ values, and 1 variance). Using the *t*-rule to determine our model identification, one finds that there are $7(8)/2 = 28$ nonredundant elements in the variance–covariance matrix. Thus, our degrees of freedom are $28 - 10$ or 18 df, which results in an over-identified model suitable for model estimation.

Model estimation

In any SEM model, paths or *parameters* are estimated in such a way that the model becomes capable of "*emulating*" the observed sample variance–covariance (or correlation) matrix. The proposed theoretical model represented by the path diagram and equations makes certain assumptions about the relationships between the involved variables, and hence has specific implications for their variances and covariances. It turns out that these implications can be worked out using a few simple relations that govern the variances and covariances of linear combinations of variables. These relations are illustrated below (for further details, see Raykov & Marcoulides, 2000, p. 19).

Let us denote variance by the letters Var and covariance by the letters Cov. For variable y (e.g., physical similarities) the first relation is stated as follows:

- Relation 1: $\mathrm{Cov}(y, y) = \mathrm{Var}(y)$.

This relation simply states that the covariance of a variable with itself is equal to that variable's variance.

- Relation 2: $\mathrm{Cov}(ax + by, cz + du) = ac\,\mathrm{Cov}(x, z) + ad\,\mathrm{Cov}(x, u) + bc\,\mathrm{Cov}(y, z) + bd\,\mathrm{Cov}(y, u)$.

The second relation allows one to find out the covariance of two linear combinations of variables. Suppose that a, b, c, and d are four constants and assume that x, y, z, and u are four variables, e.g., those denoting the scores on tests of physical similarities, behavioral similarities, physical proximity, and age. The relation is obtained according to the product of the constants with the attached covariance of each combination of variables.

- Relation 3: $$\begin{aligned}\mathrm{Var}(ax + by) &= \mathrm{Cov}(ax + by, ax + by)\\ &= a^2\mathrm{Cov}(x, x) + b^2\mathrm{Cov}(y, y)\\ &\quad + ab\mathrm{Cov}\,(x, y) + ab\mathrm{Cov}\,(x, y),\end{aligned}$$

which, on the basis of Relation 1, leads to $a^2\mathrm{Var}(x) + b^2\mathrm{Var}(y) + 2ab\,\mathrm{Cov}(x, y)$.

This relation simply states that the variance of a linear combination of variables is equal to their covariance (e.g., see Relation 1). And in the case that variables x and y are uncorrelated (i.e., $\mathrm{Cov}(x, y) = 0$), leads to $\mathrm{Var}(ax + by) = a^2\mathrm{Var}(x) + b^2\mathrm{Var}(y)$.

Any proposed theoretical model has certain implications for the variances and covariances (and the means if considered) of the involved observed variables. In order to see these implications, the above three relations are generally used. For example, consider the first two manifest variables y_1 and y_2 presented in Figure 1.3. Because both variables load on the same latent factors η_1 and η_2 we obtain directly from Relations 1 and 2 the following equality:

$$\begin{aligned}
\mathrm{Cov}(y_1, y_2) &= ((\lambda_{11} \times \eta_1) + (\lambda_{12} \times \eta_2) + (1 \times \eta_5) + \varepsilon_1, (\lambda_{21} \times \eta_1) \\
&\quad + (\lambda_{22} \times \eta_2) + (1 \times \eta_6) + \varepsilon_2) \\
&= (\lambda_{11} \times \eta_1)(\lambda_{21} \times \eta_1) + (\lambda_{12} \times \eta_2)(\lambda_{21} \times \eta_1) \\
&\quad + (\eta_5)(\lambda_{21} \times \eta_1) + (\lambda_{11} \times \eta_1)(\lambda_{22} \times \eta_2) \\
&\quad + (\lambda_{12} \times \eta_2)(\lambda_{22} \times \eta_2) + (\eta_5)(\lambda_{22} \times \eta_2) \\
&\quad + (\lambda_{11} \times \eta_1)(\eta_6) + (\lambda_{12} \times \eta_2)(\eta_6) + (\eta_5)(\eta_6) \\
&\quad + \varepsilon_1((\lambda_{21} \times \eta_1) + (\lambda_{22} \times \eta_2) + \eta_6) \\
&\quad + \varepsilon_2((\lambda_{11} \times \eta_1) + (\lambda_{12} \times \eta_2) + \eta_5) + (\varepsilon_1, \varepsilon_2) \\
&= (\lambda_{11}\lambda_{21} \times \mathrm{Var}(\eta_1)) + (\lambda_{12}\lambda_{22} \times \mathrm{Cov}(\eta_1, \eta_2)) \\
&\quad + (\lambda_{21} \times \mathrm{Cov}(\eta_1, \eta_5)) + (\lambda_{11}\lambda_{22} \times \mathrm{Cov}(\eta_2, \eta_1)) \\
&\quad + (\lambda_{12}\lambda_{22} \times \mathrm{Var}(\eta_2)) + (\lambda_{22} \times \mathrm{Cov}(\eta_2, \eta_5)) \\
&\quad + (\lambda_{11} \times \mathrm{Cov}(\eta_1, \eta_6)) + (\lambda_{22} \times \mathrm{Cov}(\eta_2, \eta_6)) \\
&\quad + \mathrm{Cov}(\eta_5, \eta_6) + \mathrm{Cov}(\varepsilon_1, \lambda_{21}) + \mathrm{Cov}(\varepsilon_1, \lambda_{22}) \\
&\quad + \mathrm{Cov}(\varepsilon_1, \eta_6) + \mathrm{Cov}(\varepsilon_2, \lambda_{11}) + \mathrm{Cov}(\varepsilon_2, \lambda_{12}) \\
&\quad + \mathrm{Cov}(\varepsilon_2, \eta_5) + \mathrm{Cov}(\varepsilon_1, \varepsilon_2).
\end{aligned}$$

However, considerable simplification of the above expression is possible, since $\mathrm{Var}(\eta_1) = \mathrm{Var}(\eta_2) = 1$,

$$\begin{aligned}
\mathrm{Cov}(\eta_1, \eta_2) &= \mathrm{Cov}(\eta_1, \eta_5) = \mathrm{Cov}(\eta_2, \eta_5) = \mathrm{Cov}(\eta_1, \eta_6) \\
&= \mathrm{Cov}(\eta_2, \eta_6) = \mathrm{Cov}(\eta_5, \eta_6) = 0,
\end{aligned}$$

and

$$\begin{aligned}
&\mathrm{Cov}(\varepsilon_1, \lambda_{21}) + \mathrm{Cov}(\varepsilon_1, \lambda_{22}) + \mathrm{Cov}(\varepsilon_1, \eta_6) + \mathrm{Cov}(\varepsilon_2, \lambda_{11}) \\
&\qquad + \mathrm{Cov}(\varepsilon_2, \lambda_{12}) + \mathrm{Cov}(\varepsilon_2, \eta_5) + \mathrm{Cov}(\varepsilon_1, \varepsilon_2) = 0.
\end{aligned}$$

Therefore:

$$\mathrm{Cov}(y_1, y_2) = \lambda_{11}\lambda_{21} + \lambda_{12}\lambda_{22}.$$

If this process were continued for every combination of p observed variables (i.e., y_1 to y_6 and x_1), the result would be the determination of every element of a model-implied variance–covariance matrix. This matrix can be denoted by $\mathbf{\Sigma}$ (the capital Greek letter sigma) and is generally referred to as the *reproduced* (or *model implied*) *covariance matrix*. For the proposed model in Figure 1.3, the reproduced covariance matrix in Table 1.1 is determined.

Table 1.1. *Reproduced covariance matrix of y_1 through y_6 and x_1 for MZ twins*

$$\begin{bmatrix}
\lambda_{11}^2 + \lambda_{12}^2 + \mathrm{Var}(\eta_5) + \varepsilon_1 & & & \\
\lambda_{11} \times \lambda_{21} + \lambda_{12} \times \lambda_{22} + \gamma_1\gamma_2 \times \mathrm{Var}(\xi_1) & \lambda_{21}^2 + \lambda_{22}^2 + \mathrm{Var}(\eta_6) + \varepsilon_2 & & \\
\lambda_{11} \times \lambda_{31} + \lambda_{12} \times \lambda_{32} + \gamma_1\gamma_3 \times \mathrm{Var}(\xi_1) & \lambda_{21} \times \lambda_{31} + \lambda_{22} \times \lambda_{32} + \gamma_2\gamma_3 \times \mathrm{Var}(\xi_1) & \lambda_{31}^2 + \lambda_{32}^2 + \mathrm{Var}(\eta_7) + \varepsilon_3 & \\
\lambda_{11} \times \lambda_{43} + \lambda_{12} \times \lambda_{44} + \gamma_1\gamma_4 \times \mathrm{Var}(\xi_1) & \lambda_{21} \times \lambda_{43} + \lambda_{22} \times \lambda_{44} + \gamma_2\gamma_4 \times \mathrm{Var}(\xi_1) & \lambda_{31} \times \lambda_{43} + \lambda_{32} \times \lambda_{44} + \gamma_3\gamma_4 \times \mathrm{Var}(\xi_1) & \lambda_{43}^2 + \lambda_{44}^2 + \mathrm{Var}(\eta_8) + \varepsilon_4 \\
\lambda_{11} \times \lambda_{53} + \lambda_{12} \times \lambda_{54} + \gamma_1\gamma_5 \times \mathrm{Var}(\xi_1) & \lambda_{21} \times \lambda_{53} + \lambda_{22} \times \lambda_{54} + \gamma_2\gamma_5 \times \mathrm{Var}(\xi_1) & \lambda_{31} \times \lambda_{53} + \lambda_{32} \times \lambda_{54} + \gamma_3\gamma_5 \times \mathrm{Var}(\xi_1) & \\
\quad \lambda_{43} \times \lambda_{53} + \lambda_{44} \times \lambda_{54} + \gamma_4\gamma_5 \times \mathrm{Var}(\xi_1) & \lambda_{53}^2 + \lambda_{54}^2 + \mathrm{Var}(\eta_9) + \varepsilon_5 & & \\
\lambda_{11} \times \lambda_{63} + \lambda_{12} \times \lambda_{64} + \gamma_1\gamma_6 \times \mathrm{Var}(\xi_1) & \lambda_{21} \times \lambda_{63} + \lambda_{22} \times \lambda_{64} + \gamma_2\gamma_6 \times \mathrm{Var}(\xi_1) & \lambda_{31} \times \lambda_{63} + \lambda_{32} \times \lambda_{64} + \gamma_3\gamma_6 \times \mathrm{Var}(\xi_1) & \\
\quad \lambda_{43} \times \lambda_{63} + \lambda_{44} \times \lambda_{64} + \gamma_4\gamma_6 \times \mathrm{Var}(\xi_1) & \lambda_{53} \times \lambda_{63} + \lambda_{54} \times \lambda_{64} + \gamma_5\gamma_6 \times \mathrm{Var}(\xi_1) & \lambda_{63}^2 + \lambda_{64}^2 + \mathrm{Var}(\eta_{10}) + \varepsilon_6 & \\
\gamma_1 \times \mathrm{Var}(\xi_1) \quad \gamma_4 \times \mathrm{Var}(\xi_1) \quad \xi_1 & \gamma_2 \times \mathrm{Var}(\xi_1) \quad \gamma_5 \times \mathrm{Var}(\xi_1) & \gamma_3 \times \mathrm{Var}(\xi_1) \quad \gamma_6 \times \mathrm{Var}(\xi_1) &
\end{bmatrix}$$

Note: Rows 6 and 8 are continuations of rows 5 and 7 respectively. Row 9 contains all the elements but because of spatial restrictions does not align with the elements in rows 5–9.

It is important to note that the elements of Σ are all *functions of model parameters.* In addition, each element of Σ has as a counterpart a corresponding numerical element (entry) in the observed sample covariance matrix obtained for the seven observed variables considered (i.e., y_1 to y_6 and x_1). Assume that the observed covariance matrix (denoted by **S**) was as follows:

$$\begin{bmatrix} 16.51 \\ 4.26 & 28.12 \\ -2.10 & 3.38 & 3.35 \\ 10.04 & 3.40 & -2.50 & 16.51 \\ 3.40 & 10.35 & 1.74 & 4.36 & 28.12 \\ -2.50 & 1.74 & 1.12 & -2.10 & 3.38 & 3.35 \\ 2.44 & 3.18 & 1.10 & 2.44 & 3.18 & 1.10 & 4.00 \end{bmatrix}.$$

For example, the top element of **S** (i.e., 16.51) corresponds to $\lambda_{11}^2 + \lambda_{12}^2 + \mathrm{Var}(\eta_5) + \varepsilon_1$ in the reproduced matrix Σ. Now imagine setting the counterpart elements of **S** and Σ equal to one another. That is, according to the proposed model displayed in Figure 1.3, set $16.51 = \lambda_{11}^2 + \lambda_{12}^2 + \mathrm{Var}(\eta_5) + \varepsilon_1$, then $4.26 = \lambda_{11} \times \lambda_{21} + \lambda_{12} \times \lambda_{22} + \gamma_1\gamma_2 \times \mathrm{Var}(\xi_1)$, and so on until, for the last element of **S**, $4.00 = \xi_1$ is set. As a result of this equality setting, a system of 28 equations (i.e., the number of nonredundant elements, with 21 covariances and 7 variances) is generated. Thus one can conceive of the process of fitting a structural equation model as a way of solving a system of equations. For each equation, its left-hand side is a subsequent numerical entry of the sample observed variance–covariance matrix **S** while its right-hand side is the corresponding expression of model parameters defined in the Σ matrix. Hence, fitting a structural equation model is conceptually "equivalent" to solving in an optimal way (discussed in the next section) this system of equations obtained according to the proposed model. This discussion also demonstrates that the model presented in Figure 1.3, like any structural equation model, *implies* a specific structuring of the elements of the covariance matrix reproduced by the model in terms of specific expressions (functions) of unknown model parameters. Therefore, if certain values for the parameters were entered into these functions, one would obtain a covariance matrix that has numbers as elements. In fact, the process of fitting a model to data with SEM programs can be thought of as repeated "insertion" of appropriate values for the parameters in the matrix Σ until a certain optimality criterion (discussed in the next section) in terms of its proximity to the matrix **S** is satisfied. Every available SEM program has built into its "memory" the exact way in which these functions of model parameters in Σ

can be obtained. Although for ease of computation most programs make use of matrix algebra, the programs in effect determine each of the expressions presented in the above-mentioned 28 equations (for further discussion, see Marcoulides & Hershberger, 1997). Fortunately, this occurs quite automatically once the user has communicated to the program the model parameters.

Model assessment and fit

The previous section illustrated how a proposed SEM model leads to the reproduction of a variance–covariance matrix **Σ** that is then fit to the observed sample variance–covariance matrix **S**. Now it would seem that the next logical question is "How can one measure or evaluate the extent to which the matrices **S** and **Σ** differ?". As it turns out, this question is particularly important in SEM because it actually permits one to evaluate the goodness of fit of the model. Indeed, if the difference between **S** and **Σ** is negligible, then one can conclude that the model represents the observed data reasonably well. On the other hand, if the difference is large, one can conclude that the proposed model is not consistent with the observed data. There are at least two reasons for such inconsistencies: (1) the proposed model may be deficient, in the sense that it is not capable of "emulating" the analyzed matrix even with most favorable parameter values; and/or (2) the data may not be good. Thus, in order to proceed with assessing model fit, we need a method for evaluating the degree to which the reproduced matrix **Σ** differs from the sample covariance matrix **S**.

In order to clarify this method, a new concept is introduced, that of distance between matrices. Obviously, if the values to be compared were scalars (single numbers) a simple subtraction of one from the other (and possibly taking the absolute value of the resulting difference) would suffice to evaluate the distance between them. However, this cannot be done directly with the two matrices **S** and **Σ**. Subtracting the matrix **S** from the matrix **Σ** does not result in a single number. Rather, a matrix of differences is obtained.

Fortunately, there are some meaningful ways to assess the distance between two matrices and, interestingly, the resulting distance measure is a single number that is easier to interpret. Perhaps the simplest way to obtain this single number involves taking the sum of squares of the differences between the corresponding elements of the two matrices. Other more complicated ways involve a multiplication of these squares with some appropriately chosen weights and then taking their sum. Perhaps the most commonly used weight is based on maximum likelihood estimation. In either case, the single number represents a sort of generalized distance measure between the two matrices considered. The bigger the number, the more different are the matrices, while the smaller the number, the more similar are the matrices.

Because in SEM this number results after comparison of the elements of **S** with those of the model-implied covariance matrix **Σ**, the generalized distance is a function of the model parameters as well as the elements of the observed variances and covariances. Therefore it is customary to refer to the relationship between the matrix distance, on the one hand, and the model parameters and **S** on the other, as a *fit function* that is typically denoted by *F*. Since it equals the distance between two matrices, *F* is always equal to a positive value or 0. Whenever the value of *F* is 0, then the two matrices considered are identical.

Before particular measures of model fit are discussed, a word of warning is in order. Even if all possible fit indices point to an acceptable model, one can *never* claim to have found the *true* model that has generated the analyzed data (of course, we exclude from consideration the cases where data are simulated according to a preset known model). SEM is most concerned with finding a model that does not contradict the data. That is to say, in an empirical session of SEM, one is typically interested in retaining the proposed model whose validity is the essence of the null hypothesis. Statistically speaking, when using SEM methodology, one is usually interested in not rejecting the null hypothesis (Raykov & Marcoulides, 2000, p. 34).

When testing a model for fit, the complete fit of the model as well as the individual parameters should be examined. Typically, choosing the appropriate fit statistic is difficult for many researchers. One of the most widely used statistics for assessing the fit of a model is the χ^2 (chi-square) goodness-of-fit statistic. This statistic is an assessment of the magnitude of difference between the initial observed covariance matrix and the reproduced matrix. The probability level that is associated with this statistic indicates whether the difference between the reproduced matrix and the original data is significant or not. A significant χ^2 test states that the difference between the two matrices is due to sampling error or variation. Typically, researchers are most interested in a nonsignificant χ^2 test. This indicates that the observed matrix and the reproduced matrix are not statistically different, therefore indicating a good fit of the model to the data. However, the χ^2 test suffers from several weaknesses, including a dependence on sample size, and vulnerability to departures of the data from multivariate normality. Thus it is suggested that a researcher should examine a number of fit criteria in addition to the χ^2 value to assess the fit of the proposed model (Raykov & Marcoulides, 2000).

To assist in the process of assessing model fit, there are many other descriptive fit statistics that are typically formulated in values that range from

1 (perfect fit) to zero (no fit). One of the more popular fit indices is the goodness-of-fit index (GFI), which can loosely be considered as a measure of the proportion of variance and covariance that the proposed model is able to explain. If the number of parameters is also taken into account then the resulting index is the adjusted goodness of fit (AGFI) (Raykov & Marcoulides, 2000, p. 38). Unfortunately, there is not a strict norm for these indices. As a rough guide, it is currently viewed that a model with a GFI or AGFI of 0.95 or above may well represent a reasonably good approximation of the data (Hu & Bentler, 1999). Quite a few other indices of model fit have been developed, each with its own strengths and weaknesses. For more comprehensive discussions of evaluating model fit, see Bollen & Long (1993) or Marsh *et al.* (1996).

The fit indices proposed above were concerned with evaluating the fit of the entire model. Although this is certainly useful to have, one should also be interested in how well various parts of the model fit. It is entirely possible for the model as a whole to fit well, but for individual sections not to fit well. Aside from this, if a model does not fit well, it is of considerable value to determine which parts of the model are contributing to model misfit. Perhaps the most useful way to determine the fit of specific sections of the model is to examine the residual matrix (Bollen, 1989). The residual matrix results from the difference between the $\mathbf{S}$ and $\mathbf{\Sigma}$ matrices. The individual residual covariances (or correlations) are $(s_{ij} - \sigma_{ij})$ where s_{ij} is the *ij*-th element of $\mathbf{S}$ and σ_{ij} is the corresponding element in $\mathbf{\Sigma}$. A positive residual means that the model underpredicts the covariance between two variables, whereas a negative one means that the model overpredicts the covariance. Of course it can be difficult to interpret the absolute magnitude of the residuals, since the magnitude of a residual is in part a function of the scaling of the two variables. Thus, examining the *correlation residuals* or the *normalized residuals* can frequently better convey a sense of the fit of a specific part of a model (Jöreskog & Sörbom, 1996).

Model modification

The requirement for SEM is that the details of the proposed model be known before the model is fit and tested with data (Marcoulides & Drezner, 2001). Often, however, theories are poorly developed and require changes or adjustments throughout the testing process. Jöreskog & Sörbom (1996) have addressed three types of situation that concern model fitting and testing. The first situation is the *strictly confirmatory* notion in which the initial model is tested against empirical data and is either accepted or rejected. The second type is the *competing or alternative model* situation. This procedure entails several proposed models that are then assessed and selected on the basis of which model more appropriately fits the observed data. The final situation is the *model generating* technique in which the scientist repeatedly modifies the proposed model until some level of fit is acquired. The decision as to which procedure will be utilized is based on the initial theory. A researcher who is firmly entrenched in his or her theory or hypotheses will conduct SEM differently from a scientist who is unsure of the interrelationships between the observed and latent variables. No matter how SEM is conducted, however, once a researcher attempts to respecify an initial model after it has been rejected by the data, the process of confirmation is over. Now SEM enters into an *exploratory* mode, in which the researcher searches for revisions to the model that will most significantly increase its fit to the data. These revisions usually entail freeing a previously fixed parameter and/or fixing a previously free parameter. Such a process of exploration is generally referred to as a *specification search* (Leamer, 1978).

All SEM computer programs come equipped with various statistics to assist in the specification search. Two of the most popular statistics are the *modification index* (MI) and the *t-ratio* (Jöreskog & Sörbom, 1996). The MI is used to determine which parameter, if freed, would contribute most to an increase in model fit and indicates the amount the χ^2 goodness-of-fit statistic would decrease if in fact the parameter were specified in the model. (Recall that with 1 df, a single parameter would significantly improve the fit of a model if it decreased the goodness of fit χ^2 by at least 8.841 points, $p < 0.05$.) On the other hand, the t-ratio assesses the significance of the individual parameters in a specified model; t-ratios of less than 2 are generally considered nonsignificant, $p = 0.05$. Presumably, those parameters which are not significant may be removed from the model without causing the model to fit significantly more badly (i.e., without causing the χ^2 goodness-of-fit statistic to increase significantly). Generally, the best strategy is first to determine which parameters should be added to the model by examining their individual MIs; then, once the list of significant MIs has been exhausted, the t-ratios should be examined to decide which parameters should be deleted from the model (Marcoulides & Hershberger, 1997). Marcoulides & Drezner (2001) have also proposed automated specification search procedures based on genetic algorithms and Tabu search procedures.

On the surface, the availability of MIs, t-ratios, similar indices, and automated specification searches may appear to be of tremendous benefit to the process of model respecification. However, certain cautionary remarks are in order. First, parameters should be added (or deleted) to the model one at a time, each time the model is re-evaluated and the indices recalculated because changes in the model result (sometimes) in dramatic changes in the values of the indices. In other words, with one parameter not in the model, another parameter may appear to be potentially significant on the basis of its MI, but, with the addition of the first parameter, the significance of the second parameter's MI disappears (fortunately, this issue is addressed in the automated specification searches proposed by Marcoulides and his colleagues). Second, as can well be imagined, even covariance matrices of moderate size (for instance, our example of a 7×7 covariance matrix) may make possible the specification of hundreds of free parameters in a model. Leaving aside the desirability of any one of these parameters, the possibility of Type I errors looms (Green *et al.*, 1998). Green *et al.* (1999) have proposed methods for controlling Type I errors during SEM specification searches. Third, even though adding a parameter may cause the model to finally fit, if the parameter is theoretically meaningless or statistically suspect, it should be avoided. Similarly even though a parameter may appear to be nonsignificant as indicated by its small t-value, it should *not* be removed from a model if it is considered theoretically or logically important.

Multisample models

Before we introduce the LISREL program and its approach to the evaluation of the model in Figure 1.3, a final, critical issue must be addressed. Although we stated earlier that according to the t-rule, the model was identified, this is in fact not so. The t-rule is a necessary but not sufficient criterion for model identification. Rather than delve into a complex discussion as to why the model is not identified, or introduce alternative sufficient criteria for model identification, a simple demonstration will suffice to show why this model in under-identified. Remember that the primary reason for solving this behavioral genetic model is to estimate genetic and environmental influences on the observed variables. Recall also that we used one type of family relation to do this, monozygotic or MZ twins. Since MZs share *all* of their genes and *all* of their common environments, we can express the covariance between MZs for an observed variables as

$$\mathrm{Cov(MZ)} = \mathrm{Var(G)} + \mathrm{Var(E)} + 2\mathrm{Cov(G, E)},$$

where G denotes genotype and E environment. However, our model stipulates no covariance between G and E, so the MZ covariance simplifies to

$$\mathrm{Cov(MZ)} = \mathrm{Var(G)} + \mathrm{Var(E)}.$$

Astute readers will no doubt question how we are to solve for two unknowns, the variance of G and the variance of E, with only one observed statistic – the covariance between MZ twins. The answer is that we cannot. Including the variance of the observed variables is of no help, for its expression is identical to that for Cov(MZ):

$$\text{Var(MZ)} = \text{Var(G)} + \text{Var(E)}.$$

How then are we to identify this model?

The solution is actually quite simple, and pre-dates the existence of SEM methodology. If we also include family members of genetic relatedness differing from that of MZs, we are now able to solve for the genetic and environmental variances of the observed variables. Traditionally, dizygotic, or DZ, twins have been used in conjunction with MZ twins, to solve for the values of these variances. This method is referred to as the "classical twin method", first used by Galton in the 1870s (Eaves *et al.*, 1989). DZ twins are a useful group to compare with MZ twins, since DZs share on average only half their genes but all of their common environments. Thus

$$\text{Cov(DZ)} = 0.5 \times \text{Var(G)} + \text{Var(E)}.$$

In fact, if we had only a single observed variable, the genetic variance of that variable[3] would be expressed as

$$\text{Var(G)} = 2 \times (\text{Cov(MZ)} - \text{Cov(DZ)}).$$

Therefore, in order to solve for the genetic and environmental variances of our model, we include a sample of MZ twins as well as one of DZ twins. For our model, the **S** (observed covariance matrix) for the DZ twins is

$$\begin{bmatrix} 16.51 & & & & & & \\ 4.26 & 28.12 & & & & & \\ -2.10 & 3.38 & 3.35 & & & & \\ 5.02 & 1.70 & -1.25 & 16.51 & & & \\ 1.70 & 5.18 & 0.87 & 4.36 & 28.12 & & \\ -1.25 & 0.87 & 0.56 & -2.10 & 3.38 & 3.35 & \\ 2.44 & 3.18 & 1.10 & 2.44 & 3.18 & 1.10 & 4.00 \end{bmatrix}$$

[3] Both the SEM model and the simple solution of the classical twin method (i.e., 2 × (Cov(MZ) – Cov(DZ)) rely on the validity of certain assumptions. If these assumptions are incorrect, then the estimate of genetic variance will be inaccurate. Among these assumptions are: (1) all genetic effects are additive (i.e., linear), (2) the covariance between genetic and environmental effects is zero, and (3) there is no assortative mating for the observed variable. For an extended discussion of the meaning and likelihood of these assumptions being met, see Eaves *et al.* (1989).

The model-implied covariance matrix ($\boldsymbol{\Sigma}$) for DZ twins is given in Table 1.2. For the most part, $\boldsymbol{\Sigma}$ is the same for MZ and DZ twins, with one critical difference: all expressions for the genetic covariance between DZ twins is weighted by 0.5; e.g., the covariance between DZs for physical similarities (i.e., Cov(y_1, y_2)) is

$$0.5(\lambda_{11} \times \lambda_{43}) + (\lambda_{12} \times \lambda_{44}) + (\gamma_2\gamma_4 \times \text{Var}(\xi_1)),$$

whereas the same covariance expression for MZs is

$$(\lambda_{11} \times \lambda_{43}) + (\lambda_{12} \times \lambda_{44}) + (\gamma_2\gamma_4 \times \text{Var}(\xi_1)).$$

Figure 1.4 presents the full, two-sample model that will be evaluated using SEM. Note that the value of the two-way arrow between the ζ values of the DZ twins' genetic latent variables is now 0.5. Most SEM programs are capable of analyzing models that require multiple samples. Such multisample analyses can usually be performed very simply, with the SEM program allowing the same parameters in the multiple groups to be estimated at the same value, or at different values. Because we would expect that each of the parameters in our model would be the same for MZ and DZ twins, we will constrain all parameters to be equal across the two groups. Thus, since 28 additional, unique variance–covariance statistics are added to the model, the total number of observed statistics becomes 56. Estimating 13 free parameters, the degrees of freedom of the model are now 43.

The LISREL model and program

Several computer programs are currently available to implement SEM models. One of the most popular and oldest programs is called LISREL (Jöreskog & Sörbom, 1993), and is based on the Keesling–Wiley–Jöreskog LInear Structural RELations model. The LISREL model consists of two basic components: a structural model,

$$\boldsymbol{\eta} = \mathbf{B}\boldsymbol{\eta} + \boldsymbol{\Gamma}\boldsymbol{\xi} + \boldsymbol{\zeta},$$

and two measurement models,

$$\begin{aligned} \boldsymbol{y} &= \boldsymbol{\Lambda}_y\boldsymbol{\eta} + \boldsymbol{\varepsilon} \\ \boldsymbol{x} &= \boldsymbol{\Lambda}_x\boldsymbol{\xi} + \boldsymbol{\delta} \end{aligned}$$

where $\boldsymbol{\eta}$ and $\boldsymbol{\xi}$ are vectors of latent variables; $\boldsymbol{y}$ and $\boldsymbol{x}$ are vectors of observed variables; $\boldsymbol{\varepsilon}$ and $\boldsymbol{\delta}$ are vectors of measurement errors; and $\boldsymbol{\zeta}$ is a vector of structural errors. $\mathbf{B}$ and $\boldsymbol{\Gamma}$ are explained later. Of these, $\boldsymbol{\eta}$ and $\boldsymbol{y}$ are endogenous (dependent) variables; $\boldsymbol{\xi}$ and $\boldsymbol{x}$ are exogenous (independent)

Table 1.2. *Reproduced covariance matrix of y_1 through y_6 and x_1 for DZ twins*

$$
\left[\begin{array}{llll}
\lambda_{11}^2 + \lambda_{12}^2 + \mathrm{Var}(\eta_5) + \varepsilon_1 & & & \\
\lambda_{11} \times \lambda_{21} + \lambda_{12} \times \lambda_{22} + \gamma_1\gamma_2 \times \mathrm{Var}(\xi_1) & \lambda_{21}^2 + \lambda_{22}^2 + \mathrm{Var}(\eta_6) + \varepsilon_2 & & \\
\lambda_{11} \times \lambda_{31} + \lambda_{12} \times \lambda_{32} + \gamma_1\gamma_3 \times \mathrm{Var}(\xi_1) & \lambda_{21} \times \lambda_{31} + \lambda_{22} \times \lambda_{32} + \gamma_2\gamma_3 \times \mathrm{Var}(\xi_1) & \lambda_{31}^2 + \lambda_{32}^2 + \mathrm{Var}(\eta_7) + \varepsilon_3 & \\
.5(\lambda_{11} \times \lambda_{43}) + \lambda_{12} \times \lambda_{44} + \gamma_1\gamma_4 \times \mathrm{Var}(\xi_1) & .5(\lambda_{21} \times \lambda_{43}) + \lambda_{22} \times \lambda_{44} + \gamma_2\gamma_4 \times \mathrm{Var}(\xi_1) & .5(\lambda_{31} \times \lambda_{43}) + \lambda_{32} \times \lambda_{44} + \gamma_3\gamma_4 \times \mathrm{Var}(\xi_1) & \lambda_{43}^2 + \lambda_{44}^2 + \mathrm{Var}(\eta_8) + \varepsilon_4 \\
.5(\lambda_{11} \times \lambda_{53}) + \lambda_{12} \times \lambda_{54} + \gamma_1\gamma_5 \times \mathrm{Var}(\xi_1) & .5(\lambda_{21} \times \lambda_{53}) + \lambda_{22} \times \lambda_{54} + \gamma_2\gamma_5 \times \mathrm{Var}(\xi_1) & .5(\lambda_{31} \times \lambda_{53}) + \lambda_{32} \times \lambda_{54} + \gamma_3\gamma_5 \times \mathrm{Var}(\xi_1) & \\
\quad \lambda_{43} \times \lambda_{53} + \lambda_{44} \times \lambda_{54} + \gamma_4\gamma_5 \times \mathrm{Var}(\xi_1) & \quad \lambda_{53}^2 + \lambda_{54}^2 + \mathrm{Var}(\eta_9) + \varepsilon_5 & & \\
.5(\lambda_{11} \times \lambda_{63}) + \lambda_{12} \times \lambda_{64} + \gamma_1\gamma_6 \times \mathrm{Var}(\xi_1) & .5(\lambda_{21} \times \lambda_{63}) + \lambda_{22} \times \lambda_{64} + \gamma_2\gamma_6 \times \mathrm{Var}(\xi_1) & .5(\lambda_{31} \times \lambda_{63}) + \lambda_{32} \times \lambda_{64} + \gamma_3\gamma_6 \times \mathrm{Var}(\xi_1) & \\
\quad \lambda_{43} \times \lambda_{63} + \lambda_{44} \times \lambda_{64} + \gamma_4\gamma_6 \times \mathrm{Var}(\xi_1) & \quad \lambda_{53} \times \lambda_{63} + \lambda_{54} \times \lambda_{64} + \gamma_5\gamma_6 \times \mathrm{Var}(\xi_1) & \quad \lambda_{63}^2 + \lambda_{64}^2 + \mathrm{Var}(\eta_{10}) + \varepsilon_6 & \\
\gamma_1 \times \mathrm{Var}(\xi_1) \qquad \gamma_4 \times \mathrm{Var}(\xi_1) \qquad \xi_1 & \gamma_2 \times \mathrm{Var}(\xi_1) \qquad \gamma_5 \times \mathrm{Var}(\xi_1) & \gamma_3 \times \mathrm{Var}(\xi_1) \qquad \gamma_6 \times \mathrm{Var}(\xi_1) &
\end{array}\right]
$$

Note: Rows 6 and 8 are continuations of rows 5 and 7 respectively. Row 9 contains all the elements but because of spatial restrictions does not align with the elements in rows 5–9.

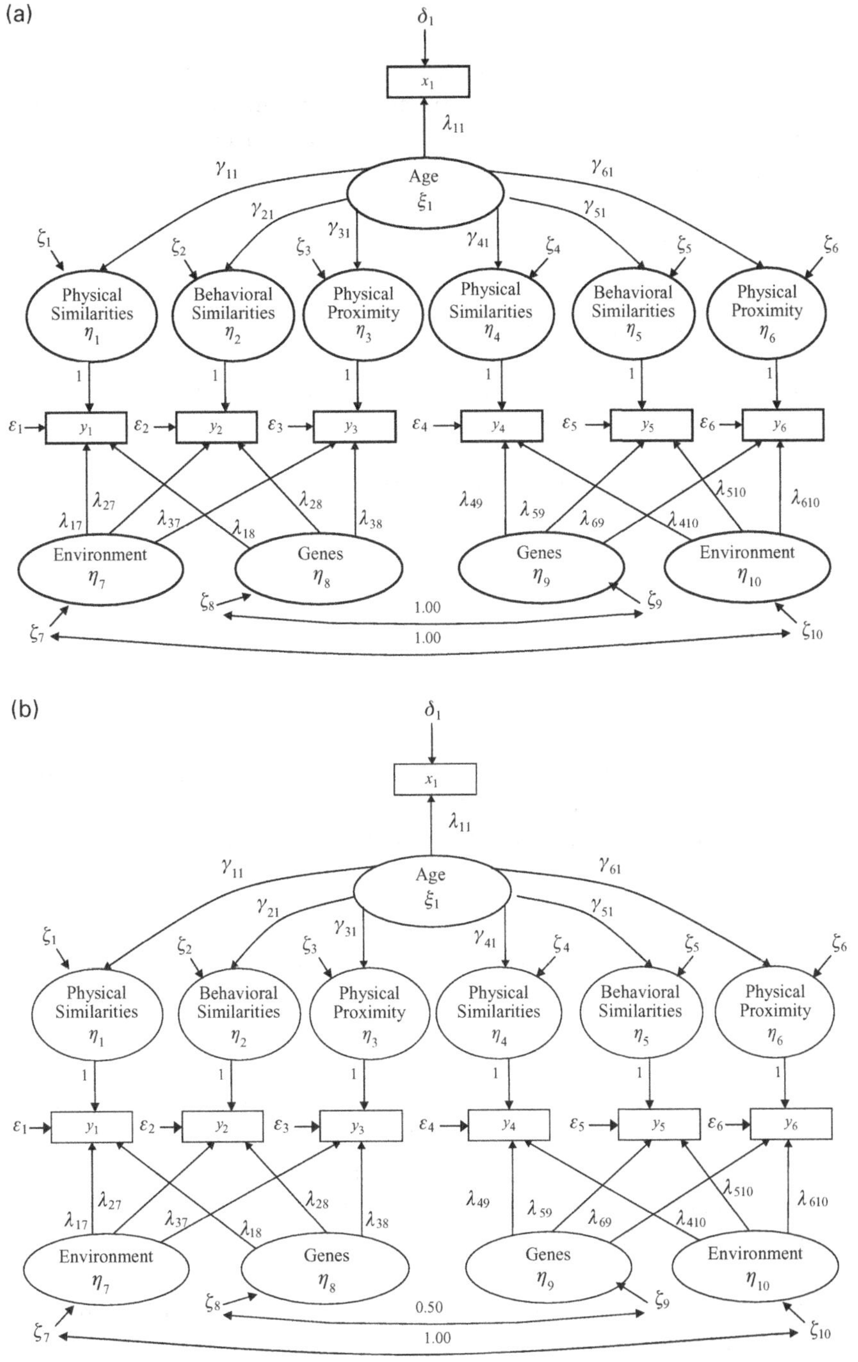

Figure 1.4. The model of sibling relatedness specified for (a) monozygotic twins and (b) dizygotic twins. Symbols are explained in the text.

variables; and, as error variables, ε, δ, and ζ are exogenous as well. All variables for now will be assumed to have zero means, an assumption that can be relaxed. (In Chapter 10, Rovine and Molenaar provide examples of evaluating the mean structure of a model). LISREL makes the following five assumptions:

1. ζ is uncorrelated with ξ.
2. ε is uncorrelated with η.
3. δ is uncorrelated with ξ.
4. ζ is uncorrelated with ε and δ.
5. $\mathbf{I} - \mathbf{B}$ is nonsingular (where $\mathbf{I}$ is the identity matrix).

Imposing these five assumptions leads to the following expression for the reproduced covariance structure (Σ) between y and $\mathbf{x}$:

$$\begin{aligned}
\Sigma &= (y, x)(y, x)^{\mathrm{T}} \\
&= \begin{bmatrix} yy^{\mathrm{T}} & yx^{\mathrm{T}} \\ xy^{\mathrm{T}} & xx^{\mathrm{T}} \end{bmatrix} \\
&= \begin{bmatrix} \Lambda_y(\mathbf{I}-\mathbf{B})^{-1}(\Gamma\Phi\Gamma + \Psi)(\mathbf{I}-\mathbf{B})^{-1}\Lambda_y^{\mathrm{T}} + \Theta_\varepsilon & \Lambda_y(\mathbf{I}-\mathbf{B})^{-1}\Gamma\Phi\Lambda_x^{\mathrm{T}} + \Theta_{\delta,\varepsilon} \\ \Lambda_x\Phi\Gamma^{\mathrm{T}}(\mathbf{I}-\mathbf{B}^{\mathrm{T}})^{-1}\Lambda_y^{\mathrm{T}} + \Theta_{\delta,\varepsilon} & \Lambda_x\Phi\Lambda_x^{\mathrm{T}} + \Theta_\delta \end{bmatrix}.
\end{aligned}$$

The reproduced covariance structure between y and x is therefore a function of nine parameter matrices. With p observed y variables, q observed x variables, m latent endogenous η variables, and n latent exogenous ξ variables, the nine parameter matrices are (including, in parentheses, their LISREL abbreviation):

- Λ_y (LAMBDA-y, LY) = a matrix of factor loadings between y and $\eta (p \times m)$;
- Λ_x (LAMBDA-x, LX) = a matrix of factor loadings between x and $\xi (q \times n)$;
- Θ_ε (THETA-epsilon, TE) = a variance–covariance matrix among the measurement errors of y $(p \times p)$;
- Θ_δ (THETA-delta, TD) = a variance–covariance matrix among the measurement errors of x $(q \times q)$;
- $\Theta_{\delta,\varepsilon}$ (THETA, TH) = a matrix of covariances between the measurement errors of x and the measurement errors of y $(q \times p)$;
- $\mathbf{B}$ (BETA, BE) = a matrix of regression coefficients among the $\eta (m \times m)$;
- Γ (GAMMA, GA) = a matrix of regression coefficients between the η and $\xi (m \times n)$;

- $\mathbf{\Phi}$ (PHI, PH) = a variance–covariance matrix of the $\xi\,(n \times n)$;
- $\mathbf{\Psi}$ (PSI, PS) = a variance–covariance matrix of the structural errors ζ of the $\eta(m \times m)$.

These nine matrices describe the variance–covariance structure among the η in terms of the variances and covariances among the ξ, between ξ and η, between ξ and x, and between η and y. Note then that there is no special 10th matrix describing the variances and covariances among the η; this matrix would be completely redundant (a linear function of at least four of the other nine matrices):

$$\eta\eta^{\mathrm{T}} = (\mathbf{I} - \mathbf{B})^{-1}(\mathbf{\Gamma}\mathbf{\Phi}\mathbf{\Gamma}^{\mathrm{T}} + \mathbf{\Psi})(\mathbf{I} - \mathbf{B})^{-1}.$$

The redundancy of this "10th matrix" underscores the primary purpose of SEM: to evaluate models describing the variance–covariance structure among the endogenous latent variables, or to explain why the η may be significantly intercorrelated. In the next section, we will apply the LISREL model to the variances and covariances among a set of variables in order to evaluate the model in Figure 1.4.

Example of LISREL analysis

The first and most critical step in SEM is communicating the model to the program. As explained previously, virtually any linear model may be communicated to LISREL using its nine parameter matrices. Most models will not require all nine matrices; some models will require only a few. Thus, in order for the behavioral genetic model to be analyzed using LISREL, the model must be formulated in terms of these nine (or fewer) matrices.

Let us first consider the LAMBDA-y ($\mathbf{\Lambda}_y$) matrix. In the model in Figure 1.4, there are six y variables (p) and 10 η variables (m). Thus $\mathbf{\Lambda}_y$ is a 6×10 matrix having the following form:

$$\begin{array}{c} \\ \begin{bmatrix} y_1 \\ y_2 \\ y_3 \\ y_4 \\ y_5 \\ y_6 \end{bmatrix} \end{array} \begin{array}{c} [\;\; \mathrm{G}_1 \quad \mathrm{E}_1 \quad \mathrm{G}_2 \quad \mathrm{E}_2 \quad y_1 \; y_2 \; y_3 \; y_4 \; y_5 \; y_6 \;\;] \\ \begin{bmatrix} \lambda_{11}\eta_1\ (1) & \lambda_{12}\eta_2\ (2) & 0 & 0 & 1 & 0 & 0 & 0 & 0 & 0 \\ \lambda_{21}\eta_1\ (3) & \lambda_{22}\eta_2\ (4) & 0 & 0 & 0 & 1 & 0 & 0 & 0 & 0 \\ \lambda_{31}\eta_1\ (5) & \lambda_{32}\eta_2\ (6) & 0 & 0 & 0 & 0 & 1 & 0 & 0 & 0 \\ 0 & 0 & \lambda_{43}\eta_3\ (1) & \lambda_{44}\eta_4\ (2) & 0 & 0 & 0 & 1 & 0 & 0 \\ 0 & 0 & \lambda_{53}\eta_3\ (3) & \lambda_{54}\eta_4\ (4) & 0 & 0 & 0 & 0 & 1 & 0 \\ 0 & 0 & \lambda_{63}\eta_3\ (5) & \lambda_{54}\eta_4\ (6) & 0 & 0 & 0 & 0 & 0 & 1 \end{bmatrix} \end{array}.$$

In the $\mathbf{\Lambda}_y$ matrix, and in all subsequent matrices, the following notational conventions are used. First, whenever a Greek symbol appears as an element, that parameter will be estimated. Second, each of the free parameters is numbered; for example $\lambda_{11}\eta_1$ is parameter number 1 as indicated by the "1" in parentheses. Third, different parameters given the same number are constrained as equal during estimation; thus $\lambda_{11}\eta_1 = \lambda_{43}\eta_3$, since both parameter are denoted as "(1)". And fourth, whenever a number appears as an element, that parameter is fixed to that value during estimation; for example, $\lambda_{41}\eta_1 = 0$ (implying that this particular parameter is not estimated at all; e.g., is not in the model); $\lambda_{15}\eta_5 = 1$ (implying that for identification purposes, the scale of η_5 has been equated to the scale of y_1 fixing a "1" as that element's value), etc.

For the LAMBDA-x ($\mathbf{\Lambda}_x$) matrix, since $q = 1$ (1 observed x variable) and $n = 1$ (1 latent independent variable), Λ_x (1×1):

$$\begin{array}{cc} & [\text{AGE}] \\ [x_1] & [1]\,. \end{array}$$

For the THETA-epsilon ($\mathbf{\Theta}_\varepsilon$) matrix, since there are 6 p (observed y variables), $\mathbf{\Theta}_\varepsilon$ is (6×6):

$$\begin{array}{c} \\ \begin{bmatrix} y_1 \\ y_2 \\ y_3 \\ y_4 \\ y_5 \\ y_6 \end{bmatrix} \end{array} \begin{array}{c} [\,y_1 \quad\quad y_2 \quad\quad y_3 \quad\quad y_4 \quad\quad y_5 \quad\quad y_6\,] \\ \begin{bmatrix} \varepsilon_{11}\,(7) & 0 & 0 & 0 & 0 & 0 \\ 0 & \varepsilon_{22}\,(8) & 0 & 0 & 0 & 0 \\ 0 & 0 & \varepsilon_{33}\,(9) & 0 & 0 & 0 \\ 0 & 0 & 0 & \varepsilon_{44}\,(7) & 0 & 0 \\ 0 & 0 & 0 & 0 & \varepsilon_{55}\,(8) & 0 \\ 0 & 0 & 0 & 0 & 0 & \varepsilon_{66}\,(9) \end{bmatrix} \end{array}.$$

For the THETA-delta ($\mathbf{\Theta}_\delta$) matrix, since $q = 1$ (1 observed x variable), $\mathbf{\Theta}_\delta$ is (1×1). However, the model in Figure 1.4 does not stipulate this parameter, due to difficulty with its identification. So Θ_δ is

$$\begin{array}{cc} & [x_1] \\ [x_1] & [0]\,. \end{array}$$

For the THETA ($\mathbf{\Theta}_{\delta,\varepsilon}$) matrix, although there are $p = 6$ observed y variables and $q = 1$ observed x variables, making $\mathbf{\Theta}_{\delta,\varepsilon}$ a (1×6) matrix of cross-error covariances, none of these covariances is stipulated in the model shown in Figure 1.4. Thus $\mathbf{\Theta}_{\delta,\varepsilon}$ is

$$\begin{array}{cc} & [y_1 \quad y_2 \quad y_3 \quad y_4 \quad y_5 \quad y_6] \\ [x_1] & [0 \quad 0 \quad 0 \quad 0 \quad 0 \quad 0]. \end{array}$$

Similarly, for the BETA (**B**) matrix, no directional paths have been specified between the 10 latent dependent variables. Thus **B** (10 × 10) is

$$\begin{matrix} & \begin{bmatrix} G_1 & E_1 & G_2 & E_2 & y_1 & y_2 & y_3 & y_4 & y_5 & y_6 \end{bmatrix} \\ \begin{bmatrix} G_1 \\ E_1 \\ G_2 \\ E_2 \\ y_1 \\ y_2 \\ y_3 \\ y_4 \\ y_5 \\ y_6 \end{bmatrix} & \begin{bmatrix} 0 & 0 & 0 & 0 & 0 & 0 & 0 & 0 & 0 & 0 \\ 0 & 0 & 0 & 0 & 0 & 0 & 0 & 0 & 0 & 0 \\ 0 & 0 & 0 & 0 & 0 & 0 & 0 & 0 & 0 & 0 \\ 0 & 0 & 0 & 0 & 0 & 0 & 0 & 0 & 0 & 0 \\ 0 & 0 & 0 & 0 & 0 & 0 & 0 & 0 & 0 & 0 \\ 0 & 0 & 0 & 0 & 0 & 0 & 0 & 0 & 0 & 0 \\ 0 & 0 & 0 & 0 & 0 & 0 & 0 & 0 & 0 & 0 \\ 0 & 0 & 0 & 0 & 0 & 0 & 0 & 0 & 0 & 0 \\ 0 & 0 & 0 & 0 & 0 & 0 & 0 & 0 & 0 & 0 \\ 0 & 0 & 0 & 0 & 0 & 0 & 0 & 0 & 0 & 0 \end{bmatrix} \end{matrix} .$$

For the GAMMA ($\boldsymbol{\Gamma}$) matrix, since there are $m = 10$ latent dependent variables and there are $n = 1$ latent independent variables, $\boldsymbol{\Gamma}$ is a (10 × 1) matrix having the following form:

$$\begin{matrix} & [\text{AGE}] \\ \begin{bmatrix} G_1 \\ E_1 \\ G_2 \\ E_2 \\ y_1 \\ y_2 \\ y_3 \\ y_4 \\ y_5 \\ y_6 \end{bmatrix} & \begin{bmatrix} 0 & \\ 0 & \\ 0 & \\ 0 & \\ \gamma_{51} & (10) \\ \gamma_{61} & (11) \\ \gamma_{71} & (12) \\ \gamma_{81} & (10) \\ \gamma_{91} & (11) \\ \gamma_{10,1} & (12) \end{bmatrix} \end{matrix} .$$

For PHI ($\boldsymbol{\Phi}$), since $n = 1$, $\boldsymbol{\Phi}$ is (1 × 1):

$$\begin{matrix} & [\text{AGE}] \\ [\text{AGE}] & [\theta_{11} \quad (13)]. \end{matrix}$$

The last and ninth matrix is PSI ($\boldsymbol{\Psi}$). Since there are $q = 10$ latent dependent variables, $\boldsymbol{\Psi}$ is a (10 × 10) variance–covariance matrix. However, it differs slightly in its parameter specifications for MZ and DZ twins. For

MZ twins, PSI is:

$$[\,G_1 \quad E_1 \quad G_2 \quad E_2 \quad y_1 \quad y_2 \quad y_3 \quad y_4 \quad y_5 \quad y_6\,]$$

$$\begin{bmatrix} G_1 \\ E_1 \\ G_2 \\ E_2 \\ y_1 \\ y_2 \\ y_3 \\ y_4 \\ y_5 \\ y_6 \end{bmatrix} \begin{bmatrix} 1 & 0 & 0 & 0 & 0 & 0 & 0 & 0 & 0 & 0 \\ 0 & 1 & 0 & 0 & 0 & 0 & 0 & 0 & 0 & 0 \\ 1 & 0 & 1 & 0 & 0 & 0 & 0 & 0 & 0 & 0 \\ 0 & 1 & 0 & 1 & 0 & 0 & 0 & 0 & 0 & 0 \\ 0 & 0 & 0 & 0 & 0 & 0 & 0 & 0 & 0 & 0 \\ 0 & 0 & 0 & 0 & 0 & 0 & 0 & 0 & 0 & 0 \\ 0 & 0 & 0 & 0 & 0 & 0 & 0 & 0 & 0 & 0 \\ 0 & 0 & 0 & 0 & 0 & 0 & 0 & 0 & 0 & 0 \\ 0 & 0 & 0 & 0 & 0 & 0 & 0 & 0 & 0 & 0 \\ 0 & 0 & 0 & 0 & 0 & 0 & 0 & 0 & 0 & 0 \end{bmatrix},$$

and for DZ twins, PSI is:

$$[\,G_1 \quad E_1 \quad G_2 \quad E_2 \quad y_1 \quad y_2 \quad y_3 \quad y_4 \quad y_5 \quad y_6\,]$$

$$\begin{bmatrix} G_1 \\ E_1 \\ G_2 \\ E_2 \\ y_1 \\ y_2 \\ y_3 \\ y_4 \\ y_5 \\ y_6 \end{bmatrix} \begin{bmatrix} 1 & 0 & 0 & 0 & 0 & 0 & 0 & 0 & 0 & 0 \\ 0 & 1 & 0 & 0 & 0 & 0 & 0 & 0 & 0 & 0 \\ 0.5 & 0 & 1 & 0 & 0 & 0 & 0 & 0 & 0 & 0 \\ 0 & 1 & 0 & 1 & 0 & 0 & 0 & 0 & 0 & 0 \\ 0 & 0 & 0 & 0 & 0 & 0 & 0 & 0 & 0 & 0 \\ 0 & 0 & 0 & 0 & 0 & 0 & 0 & 0 & 0 & 0 \\ 0 & 0 & 0 & 0 & 0 & 0 & 0 & 0 & 0 & 0 \\ 0 & 0 & 0 & 0 & 0 & 0 & 0 & 0 & 0 & 0 \\ 0 & 0 & 0 & 0 & 0 & 0 & 0 & 0 & 0 & 0 \\ 0 & 0 & 0 & 0 & 0 & 0 & 0 & 0 & 0 & 0 \end{bmatrix}.$$

Note that the only element which differs between the two $\boldsymbol{\Psi}$ matrices is the fixed genetic covariance (i.e., $\text{Cov}(G_1, G_2)$) for the twins. In the case of MZs, $\text{Cov}(G_1, G_2) = 1$, whereas for DZs $\text{Cov}(G_1,G_2) = 0.5$. All other parameter matrices are identical for the two twin groups.

Table 1.3 presents the LISREL program used to evaluate the model shown in Figure 1.4. Although readers unfamiliar with LISREL should consult the program's documentation (Jöreskog & Sörbom, 1993) for a complete explanation of the syntax used, a brief overview of the program statements used in Table 1.3 would be helpful at this point. First, it is advisable that the first line be a title for the program, for later, easy reference. The title tells us that the first group to be analyzed will be MZ twins. The third

Table 1.3. *LISREL program for behavioral genetic model*

```
LISREL PROGRAM FOR BEHAVIORAL GENETIC MODEL
MZ TWINS
DA NG=2 NI=7 NO=200 MA=CM
LA
T1_PS T1_BS T1_PP T2_PS T2_BS T2_PP AGE
CM
16.51
4.26 28.12
-2.10 3.38 3.35
10.04 3.40 -2.50 16.51
3.40 10.35 1.74 4.36 28.12
-2.50 1.74 1.12 -2.10 3.38 3.35
2.44 3.18 1.10 2.44 3.18 1.10 4.00
MO NY=6 NX=1 NE=10 NK=1 LX=FU,FI LY=FU,FI TD=ZE TE=DI c
  GA=FU,FI PH=SY,FR BE=ZE PS=SY,FI TH=ZE
LK
AGE
LE
T1_G T1_E T2_G T2_E T1_PS T1_BS T1_PP T2_PS T2_BS T2_PP
VA 1.0 LX (1,1)
FR LY(1,1) LY(2,1) LY(3,1) LY(1,2) LY(2,2) LY(3,2)
EQ LY(1,1) LY(4,3)
EQ LY(2,1) LY(5,3)
EQ LY(3,1) LY(6,3)
EQ LY(1,2) LY(4,4)
EQ LY(2,2) LY(5,4)
EQ LY(3,2) LY(6,4)
VA 1.00 LY(1,5) LY(2,6) LY(3,7) LY(4,8) LY(5,9) LY(6,10)
EQ TE(1,1) TE(4,4)
EQ TE(2,2) TE(5,5)
EQ TE(3,3) TE(6,6)
FR GA(5,1) GA(6,1) GA(7,1)
EQ GA(5,1) GA(8,1)
EQ GA(6,1) GA(9,1)
EQ GA(7,1) GA(10,1)
VA 1.00 PS(1,1) PS(2,2) PS(3,3) PS(4,4)
VA 1.00 PS(1,3)
VA 1.00 PS(2,4)
FR PH(1,1)
ST 5 ALL
OU NS AD=OFF IT=500 MI RS
```

Table 1.3. (*cont.*)

```
DZ TWINS
DA NI=7 NO=200 MA=CM
LA
T1_PS T1_BS T1_PP T2_PS T2_BS T2_PP AGE
CM
16.51
4.26 28.12
-2.10 3.38 3.35
5.02 1.70 -1.25 16.51
1.70 5.18 0.87 4.36 28.12
-1.25 0.87 0.56 -2.10 3.38 3.35
2.44 3.18 1.10 2.44 3.18 1.10 4.00
MO LX=IN LY=IN TD=IN TE=IN GA=IN PH=IN PS=SY,FI BE=IN TH=IN
LK
AGE
LE
T1_G T1_E T2_G T2_ E T1_PS T1_BS T1_PP T2_PS T2_BS T2_PP
VA 1.00 PS(1,1) PS(2,2) PS(3,3) PS(4,4)
VA 0.5 PS(3,1)
VA 1.00 PS(4,2)
ST 5 ALL
OU NS AD=OFF MI RS
```

c, indicates continuation of line in LISREL notation.

line is the DA (data line), which provides information as to the number of groups or samples in the model (NG = 2); the total number of observed variables (NI = 7 = $y + x$); the number of observations in the first group (NO = 200); and the type of data matrix one would like to analyze (MA = CM) – in this case a covariance matrix. Next, labels (LA) are assigned to each of the seven observed variables in the order they appear in the input data (e.g., T1_PS = physical similarities for twin 1; T1_BS = behavioral similarities for twin 1; etc.). Then the actual input data matrix is given, first denoted by a CM and then followed by the covariance matrix itself. Following the covariance matrix is the MO (or Model line), which provides a preliminary description of the model for MZ twins: the number of y variables (NY = 6); the number of x variables (NX = 1); the number of η (latent dependent) variables (NE = 10); the number of ξ (latent independent) variables (NK = 1); and the initial specification of the LISREL parameter matrices. By initial specification, we mean the overall structure of the matrix (i.e., FU = full or asymmetric (typically a regression matrix); SY = symmetric (typically a variance–covariance matrix); ZE = zero or no free parameters; DI = diagonal or only diagonal parameters are free), and whether all of the matrix parameters are initially fixed (FI) or free (FR). Of course, unless the matrix is designated as ZE, some modification to the initial entirely fixed or entirely free status must be made to completely specify the model. Proceeding through the MO line, these modifications are specified. For example, although the $\mathbf{\Lambda}_y$ (LY) matrix is initially specified as fixed, the parameters λ_{11}, λ_{21}, λ_{31}, λ_{12}, λ_{22}, λ_{23} are subsequently freed by using a FREE statement (i.e., FR LY(1,1) LY(2,1) LY(3,1) LY(1,2) LY(2,2) LY(3,2)). Further, these newly freed parameters are constrained to be equal to other $\mathbf{\Lambda}_y$ parameters by using an EQUAL (EQ) statement. For example, the genetic loading LY(1,1) for twin 1 is equated to the same genetic loading LY(4,3) for twin 2 during estimation. Other parameters may have their status changed by assigning them a fixed value during estimation other than zero. For example, by using the value (VA) statement, the parameter LY(1,5) is fixed to the value of 1 – it is thus not estimated. As another example, the genetic covariance of 1 between MZ twins is specified by fixing PS(1,3) to 1. Optional but useful are the LK and LE commands, which assign, respectively, labels to the latent independent and latent dependent variables. Lastly, each group must end with an output (OU) line. LISREL provides a number of optional commands for the output line. Here, "NS" refers to no starting values. Unless otherwise instructed, LISREL will provide its own starting values for parameter estimation. If NS is specified, the user must provide starting values, given on the ST line directly above the output line. A starting value of "5" is assigned to *all* free parameters. A number of potential reasons exist as to why a user might prefer to provide starting values rather than have the program do so. Probably the most common reason is that, despite the sophistication of LISREL's method for determining starting values, occasionally a model is so complex or the data are so inconsistent with the model that the model never converges (final, best estimates of the parameters are never achieved) with the starting values LISREL provides. In this case, it is up to the user to use his or her knowledge of the domain being modeled, or simply to engage in trial and error, to determine adequate starting values. For behavioral genetic models, it is usually necessary for the user to provide starting values for a different reason. There are several peculiarities in behavioral genetic models that LISREL finds troublesome. For one, the $\mathbf{\Psi}$ matrices for MZ and DZ twins are ill-conditioned owing to the presence of one or more 1s off the main diagonal. Second, behavioral genetic models are unusual due to the specification of more latent than observed variables. Recall that in the present model (Figure 1.4), there are 11 latent variables, while there are only 7 observed variables. Although LISREL will still estimate these models, the starting values it provides are usually not good enough to achieve convergence, or, if convergence occurs, inaccurate warning messages concerning identification problems appear along with the absence of such important statistics as modification indices. ADD=OFF on the output line requests that LISREL not check the statistical admissibility of the model. Behavioral genetic models will fail this admissibility check, and the program will cease running, owing to the ill-conditioned status of the $\mathbf{\Psi}$ matrix described above. As a default, LISREL will iterate three times the number of free parameters in order to find the best parameter estimates; IT = 500 overrides that default by requesting that LISREL iterate 500 or fewer times to find the best parameter estimates. Lastly, MI and RS, respectively, request that LISREL provide modification indices and residual statistics to assist in the assessment of model fit.

Following the output line of the first group is the title of the second group, "DZ TWINS". Largely the same syntax format is used for the second group, with an important exception. On the model line, note that eight of the nine parameters matrices are denoted by "IN", with PS = SY,FI. "IN" refers to invariant parameter estimation, informing LISREL that these parameters should have the same value in MZ and DZ twins. However, the specification of the $\mathbf{\Psi}$ matrix differs between MZ and DZ twins, and thus must be separately specified in both groups. (Recall that the genetic covariance, fixed in the $\mathbf{\Psi}$ matrix, is 1 for MZs and 0.5 for DZs.)

After the LISREL program has been run, using maximum likelihood estimation (the default in LISREL) the overall fit of the model is first assessed. LISREL provides more than two dozen fit indices for the whole model. We will report only two, the χ^2 goodness-of-fit statistic and the goodness-of-fit (GFI) index. For the model shown in Figure 1.4, $\chi^2 = 108.89$, df $= 43$, $p = 0.000$, and GFI $= 0.90$. Both fit indices essentially agree that the model as a whole does not fit the data: χ^2 is rejected (the discrepancy between the observed and reproduced covariance matrices is significantly large), and the GFI is below 0.95. Incidentally, both fit measures are reported for each group separately as well as for both groups combined (the numbers given above). For MZs, $\chi^2 = 43.26$ and GFI $= 0.94$; for DZs, $\chi^2 = 86.63$ and the GFI $= 0.90$. Apparently, the model fit the MZ data slightly better than the DZ data.

A specification search is now conducted to evaluate how the model may best be improved. Examining the modification indices, it is noted that those parameters that represent the covariance between the errors of the same observed variable for twins 1 and 2 would significantly add to model fit if freed. For example, the MI for the error covariance between the physical proximity measure for twins 1 and 2 is 11.04. Freeing these correlated errors is certainly substantively meaningful: one would predict that some of those errors which influence one twin's score on a variable would influence the other twin's score on the same variable.

Thus three error covariances (one for each of the y variables for a twin) were added to the model, as shown in Figure 1.5. We appear to be violating our own advice given earlier that one should re-evaluate a model after freeing only one parameter at a time; however, theoretically, it makes sense to free all three possible error covariances even if only one turns out to significantly improve the fit of the model. It must be remembered, though, that we lose three degrees of freedom by freeing these three parameters. The program for the model shown in Figure 1.5 appears in Table 1.4. Note the modification that has been made to the specification of the TE matrix. Now, TE = SY,FI instead of TE = DI. Therefore we must explicitly inform LISREL which of the TE parameters are to be estimated (freed). The model as a whole now fits the data well: $\chi^2 = 52.39$, df$= 40$, $p = 0.091$; for MZs, $\chi^2 = 38.93$ and GFI $= 0.97$; for DZs, $\chi^2 = 31.98$ and GFI $= 0.95$. The parameter estimates are provided in Table 1.5.

The first three parameter estimates in Table 1.5 refer to the genetic loadings of physical similarities, behavioral similarities, and physical proximity and the genetic latent variable. All three genetic loadings are significant. If one wishes to determine how much of the variance in these three measures each is accounted for by genetic effects, the division of the sum of squared genetic loading by the total variance of the observed variable is computed: for physical similarities, $(3.16)^2/16.51 = 60\%$; for behavioral similarities, $(1.41)^2/28.12 = 7\%$; for physical proximity, $(-0.43)^2/3.35 = 6\%$. Each of these percentages is referred to as a *heritability* – the heritability of trait is the proportion of its variance due to genetic effects. Apparently, physical similarity has the largest heritability. Conversely, the *environmentality* of a trait is defined as the proportion of its variance due to environmental effects. Thus each of the variable loadings on the environmental latent variable (parameters 4 through 6) is squared and divided by the total variance of the variable: for physical similarities, $(-1.33)^2/16.51 = 11\%$; for behavioral similarities, $(1.81)^2/28.12 = 12\%$; for physical proximity, $(1.10)^2/3.35 = 36\%$. In this case, physical proximity has the largest environmentality, although all three environmentalities are significant.

According to the model shown in Figure 1.5, each observed variable's variance is due to genetic, environmental, age, and error effects. All

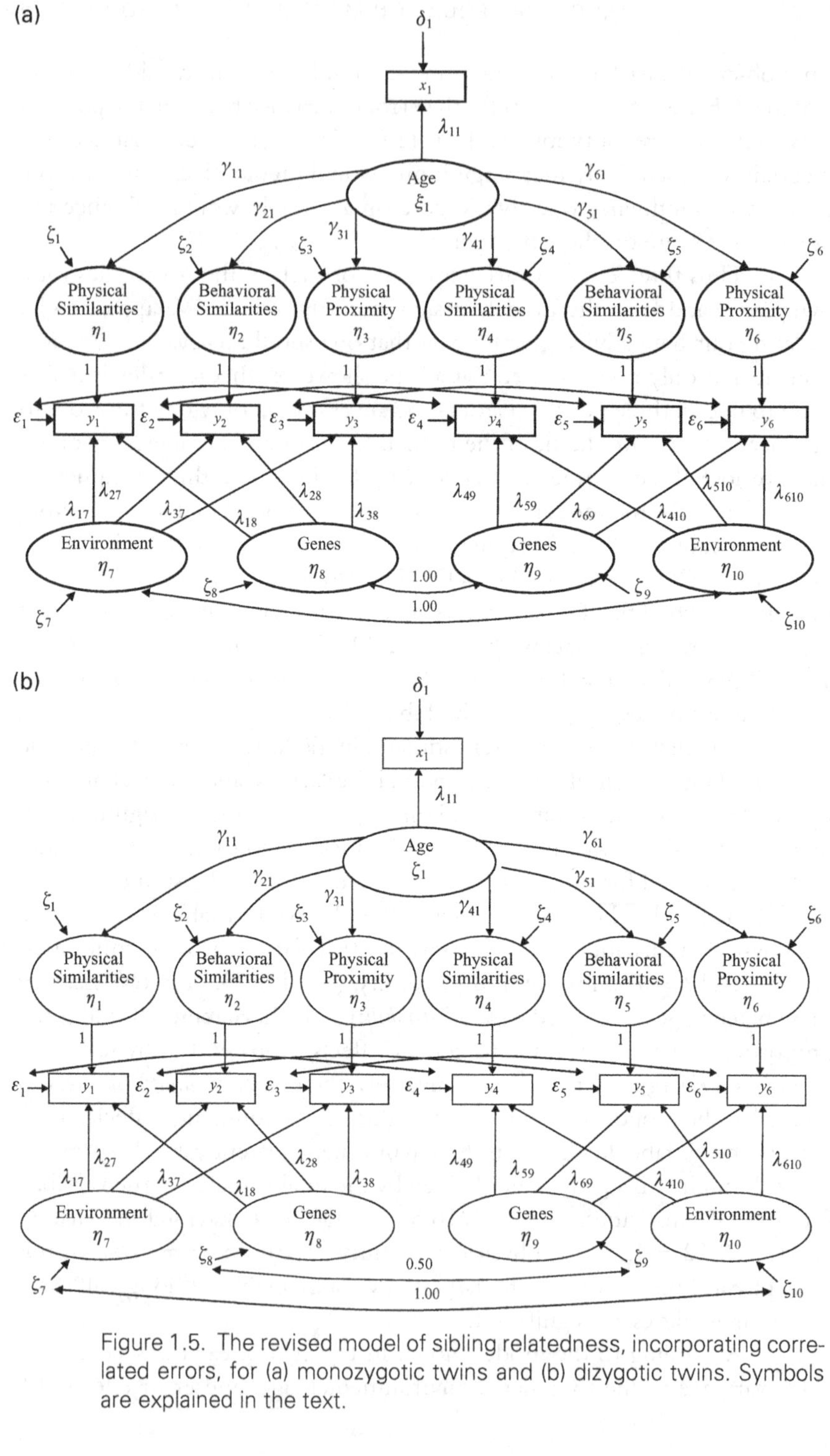

Figure 1.5. The revised model of sibling relatedness, incorporating correlated errors, for (a) monozygotic twins and (b) dizygotic twins. Symbols are explained in the text.

Table 1.4. *LISREL program for behavioral genetic model including the three error covariances*

```
LISREL PROGRAM FOR BEHAVIORAL GENETIC MODEL
INCLUDING THE THREE ERROR COVARIANCES
MZ TWINS
DA NG=2 NI=7 NO=200 MA=CM
LA
T1_PS T1_BS T1_PP T2_PS T2_BS T2_PP AGE
CM
16.51
4.26 28.12
-2.10 3.38 3.35
10.04 3.40 -2.50 16.51
3.40 10.35 1.74 4.36 28.12
-2.50 1.74 1.12 -2.10 3.38 3.35
2.44 3.18 1.10 2.44 3.18 1.10 4.00
MO NY=6 NX=1 NE=10 NK=1 LX=FU,FI LY=FU,FI TD=ZE TE=SY,FI c
  GA=FU,FI PH=SY,FR BE=ZE PS=SY,FI TH=ZE
LK
AGE
LE
T1_G T1_E T2_G T2_E T1_PS T1_BS T1_PP T2_PS T2_BS T2_PP
VA 1.0 LX (1,1)
FR LY(1,1) LY(2,1) LY(3,1) LY(1,2) LY(2,2) LY(3,2)
EQ LY(1,1) LY(4,3)
EQ LY(2,1) LY(5,3)
EQ LY(3,1) LY(6,3)
EQ LY(1,2) LY(4,4)
EQ LY(2,2) LY(5,4)
EQ LY(3,2) LY(6,4)
VA 1.00 LY(1,5) LY(2,6) LY(3,7) LY(4,8) LY(5,9) LY(6,10)
FR TE(1,1) TE(2,2) TE(3,3)
EQ TE(1,1) TE(4,4)
EQ TE(2,2) TE(5,5)
EQ TE(3,3) TE(6,6)
FR TE(1,4) TE(2,5) TE(2,6)
FR GA(5,1) GA(6,1) GA(7,1)
EQ GA(5,1) GA(8,1)
EQ GA(6,1) GA(9,1)
EQ GA(7,1) GA(10,1)
```

Table 1.4. (*cont.*)

```
VA 1.00 PS(1,1) PS(2,2) PS(3,3) PS(4,4)
VA 1.00 PS(1,3)
VA 1.00 PS(2,4)
FR PH(1,1)
ST 5 ALL
OU NS AD=OFF IT=500 MI RS
DZ TWINS
DA NI=7 NO=200 MA=CM
LA
T1_PS T1_BS T1_PP T2_PS T2_BS T2_PP AGE
CM
16.51
4.26 28.12
-2.10 3.38 3.35
5.02 1.70 -1.25 16.51
1.70 5.18 0.87 4.36 28.12
-1.25 0.87 0.56 -2.10 3.38 3.35
2.44 3.18 1.10 2.44 3.18 1.10 4.00
MO LX=IN LY=IN TD=IN TE=IN GA=IN PH=IN PS=SY,FI BE=IN TH=IN
LK
AGE
LE
T1_G T1_E T2_G T2_E T1_PS T1_BS T1_PP T2_PS T2_BS T2_PP
VA 1.00 PS(1,1) PS(2,2) PS(3,3) PS(4,4)
VA 0.5 PS(3,1)
VA 1.00 PS(4,2)
ST 5 ALL
OU NS AD=OFF MI RS
```

c, indicates continuation of line in LISREL notation.

four influences combined should account for 100% of the variance of a variable. As an example, let us partition the variance of physical similarities into these four influences. The expression for the variance of physical similarities is:

$$\begin{aligned}\text{Var(PS)} &= \lambda_{11}^2 + \lambda_{12}^2 + \left(\gamma_{11}^2 \times \text{Var}(\xi_{11})\right) + \text{Var}(\varepsilon_{11})\\ &= (3.16)^2 + (-1.33)^2 + (0.61)^2(4.00) + (3.25)\\ &= 9.99 + 1.78 + 1.49 + 3.25\\ &= 16.51,\end{aligned}$$

Table 1.5. *Maximum likelihood parameter estimates from behavioral genetic model with correlated errors*

Number	Parameter	Estimate	Standard error	t-value
1	λ_{11} & λ_{43}	3.16	0.56	8.78
2	λ_{21} & λ_{53}	1.41	0.60	2.34
3	λ_{31} & λ_{63}	−0.43	0.20	−2.15
4	λ_{12} & λ_{44}	−1.33	0.56	−2.40
5	λ_{22} & λ_{54}	1.81	0.91	2.00
6	λ_{32} & λ_{64}	1.10	0.39	2.85
7	γ_{11} & γ_{41}	0.61	0.08	7.55
8	γ_{21} & γ_{51}	0.79	0.10	8.09
9	γ_{31} & γ_{61}	0.28	0.03	8.21
10	$\mathrm{Var}(\xi_{11})$	4.00	0.28	14.11
11	$\mathrm{Var}(\varepsilon_{11})$ & $\mathrm{Var}(\varepsilon_{44})$	3.25	2.60	1.25
12	$\mathrm{Var}(\varepsilon_{22})$ & $\mathrm{Var}(\varepsilon_{55})$	20.34	4.72	4.31
13	$\mathrm{Var}(\varepsilon_{33})$ & $\mathrm{Var}(\varepsilon_{66})$	1.65	0.81	2.03
14	$\mathrm{Cov}(\varepsilon_1, \varepsilon_4)$	−3.15	2.40	−1.31
15	$\mathrm{Cov}(\varepsilon_2, \varepsilon_5)$	0.42	4.54	0.09
16	$\mathrm{Cov}(\varepsilon_3, \varepsilon_6)$	−0.82	0.82	1.00

where λ_{11}^2 is the genetic variance; λ_{12}^2 is the environmental variance; $\gamma_{11}^2 \times \mathrm{Var}(\xi_{11})$ is the age variance; and $\mathrm{Var}(\varepsilon_{11})$ is the error variance. According to the data in Table 1.5, all three error variances and all three age variances are significant, underlining the importance of including these two covariates in the model for a valid estimation of genetic and environmental influences.

Interestingly, none of the three error covariances is significant, although all three in combination produced a significant increment in the fit of the model. One can determine exactly the incremental significance of including these three parameters by conducting a *chi-square difference test* between the model with and without the three parameters. For the model without the three parameters (referred to as the "reduced model"), $\chi^2 = 108.89$, df $= 43$; for the model with the three parameters (referred to as the "full model"), $\chi^2 = 52.36$, df $= 40$. The difference in χ^2 values is itself distributed as a χ^2 with degrees of freedom equal to the difference in degrees of freedom of the two models. Therefore: $108.89 - 52.36 = 56.53$, with df $= 43 - 40 = 3$. A χ^2 value of 56.53, with df $= 3$, is highly significant, $p < 0.0000$. Thus, freeing the three error covariances significantly reduces the χ^2 statistic, thus significantly improving the fit of the model.

References

Bollen, K. A. (1989). *Structural Equation Models with Latent Variables.* New York: Wiley.

Bollen, K. A. & Long, J. S. (eds.) (1993). *Testing Structural Equation Models.* Newbury Park, CA: Sage.

Byrne, B. M. (1989). *A Primer of LISREL: Basic Applications and Programming for Confirmatory Factor Analysis.* New York: Springer-Verlag.

Byrne, B. M. (1994). *Structural Equation Modeling with EQS and EQS/Windows: Basic Concepts, Applications, and Programming.* Thousand Oaks, CA: Sage.

Byrne, B. M. (2000). *Structural Equation Modeling with AMOS: Basic Concepts, Applications, and Programming.* Mahwah, NJ: Lawrence Erlbaum Associates.

Eaves, L. J., Eysenck, H. J. & Martin, N. G. (1989). *Genes, Culture and Personality: An Empirical Approach.* San Diego: Academic Press.

Green, S. B., Thompson, M. S. & Babyak, M. A. (1998). A Monte Carlo investigation of methods for controlling Type I errors with specification searches in structural equation modeling. *Multivariate Behavioral Research*, **33**, 365–383.

Green, S. B., Thompson, M. S. & Poirier, J. (1999). Exploratory analyses to improve model fit: errors due to misspecification and a strategy to reduce their occurrence. *Structural Equation Modeling*, **6**, 113–126.

Jöreskog, K. & Sörbom, D. (1993). *LISREL 8: User's Reference Guide.* Chicago: Scientific Software.

Hayduk, L. A. (1987). *Structural Equation Modeling with LISREL: Essentials and Advances.* Baltimore, MD: Johns Hopkins University Press.

Hayduk, L. A. (1996). *LISREL Issues, Debates, and Strategies.* Baltimore, MD: Johns Hopkins University Press.

Hoyle, R. H. (ed.) (1995). *Structural Equation Modeling: Concepts, Issues, and Applications.* Thousand Oaks, CA: Sage.

Hu, L. & Bentler, P. M. (1999). Cutoff criteria for fit indexes in covariance structure analysis: conventional criteria versus new alternatives. *Structural Equation Modeling*, **6**, 1–55.

Jöreskog, K. G. & Sörbom, D. (1996). *LISREL 8: User's Reference Guide.* Chicago: Scientific Software International.

Leamer, E. E. (1978). *Specification Searches: Ad Hoc Inference with Nonexperimental Data.* New York: Wiley.

Marcoulides, G. M. (1989). Measuring computer anxiety: the Computer Anxiety Scale. *Educational and Psychological Measurement*, **49**, 733–739.

Marcoulides, G. M. & Drezner, Z. (2001). Specification searches in structural equation modeling with a genetic algorithm. In G. A. Marcoulides & R. E. Schumacker (eds.), *New Developments and Techniques in Structural Equation Modeling*, pp. 247–268. Mahwah, NJ: Lawrence Erlbaum Associates.

Marcoulides, G. A. & Hershberger, S. L. (1997). *Multivariate Statistical Methods: A First Course.* Mahwah, NJ: Lawrence Erlbaum Associates.

Marcoulides, G. A. & Schumacker, R. E. (eds.) (1996). *Advanced Structural Equation Modeling: Issues and Techniques.* Mahwah, NJ: Lawrence Erlbaum Associates.

Marsh, H. W., Balla, J. R. & Hau, K. T. (1996). An evaluation of incremental fit indices: A clarification of mathematical and empirical properties. In G. A. Marcoulides & R. E. Schumacker (eds.), *Advanced Structural Equation Modeling: Issues and Techniques*, pp. 315–354. Mahwah, NJ: Lawrence Erlbaum Associates.

Plomin, R., DeFries, J. C., McClearn, G. E. & Rutter, M. (1997). *Behavioral Genetics*, 3rd edition. New York: W. H. Freeman & Co.

Raykov, T. & Marcoulides, G. A. (2000). *A First Course in Structural Equation Modeling.* Mahwah, NJ: Lawrence Erlbaum Associates.

Schumacker, R. E. & Lomax, R. (1996). *A Beginner's Guide to Structural Equation Modeling.* Mahwah, NJ: Lawrence Erlbaum Associates.

Schumacker, R. E. & Marcoulides, G. M. (eds.) (1998). *Interaction and Nonlinear Effects in Structural Equation Modeling.* Mahwah, NJ: Lawrence Erlbaum Associates.

Segal, N., Weisfeld, G. E. & Weisfeld, C. C. (1997). *Uniting Psychology and Biology: Integrative Perspectives on Human Development.* Washington, DC: American Psychological Association.

Shipley, B. (2000). *Cause and Correlation in Biology: A User's Guide to Path Analysis, Structural Equations and Causal Inference.* Cambridge: Cambridge University Press.

2 Concepts of structural equation modeling in biological research

Bruce H. Pugesek

Abstract

This chapter reviews, in a nonmathematical way, some of the key concepts that set structural equation modeling (SEM) apart from conventional parametric multivariate methods. The chapter is sectioned by the somewhat overlapping topics of measurement error, latent variables, hypothesis testing, and complex models. In the section dealing with measurement error, I describe some of the sources of measurement error found in nonexperimental field data and how measurement error in concert with collinearity among independent variables can bias and distort regression path coefficients, factor loadings, and other outputs of conventional statistical methods. In the next section, latent variables, I introduce the latent variable with multiple indicators and demonstrate their efficacy for dealing with collinearity and reducing measurement error. As a result, higher precision of estimates of path coefficients may be obtained with SEM analysis over conventional methods. In the following section, hypothesis testing, I demonstrate how the use of latent variables introduces a new dimension to the analysis in the form of model fitting and hypothesis testing. Finally, I provide examples of how conventional methodologies can be extended with SEM for the analysis of complex systems.

Introduction

Given the complexity of structural equation modeling (SEM), researchers will naturally ask the question, "What benefit is to be obtained in applying these methods?". The answer to this question lies in a discourse of some of the key concepts of SEM. Some of these concepts relate to mathematical aspects of SEM model formulation that are not readily apparent through simple inspection of the equations. Other aspects relate to epistemological issues of applying SEM in a research investigation. I have subdivided this chapter into the major categories, measurement error, latent variables, hypothesis testing, and modeling complex systems. Each category tends to overlap somewhat with the others; however, each provides a major focal point for understanding the benefits of SEM and how it differs from other statistical methods. A fifth category, model flexibility, might have been added to the list to describe the large number and types of SEM models that can be employed. However, the numerous examples provided in this volume serve that purpose instead.

Measurement error

Many field biologists tend to think of measurement error primarily as a mistake in the measurement of a variable. Was the distance exactly measured? Were the number of survivors accurately counted? Was the response correctly entered in the data log? These types of measurement error, known as construct reliability (Bollen, 1989), are frequently given considerable advanced thought by field biologists in an effort to minimize the impact of error on research results. The solution is often rigorous quality control. Perhaps also, measurements are taken with higher than required tolerances so that measurement error can be minimized to insignificant quantities. Weight, for example, may be measured to the nearest thousandth of a gram when only the nearest tenth of a gram is required. Indeed, one of the assumptions of linear regression and many multivariate models is that the independent variable was measured without error (Neter *et al.*, 1985). Therefore, it is incumbent upon researchers to satisfy themselves that such is the case.

Construct reliability is not the only source of measurement error. Another, construct validity (Bollen, 1989), considers the match between the measurement-level or indicator variable and the construct it seeks to measure. If the indicator variable has a less than perfect correspondence to the construct it seeks to measure, it contains measurement error. For example, an investigator who seeks to determine the impact of salinity stress on coastal freshwater marsh vegetation may choose to take a daily measure of salinity as his measure of salinity stress. We know that salinity stress on plants is a complex phenomenon that involves both level and duration of exposure to salinity and that it can be influenced by tides, wind, and rainfall. Therefore, a daily measure of salinity may tell us something about salinity stress, but this single indicator is not a perfect measure of the construct "salinity stress".

Figure 2.1 illustrates the concepts of measurement level indicator variables, dependent variables, and measurement error in regression as well as introduces the graphical display of path models (see also Hershberger *et al.*, Chapter 1). A simple linear regression (Figure 2.1a) consists of an independent variable, I1, and a dependent variable, D1. The single-headed arrow

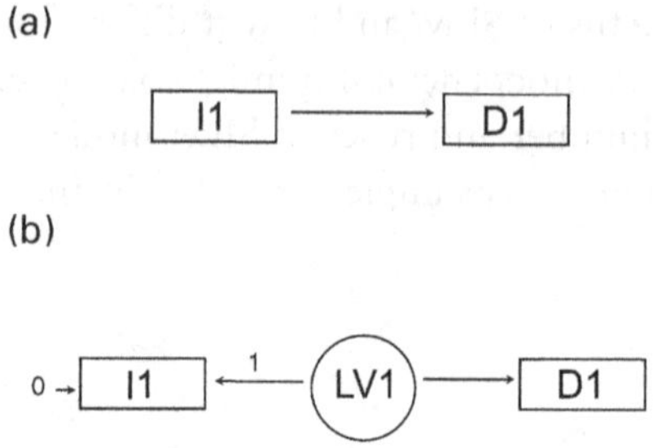

Figure 2.1. Path model diagrams. (a) A simple linear regression with independent variable (I1) and dependent variable (D1). (b) A path model depicting an unmeasured construct with a single indicator. The latent variable (LV1) acts as an independent variable that predicts the dependent variable.

between independent and dependent variables represents the regression. The arrow points towards the dependent variable. Regression coefficients are often posted on the line (see path models throughout this volume). Figure 2.1b depicts the identical model, but includes information about an unmeasured or latent construct that causes the observed values of the independent variable, as well as the assumption that the independent variable is measured without error. Variance in the independent variable, I1, is explained by an unmeasured construct, LV1, represented by a circle and measurement error. Measurement error is assumed to be zero, therefore, the path from the error term to I1 is set at zero. Since the latent variable and the indicator variable are identical, the path from LV1 to I1 is set to 1 to signify that 100% of the variance of the indicator is explained by the latent variable. When a simple regression is performed, the investigator assumes that the model in Figure 2.1b is true. Suppose a researcher theorized that the size of individuals of a species of waterfowl was related to clutch size. The researcher uses a single measure, weight, as the measure of body size, but recognizes that the independent variable is not a perfect measure of the conceptual variable, body size. This is so because the physical condition of birds may contribute to weight in addition to overall size. Thus variance in the variable weight is composed of a portion attributable to size, and a portion attributable to physical condition, where some birds are better muscled, have more body fat, etc. as compared with others. A simple linear regression of weight on clutch size would undoubtably contain measurement error in the form of reduced construct validity even if the measurement of body weight had proceeded with perfect construct reliability. Researchers have long recognized that shortcomings such as this occur with nonexperimental field data and have generally followed two approaches in an effort to surmount the problem.

One approach the researcher might choose is to utilize experimentation to control for variation due to body condition by using only birds that have remained in holding pens until body condition is constant among all experimental birds. Indeed, it has been recommended that classical controlled experiments complement nonexperimental field data studies in order to determine the true nature of relations among dependent and independent variables (James & McCulloch, 1990). One undesirable effect of controlled experimentation is that the experiment may interrupt the natural process that the researcher wishes to investigate. Results of the experiment are, therefore, difficult to interpret when taken out of their natural context. In the hypothetical experiment described here, it is likely that the natural sequence of behaviors leading up to egg-laying would be severely disrupted by an artificial holding period. Furthermore, experimentation in many instances can be impractical or unethical. Finally, controlled experiments that seek to control for all variation with the exception of that of the experimental treatment variable do not allow researchers to study the complex interactions among variables, an objective that is often at the heart of the research undertaking. In the hypothetical clutch size study, for example, body size might be related to clutch size only under conditions when food supply was limited prior to egg-laying.

A second approach, known as statistical control, involves multivariate data analysis. Here the researcher analyzes data on a number of variables in an attempt to quantify their interrelationships. The waterfowl researcher might use multiple regression to investigate the relationship of body size to clutch size. Also included in the model are the independent variables food availability and a measure of body condition. By including the additional independent variables, the researcher wishes to partition the variance explained in clutch size among several causal agents. At the same time, it is hoped that the construct validity problems of the independent variable weight will be solved by statistically controlling for factors such as body condition that may distort the relationship between body size and the dependent variable. Figure 2.2 depicts a path model for a multiple regression

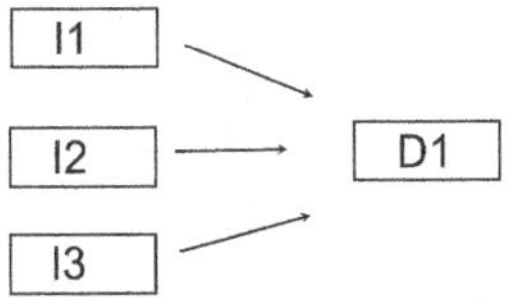

Figure 2.2. Path model for a multiple regression with three independent (I) and one dependent (D) variable.

analysis. A single dependent variable, D1, is regressed on three independent variables, I1 to I3.

Multivariate statistics are designed to maximize the amount of variation explained by the statistical model. In multiple regression, the analysis attempts to maximize the variance explained in the dependent variable. The analysis proceeds by regressing the first independent variable in the model on the dependent variable. Then each succeeding variable is regressed on the unexplained variance of the dependent variable remaining after the regressions of the proceeding variables. As a consequence, multiple regression performs well its function to maximize variance explained in the dependent variable. However, it does not perform well the task of partitioning the variance explained by each independent variable. If variables are measured with error, regression coefficients will be biased in an unpredictable fashion just as in simple linear regression. Furthermore, a new dimension is added to the problem through collinearity (covariance) among independent variables. When two independent variables covary, variance in the dependent variable that is explained by one independent variable may be transferred to a second independent variable. Thus the regression coefficient (β) of one independent variable can be deflated, while the β-value of the second independent variable is inflated. Neither regression coefficient is, of course, an accurate representation of the relations between independent and dependent variables. This feature of multiple regression analysis explains the well-known phenomenon in which coefficients derived from multiple regression change depending on their order of entry into the regression model or as a consequence of the inclusion or exclusion of independent variables in the model (Neter *et al.*, 1985).

Table 2.1 illustrates the degree to which multiple regression coefficients may undergo bias and distortion due to measurement error and collinearity. The calculated β-values are derived from multiple regression with three independent variables with measurement errors listed on the left-hand side of the table. Values generated for various levels of collinearity among independent variables noted at the top are presented in the body of the table. Each β-value should equal exactly 0.50. Results are drawn from the population rather than a sample, therefore results do not reflect random sampling variation. Such variation would tend to increase the bias and distortion observed at the population level. A detailed description of the mathematical model that formulated these results is available in Pugesek & Tomer (1995). Note from Table 2.1 that regression coefficients diverge from their true values in an unpredictable fashion, sometimes with inflated values and at other times with deflated values. In this example, regression

Table 2.1. *Value of regression coefficients β as a function of values of the simulation parameters. True values were 0.500 in all cases. Regression coefficients appear in blocks of three*

	Intercorrelations among true values				
Error variance	0	0.2	0.4	0.6	0.8
0	0.500	0.515	0.529	0.550	0.591
0	0.500	0.515	0.529	0.550	0.591
0.2	0.417	0.412	0.397	0.367	0.295
0	0.500	0.525	0.549	0.579	0.629
0	0.500	0.525	0.549	0.579	0.629
0.4	0.357	0.350	0.329	0.289	0.210
0	0.500	0.530	0.562	0.611	0.722
0.2	0.417	0.424	0.422	0.407	0.361
0.2	0.417	0.424	0.422	0.407	0.361
0	0.500	0.541	0.584	0.647	0.780
0.2	0.417	0.432	0.438	0.431	0.390
0.4	0.357	0.361	0.351	0.324	0.260
0	0.500	0.553	0.608	0.687	0.848
0.4	0.357	0.368	0.365	0.344	0.283
0.4	0.357	0.368	0.365	0.344	0.283
0.2	0.417	0.437	0.450	0.458	0.464
0.2	0.417	0.437	0.450	0.458	0.464
0.2	0.417	0.437	0.450	0.458	0.464
0.2	0.417	0.447	0.469	0.489	0.513
0.2	0.417	0.447	0.469	0.489	0.513
0.4	0.357	0.372	0.375	0.367	0.342
0.2	0.417	0.457	0.489	0.524	0.574
0.4	0.357	0.380	0.391	0.393	0.382
0.4	0.357	0.380	0.391	0.393	0.382
0.4	0.357	0.389	0.409	0.423	0.433
0.4	0.357	0.389	0.409	0.423	0.433
0.4	0.357	0.389	0.409	0.423	0.433

coefficients diverged from their true value by upwards of 70%. Also note that bias and distortion can occur even when the independent variable is perfectly measured but is correlated with variables that themselves contain measurement error.

In our example study of avian body size, let us assume that up to 25% of the variation in weight may be explained by body size. Also, weight may vary by 25% depending on physical condition. Under these circumstances, measurement error in the variable weight may be as high as 50% of the variance. It is not inconceivable, therefore, that levels of bias and distortion of the regression coefficients displayed in Table 2.1 commonly occur in nonexperimental field data.

This brief discourse outlines some of the foibles of measurement error, the difficulty it creates in our attempts to test hypotheses and quantify relationships in natural systems, and some methods employed to maximize the precision to our research results. I have restricted the discussion to simple and multiple regression; however, the principles discussed here apply equally to other parametric methods. Neither approach, controlled experimentation or conventional multivariate analysis techniques with statistical control, are completely satisfactory answers for the problem of dealing with measurement error. The controlled experiment is frequently not appropriate, especially under circumstances in which we want to know how variables behave in an uncontrolled natural setting. Conventional multivariate analysis of nonexperimental field data does not deal well with measurement error, and collinearity among indicator variables can exacerbate rather than statistically control for measurement error. The limitations of experimental and statistical methodologies discussed here may well slow progress of scientific investigation in many areas of ecology as well as contribute to controversies arising from conflicting results in field studies. Competition theory, in particular the relative importance of abiotic and biotic competitive forces, is but one example (Grace & Pugesek, 1997).

Latent variables

Latent variables, also known as factors, unmeasured constructs or latent constructs, were previously defined as conceptual variables that the researcher does not measure directly but attempts to estimate with measurement-level indicator variables (see also Hershberger *et al.*, Chapter 1; Tomer & Pugesek, Chapter 5). Conceptual variables, as depicted in Figure 2.1b, are often imperfectly represented by a single indicator. However, it is possible in SEM to construct latent variables from two or more measured variables by using confirmatory factor analysis. In this section, I will describe some of the advantages of using latent constructs with multiple indicators.

Many variables that we wish to measure in ecological research are complex constructs that cannot be adequately represented by a single

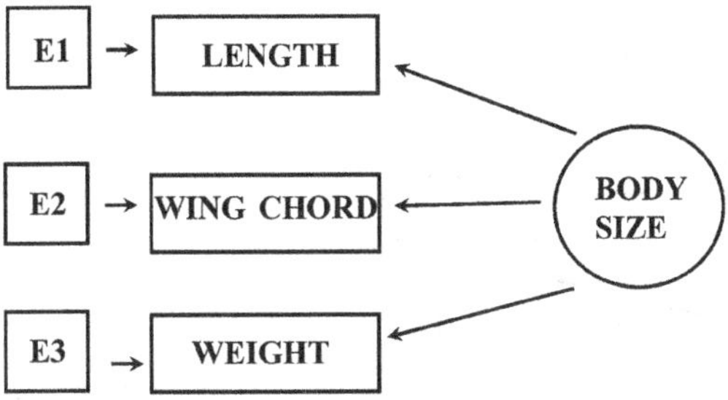

Figure 2.3. Path model of a latent variable, body size. The latent variable is measured with three indicators, body length, wing chord, and weight. The variance of each indicator is divided into two portions, that portion that contributes to the body size factor and the remaining error variance (E1 to E3).

measure. Salinity stress and avian body size, mentioned above, are but two examples. Others might include, habitat physical structure, soil fertility, and territory quality. In Figure 2.3, I recast the body size variable as a latent variable with three indicators, weight, wing chord length, and body length in a confirmatory factor analysis (CFA). The covariance among the three indicator variables formulates the latent variable, body size, which we can now regress on other variables such as clutch size.

Several advantages are obtained by measuring body size with three variables instead of a single indicator. Construct validity of the three-variable measure of body size is typically higher as compared with the single indicator construct. Large birds, in addition to having higher body weight, will also tend to have longer body lengths and longer wings. The shared covariance among these three variables is less likely to include variance in body weight due to other sources, such as body condition, that are unrelated to body size. Variance in each of the three variables that is unrelated to body size will compose the residual error variance. Variance in body weight due to body condition, for example, will become part of the residual error variance and is not used in a regression of the latent construct body size on clutch size. The resulting regression coefficient is a more accurate representation of the relationship. Note that collinearity among measures of body size is, in this instance, effectively utilized as opposed to being an impediment to the analysis. By carefully selecting variables for construction of latent variables, the researcher can utilize collinearity among selected variables to improve the precision of the measurement instrument.

In order to illustrate the increased precision of SEM estimates, I have constructed a simulation model based on the path model in Figure 2.4.

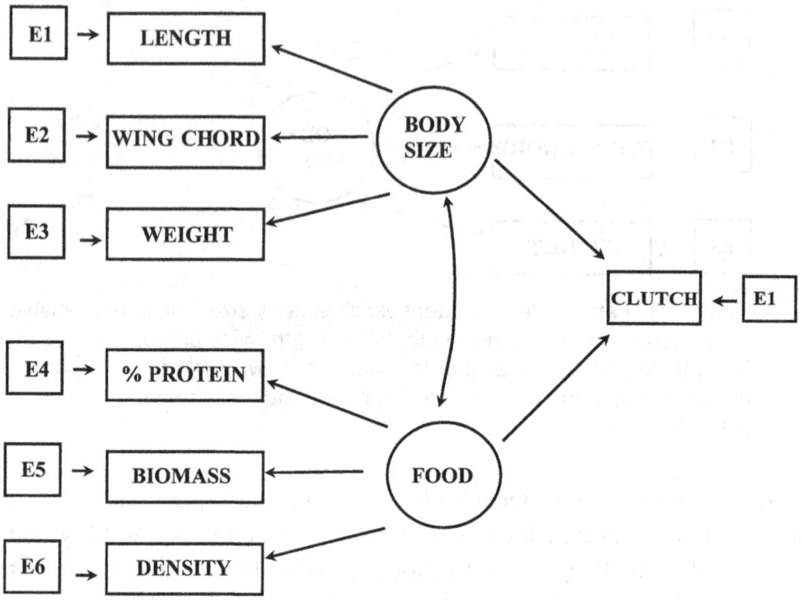

Figure 2.4. A path model for the simulation study in Appendix 2.1 and Table 2.2. Six exogenous indicator variables are used to construct two latent variables, body size and food availability. The latent variables predict one endogenous indicator, clutch size. A correlation is modeled between the latent variables.

I used EQS software (Bentler, 1992) to construct a true model that generated nine randomly generated data sets (see Appendix 2.1). Each of the nine true models were programmed with levels of error variance set at 0.10, 0.30, or 0.40 for pairs of indicators for each of the three exogenous latent variables. Variances of latent variables were set to 1.0. Covariances between latent variables were set at 0.40 in the first seven models. To examine the effects of unequal covariances between latent variables, covariances between latent variables were set at 0.70, 0.70, and 0.40 in model 8 and model 9. In all models, path coefficients between latent variables and the dependent variable equaled 0.80. The error variance of the dependent variable was set to be 0.80. The large number of cases in each output data set, $N = 10\,000$, ensured that the effects of sampling variation were minimized.

Data were analyzed with EQS using a structural model of the type in Figure 2.4, with the amount of error variance for the observed independent variables set to be free (i.e. error variances were estimated). Estimates of the path coefficients for the models were very close to 0.80 (Table 2.2).

Table 2.2. *Path coefficients and regression coefficients estimated in generated samples*

	Error variance[a]						Path coefficients			Regression coefficients					
										P1		P2		P3	
Model	V1	V2	V3	V4	V5	V6	P1	P2	P3	I1	I2	I3	I4	I5	I6
1	0.1	0.1	0.1	0.1	0.1	0.1	0.800	0.793	0.807	0.355	0.423	0.362	0.411	0.381	0.397
2	0.1	0.1	0.1	0.1	0.3	0.3	0.801	0.793	0.806	0.369	0.433	0.372	0.425	0.343	0.352
3	0.1	0.1	0.1	0.1	0.4	0.4	0.800	0.793	0.807	0.375	0.437	0.377	0.431	0.325	0.333
4	0.3	0.3	0.3	0.3	0.3	0.3	0.800	0.788	0.809	0.345	0.392	0.353	0.380	0.365	0.376
5	0.4	0.4	0.4	0.4	0.4	0.4	0.800	0.786	0.812	0.337	0.381	0.347	0.369	0.356	0.367
6	0.1	0.1	0.3	0.3	0.3	0.3	0.801	0.787	0.806	0.377	0.451	0.342	0.369	0.354	0.361
7	0.1	0.1	0.4	0.4	0.4	0.4	0.801	0.785	0.808	0.387	0.463	0.330	0.353	0.341	0.347
8[b]	0.1	0.1	0.1	0.1	0.4	0.4	0.789	0.802	0.814	0.444	0.505	0.332	0.386	0.296	0.306
9[b]	0.1	0.1	0.4	0.4	0.4	0.4	0.797	0.792	0.811	0.505	0.575	0.269	0.290	0.277	0.287

V, error variance; P, path coefficient; I, regression coefficient.

[a] Refers to first variable, second variable, etc.

[b] The correlations among factors were 0.7, 0.7, and 0.3. In all other cases the correlations were 0.4.

Data were also analyzed as a multiple regression model with six independent variables whose error variances were null. When a regression model that assumes null error variance is estimated, the path coefficients of 0.80 are split equally between the two indicator variables. Given the large sample size, we should expect regression coefficients for each of the independent variables to equal approximately 0.40. However, as can be observed in Table 2.2, considerable divergences of the multiple regression coefficients occurred. Error variances tended to produce a downwards bias on regression coefficients. Also note that substantial upwards and downwards distortion of coefficients occurred when error variances were unequal. Variables with larger error variances tended to be biased downwards, while variables with small error variances were biased upwards. More extreme levels of bias and distortion occurred in models with unequal covariances among latent variables. In models 8 and 9, which included levels of error variances of 0.10 and 0.40, regression coefficients departed from the true value of 0.40 by as much as 44%. In contrast, the greatest departure from the true value of an SEM path coefficient was 2%.

Jaccard & Wan (1995) obtained similar results in simulation studies of interaction effects. While multiple regression did a poor job of estimating interactions due to the effects of measurement error, SEM performed well.

The results presented here and in the previous section underscore the degree to which the results of just-identified (see Hershberger *et al.*, Chapter 1) regression models such as multiple regression lack precision. Measurement error, resulting from low construct reliability or low construct validity, can combine with collinearity among independent variables to skew regression coefficients in an unpredictable fashion. The use of latent variables in SEM allows researchers to produce regression coefficients that are more precise by effectively modeling measurement error and collinearity among indicator variables. In the next section, I will discuss another feature of SEM, hypothesis testing, that is facilitated through the use of latent variables.

Hypothesis testing

A SEM model using latent variables is typically over-identified, having fewer parameters to estimate than there are variances and covariances in the data matrix. Conventional parametric methods are just-identified models. The number of parameters estimated by the model equals the number of variances and covariances in the data matrix. This distinction leads to two important differences between SEM and standard parametric methods such as multiple regression, principal component analysis, multivariate analysis of variance (MANOVA), canonical correlation, and others.

Both just-identified models and under-identified models require a calculating algorithm that maximizes some aspect of the model. Multiple regression maximizes the amount of variance explained in the dependent variable. Principal components analysis maximizes the amount of intercorrelation among the set of variables in the model. The first factor is the first maximized set of intercorrelations, and each successive factor operates on the residuals remaining from the preceding factors. All models focus on maximizing covariance in one way or another, otherwise an infinite number of solutions is possible.

SEM optimizes the model–implied variance–covariance matrix $\boldsymbol{\Sigma}$. This is possible because the model is over-identified. Path coefficients are estimated through a process that solves for each model parameter in a set of simultaneous equations that includes all related variables and their direct and indirect effects. This is why, in addition to the requirement of over-identification, model identification procedures also require that every estimated parameter is identified in a SEM model (see Hershberger *et al.*, Chapter 1; Tomer & Pugesek, Chapter 5). The result is parameter estimates that were not derived through a process of cascading maximizations of variance explained on the most important variable, and likewise for the next most important variable operating on the residuals. It is fairly obvious that nature does not operate in this way and neither should our empirical estimates of nature's behavior. A Wrightian path analysis (Wright, 1921), for example, calculates indirect effects of variable A on variable C in a sequence A → B → C as a partial regression coefficient after taking into account the direct relations of A on B and B on C. An SEM model in which A, B, and C were latent variables would calculate direct and indirect effects based on parameter estimates that maximize the match between $\boldsymbol{\Sigma}$ and the sample variance–covariance matrix. Each direct and indirect effect would be derived from all the information available on those paths.

A second difference between conventional parametric methods and SEM stemming from the over-identified nature of the SEM model is the added dimension of model fit. Because the SEM model estimates the sample variance–covariance matrix with fewer pieces of information than are available, degrees of freedom are available to ask the question "Does our model explain the matrix just as well?". Other chapters deal in depth with the specifics of model fitting, significance of parameters, modification indices, confirmatory and exploratory models, and other issues that flow from the fact that SEM models are over-identified. Here I will comment on the fact that the SEM model asks more than the question, "Is my path coefficient significant?". The SEM model also asks the question, "Does my overall model fit the data?". In other words, do the data confirm or disconfirm my theory about the interrelationships among the variables in the data set? This unique feature of SEM allows investigators to address questions about the structure and function of complex systems. SEM provides the framework for investigation of systems of relationships and diagnostic tools to advance research in an iterative fashion where research data analysis proceeds with modification of theory, testing, and confirmation with additional data. All of the above occur within the context of scientific hypothesis testing with inferential statistics. This feature of SEM should not be underemphasized in our thinking. In the next section I discuss the advantages of SEM in modeling complex systems.

Modeling complex systems

The parameter matrices used to construct SEM models allow considerable flexibility in construction of models. The major constraint for SEM models lies in the requirement that models must be identified. This constraint not withstanding, a wide array of models are possible, opening the door for testing hypotheses about our understanding of complex natural systems. In this section, I will return to the example of avian body size to illustrate the types of method that have been used to investigate avian body size and what is possible using SEM.

Avian biologists have recognized the need to consider a general size factor when relating independent variables such as body weight to dependent variables such as clutch size (Lougheed *et al.*, 1991). The approach taken has been to first analyze data with principal components analysis (PCA), and extract residuals remaining from the first factor, which is considered to be a size factor. Next the dependent variable is regressed on the residuals of the independent variable in question (e.g., body weight). The idea here is to regress the dependent variable on variation in body weight that is unrelated to body size. A separate analysis regressing the dependent variable on factor scores of the body size factor can be conducted if there is an interest in relating the dependent variable to a general body size factor.

This approach is similar to an SEM approach, but a number of differences should be noted. First, factors obtained with PCA may differ from those obtained by confirmatory factor analysis (CFA) used in SEM. The body size factor obtained with PCA does not take into account measurement error (see previous sections). Also, PCA is a variance-maximizing technique which seeks to obtain a first factor that explains the maximum amount of variation among all the variables in the data set. The CFA body size factor would seek the maximum covariation among the subset of variables specified by the researcher for the body size factor. Therefore, results obtained with CFA are more likely to precisely define the body size factor with less bias and less distortion to factor loadings because measurement error of the independent variables can be made part of the model. A second difference is that with conventional PCA no assessment of model fit or hypothesis testing is possible.

The PCA method of extracting variation in body weight attributable to body size is a good start at approximating the relationship between body condition and body size and a dependent variable such as clutch size. SEM methods would improve upon this by providing more precision to estimates, a simultaneous analysis of the relationship between the size factor and body weight residuals and the dependent variable, and an assessment of model fit (Figure 2.5).

More complex models are also possible with SEM. The ultimate dependent variable of interest might be a latent variable as opposed to a single indicator such as clutch size. The latent variable, breeding success, in Figure 2.6 is created from clutch size, laying date, and egg volume. The investigator might believe here that large clutch size occurs concomitantly with early laying and high egg volume. A latent variable constructed from their covariance might provide a better indicator of overall breeding success than would clutch size alone. An additional latent variable, food availability, is added to the model to examine additionally the relationship between food prior to the egg-laying period and breeding success. The paths with

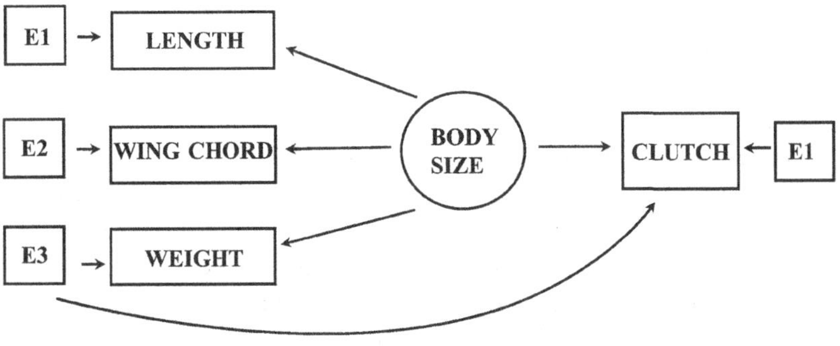

Figure 2.5. A path model depicting an SEM approach to modeling the relationship between body weight and clutch size. SEM simultaneously relates a body size factor as well as the residual variance of body weight to the dependent variable clutch size.

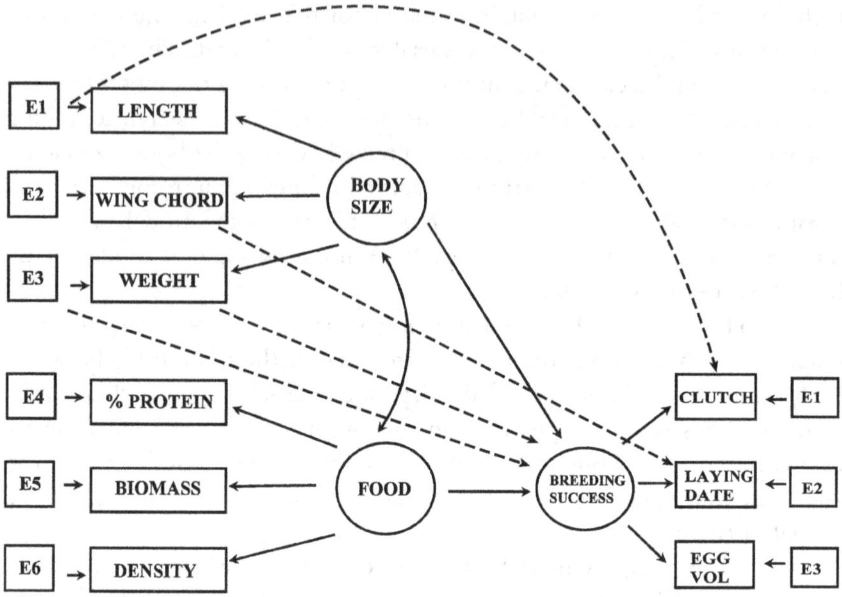

Figure 2.6. A path model depicting a more complex analysis of biological factors influencing reproductive performance. The single dependent variable clutch size is replaced by a latent variable that may have more capability in predicting ultimate reproductive success. An additional latent variable measuring food availability is included in the model. Paths with dashed lines depict some of the additional relationships that can be explored among indicators and residual error terms within latent variables.

dashed lines highlight some relationships that can be contemplated within a latent construct such as body size. Not all paths are possible in a single model owing to model identification considerations. However, each path could be incorporated singly into a model such as the one presented in Figure 2.6. Paths from any indicator or its error term can be modeled to either an endogenous latent variable or one of its indicators.

Returning to the original problem of modeling the relationship between weight, body size, and clutch size, it might be more advantageous to consider creating a latent variable, body condition, in order to model its relationship with clutch size. In Figure 2.7, a latent variable body condition is created by modeling body weight, with two additional indicators. The first, conductance, is developed from data on electrical conductance differences between body fat and muscle/organ tissue. The second variable, keel, is a field assessment of curvature of the keel ranked by level of convexity and concavity. Note that body weight loads on both the body size and body

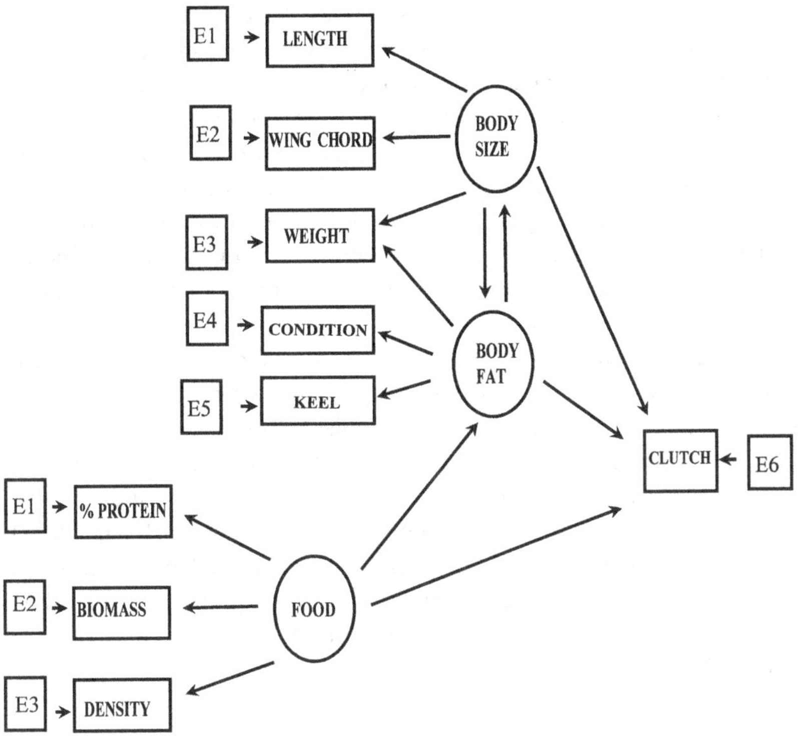

Figure 2.7. A hypothetical model for researching the relationship of food availability, body size, and body condition with clutch size. Included in the model are a biological feedback loop, and direct and indirect effects of latent variables.

condition factor. Two-way biological feedback is modeled between body size and body condition. It is unlikely that body condition would effect body size but is included in the model to demonstrate that it is possible to model biological feedback. Food availability, the single exogenous variable in the model, is modeled with indirect and direct effects on clutch size. Food availability would probably affect body condition, which in turn affects clutch size. A direct effect on clutch size independent of the relationship through body condition may also be included in the model.

Conclusions

At first glance it would appear that the collection of data on additional variables to construct latent variables would impact negatively on the potential for a successful research project. Additional data collection would inevitably increase research costs. Researchers would also have to consider, from a logistical standpoint, the tradeoffs between collecting more data on fewer subjects versus fewer data on a larger sample. Results provided in Tables 2.1 and 2.2 demonstrate that the problems with measurement error and collinearity observed in conventional multivariate methods such as multiple regression persist even when coefficients are calculated with very large samples or at the population level. Therefore, utilizing conventional methods with a larger sample size is probably not a good tradeoff for the additional expense, effort or reduction in sample size required to collect data on additional variables for a SEM study.

In addition to the increased precision of estimates, SEM affords the opportunity to model complex constructs and their interrelationships. These complex variables are often the variables of interest in an ecological study. SEM allows investigators to confirm or disconfirm complex models. Thus researchers can subject their theories about the structure and function of complex ecological systems to rigorous hypothesis testing. Finally, SEM provides considerable diagnostic capability so that model parameters can be added to, and deleted from, models as warranted by the data. The SEM approach is, therefore, an iterative procedure that assists researchers in building and refining theory about complex systems.

Appendix 2.1. The EQS program (Bentler, 1992) used to generate a random sample from a population defined by a model including three latent variables with two indicators each

```
/TITLE
 Model with 3 latent variables and fitness for study of
   meas. errors
/SPECIFICATION
 CAS=10000; VAR=7; ME=ML; ANALYSIS=COV;
/EQUATIONS
 V1=F1+E1;
 V2=1*F1+E2;
 V3=F2+E3;
 V4=1*F2+E4;
 V5=F3+E5;
 V6=1*F3+E6;
 V7=F4;
 F4=.8*F1+.8*F2 +.8*F3 +D4;
```

```
/VARIANCES
 F1=1*; F2=1*; F3=1*;
 E1=.1*; E2=.1*; E3=.1*; E4=.1*; E5=.1*; E6=.1*; D4=.80*;
/COVARIANCES
 F1 to F3=.4*;
/SIMULATION
 SEED=1443289647;
 REPLICATION=1;
 POPULATION=MODEL;
 DATA='EQS';
 SAVE=SEPARATE;
/OUTPUT
 LISTING; PARAMETER ESTIMATES; STANDARD ERRORS;
/END
```

References

Bentler, P. M. (1992). *EQS: A Structural Equation Program Manual.* Los Angeles: BMDP Statistical Software Inc.

Bollen, K. A. (1989). *Structural Equations with Latent Variables.* New York: Wiley.

Grace, J. B. & Pugesek, B. H. (1997). A structural equation model of plant species richness and its application to a coastal wetland. *American Naturalist,* **149**, 436–460.

Jaccard, J. & Wan, C. K. (1995). Measurement error in the analysis of interaction effects between continuous predictors using multiple regression: multiple indicator and structural equation approaches. *Psychological Bulletin,* **117**, 348–357.

James, F. C. & McCulloch, C. E. (1990). Multivariate analysis in ecology and systematics: panacea or Pandora's box? *Annual Review of Ecology and Systematics,* **21**, 129–166.

Lougheed, S. C., Arnold, T. W. & Bailey, R. C. (1991). Measurement error of external and skeletal variables in birds and its effect on principal components. *Auk,* **108**, 432–436.

Neter, J., Wasserman, W. & Kutner, M. H. (1985). *Applied Linear Statistical Models: Regression, Analysis of Variance, and Experimental Designs.* Homewood, IL: Richard D. Irwin.

Pugesek, B. H. & Tomer, A. (1995). Determination of selection gradients using multiple regression versus structural equation models (SEM). *Biometrical Journal,* **37**, 449–462.

Wright, S. (1921). Correlation and causation. *Journal of Agricultural Research,* **10**, 557–585.

3 Modeling a complex conceptual theory of population change in the Shiras moose: history and recasting as a structural equation model

Bruce H. Pugesek

Abstract

This chapter presents an example of how structural equation modeling (SEM) could be used to test a classic theoretical model of population dynamics of the Shiras moose (*Alces alces*). A longitudinal model is developed in which population density is measured in two waves. The change in population density between the two periods of measure is modeled in relation to a complex set of interrelationships among environmental and population level variables. Included in the model are examples of composite variables and nonzero fixed parameters. Analysis of a simulated data set demonstrates the procedures of a typical SEM study that begins first with a measurement model and proceeds with a series of exploratory and confirmatory analyses. The use and pitfalls of fit statistics, *t*-values, modification indices, and Q–Q plots as diagnostic tools are demonstrated. Two types of estimate, covariance estimates and standardized estimates, are contrasted. Examples of the calculation of total effects from direct and indirect effects are presented. Results demonstrate a significant potential for using SEM to develop expert systems and ecological models.

Introduction

A treatise on the Jackson Hole Shiras moose (Houston, 1968) provides us with an opportunity to demonstrate the capability of structural equation modeling (SEM) to develop and test theory on the structure and function of complex ecological processes. Houston (1968) collected data over a 4-year period on various biotic and abiotic variables that he believed impacted on demographic variables that determine moose population size: natality, mortality, and dispersal. He summarized his findings in a diagram that he labeled "Probable mechanisms regulating moose population size in the Jackson Hole area." The diagram, recreated here in Figure 3.1, represented his biological interpretation of the data. It was correctly labeled "probable mechanisms" because there were no methods available at the time with which to test his interpretation regarding the complex interrelationships that he envisioned, and

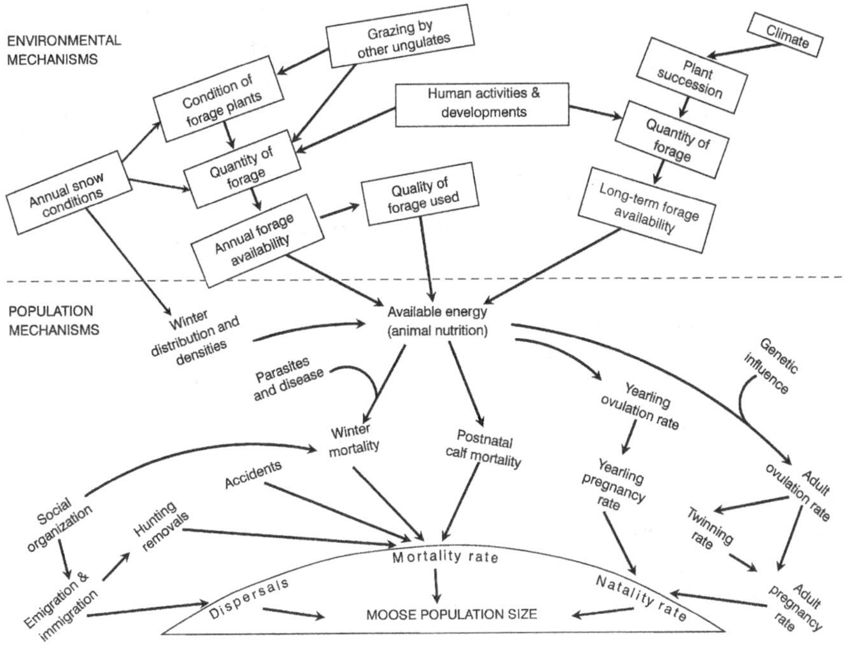

Figure 3.1. Diagrammatic representation of the conceptual model proposed by Houston (1968).

to assign values to the model paths. In this chapter I will demonstrate how these tasks may be accomplished with SEM methods, and, in doing so, introduce data simulation methods with EQS software (Bentler & Wu, 1995), some concepts of longitudinal models, composite variables, and nonzero fixed parameters. I will also discuss exploratory and confirmatory analysis, model fitting, covariance and standardized solutions, and calculation of total effects from direct and indirect effects.

Houston viewed the determinants of moose population size as a set of environmental mechanisms, consisting of major abiotic and biotic variables, that interacted to influence the food energy available to moose (Figure 3.1). The effects of variables on each other occurred as a tiered structure where abiotic factors such as snow conditions, climate, competition from other ungulates, and human development affected the quality and the quantity of forage. The energy available from forage foods was in turn the predominant feature that controlled moose population size. Low energy availability, for example, would increase overwintering mortality and decrease birth rate. Social organization affected emigrations and immigrations, which Houston collectively called dispersals. Hunting reduced winter mortality by reducing competition for forage but at the same time contributed to overall mortality. Genetic factors were also thought to play a role in selecting for adults that were tolerant of low food availability but had lower fecundity.

The Shiras moose population model

Data

Data were generated by simulation using EQS software (Bentler & Wu, 1995). Variables used in the simulation (Tables 3.1 and 3.2; Figure 3.2) and their interrelationships were created as they were conceptualized by Houston (1968), with one exception as discussed below. In most cases, Houston's variables were modeled as latent variables, for which I devised one or more indicators. I describe below each latent variable and its indicators. Latent variables, labeled F#, and indicators labeled V#, correspond to variables in the EQS (Bentler, 1995) simulation program presented in Appendix 3.1, and to the variable labels in path diagrams (Figures 3.2, 3.3, and 3.5 to 3.9). All variables are continuous and normally distributed. Moose population density is measured as the number of animals in a specified area (sampling units) that have been randomly selected from the available habitat. All other variables are sampled from within each sampling unit.

- *Snow* (F1) – Measures Houston's annual snow conditions.
  - SnowMelt (V1) – The date that snow cover dissipates.
  - MaxDepth (V2) – The maximum depth of winter snow cover.
  - SnowFall (V3) – The average depth of winter snowfalls.
- *Graze* (F2) – Measures the abundance of competing ungulates.
  - Scats (V4) – The frequency of scats of other ungulate species.
  - Tracks (V5) – The frequency of tracks of other ungulate species.
- *Human* (F3) – The impact of human disturbance caused by habitat flooding.
  - Floods (V6) – The depth of maximum flooding resulting from water control and management activities.
- *Pop98* (F4) – Moose population density in 1998.
  - Census98 (V7) – Moose population counts within each sampling unit using distance sampling methods (Buckland *et al.*, 1993).
- *Social* (F5) – A measure of social organization indicative of social stress that would ultimately result in dispersal of moose into or out

Table 3.1. *Indicator and latent variables*

Indicator	Indicator	Latent variable
V1	SnowMelt	Snow (F1)
V2	MaxDepth	Snow (F1)
V3	SnowFall	Snow (F1)
V4	Scats	Graze (F2)
V5	Tracks	Graze (F2)
V6	Floods	Human (F3)
V7	Census98	Pop98 (F4)
V8	Displays	Social (F5)
V9	Attacks	Social (F5)
V10	TagI	Hunting (F6)
V11	TagR	Hunting (F6)
V12	Assay	Genetic (F7)
V13	%Used	Cndition (F8)
V14	%Hedged	Cndition (F8)
V15	Biomass	QuantFor (F9)
V16	LAIndex	QuantFor (F9)
V17	LeadLGN	QuantFor (F9)
V18	Nitrogen	QualFor (F10)
V19	Protein	QualFor (F10)
V20	Fiber	QualFor (F10)
V21	HSI	Longterm (F11)
V22	Parasite	Disease (F13)
V23	IntHost	Disease (F13)
V24	AltHost	Disease (F13)
V25	YoungO	Ovulate (F14)
V26	AdultO	Ovulate (F14)
V27	TagMove	Disperse (F15)
V28	RadioTag	Disperse (F15)
V29	Accident	Mortal (F16)
V30	WintMort	Mortal (F16)
V31	%CafMort	Mortal (F16)
V32	APreg	Natality (F17)
V33	Twins	Natality (F17)
V34	YPreg	Natality (F17)
V35	Census99	Pop99 (F19)

Table 3.2. *Composite variables*

Latent variable	Variable name	Composite variable
F9	QuantFor	Energy (F12)
F10	QualFor	Energy (F12)
F11	Longterm	Energy (F12)
F6	Hunting	PopChang (F18)
F15	Disperse	PopChang (F18)
F16	Mortal	PopChang (F18)
F17	Natality	PopChang (F18)

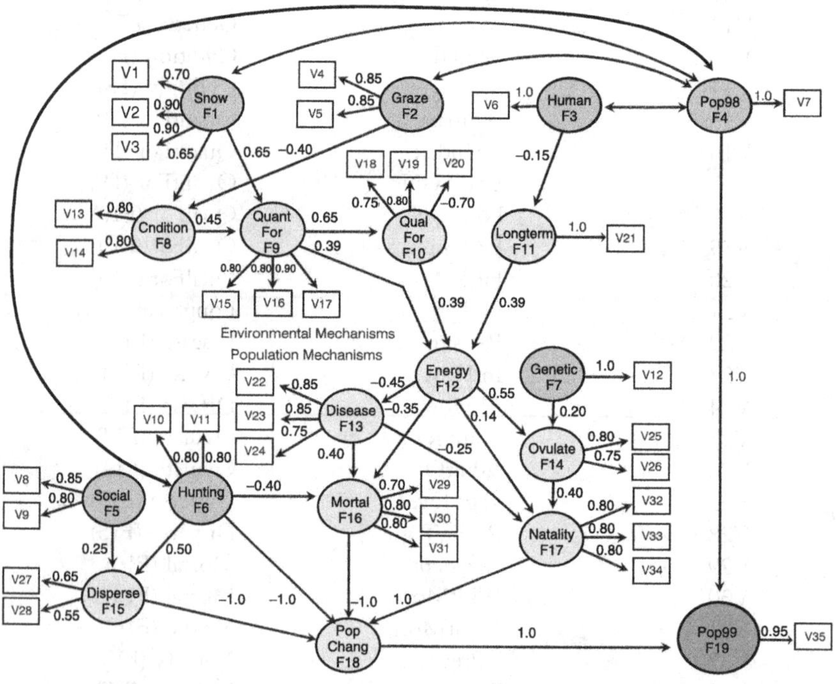

Figure 3.2. Path diagram of simulated population model. Path coefficients are the implied population covariances from which samples were generated for analysis. Latent variables with dark gray backgrounds are exogenous. Latent variables with light gray backgrounds are endogenous. Pop99, with bold border and dark gray background, is the endogenous variable that the model seeks to explain.

of sampling units. This latent variable corresponds to Houston's (1968) "social organization".

  - Displays (V8) – The frequency of agonistic displays and vocalizations.
  - Attacks (V9) – The frequency of overt agonistic behavior.

- *Hunting* (F6) – A measure of hunting pressure on the moose population.
  - TagI (V10) – The number of hunting tags issued for an area.
  - TagR (V11) – The number of tags observed at check stations.
- *Genetic* (F7) – The genetic predisposition to tolerate low food availability and to ovulate with reduced frequency.
  - Assay (V12) – Micrograms per liter of fat-catabolizing enzyme.
- *Cndition* (F8) – The condition of forage foods.
  - %Used (V13) – The percentage of plant stems used by foragers.
  - %Hedged (V14) – The percentage of the forage plant that has been browsed.
- *QuantFor* (F9) – The quantity of available forage.
  - Biomass (V15) – The biomass of forage plants available.
  - LAIndex (V16) – The leaf area index.
  - LeadLGN (V17) – The value for leader length.
- *QualFor* (F10) – The quality of forage available to moose.
  - Nitrogen (V18) – The nitrogen content per gram of forage plants.
  - Protein (V19) – The protein content per gram of forage plants.
  - Fiber (V20) – The cellulose content per gram of forage plants.
- *Longterm* (F11) – The impacts of long-term habitat change on forage availability.
  - HSI (V21) – Habitat changes as measured by habitat suitability index for forage plant quantity.
- *Energy* (F12) – Available energy within the sampling unit. This latent variable has no measurement-level indicators but is a composite of equally weighted components of the three latent variables QuantFor, QualFor, and Longterm.
- *Disease* (F13) – The level of physical threat from disease/parasitic infection.
  - Parasite (V22) – The numbers of parasites per kilogram live weight.
  - IntHost (V23) – The level of disease antibodies detected from blood serum.
  - Althost (V24) – The frequency of parasites in alternative hosts.

- *Ovulate* (F14) – The fertility rate of females.
  - YoungO (V25) – The ovulation rate of yearling females.
  - AdultO (V26) – The ovulation rate of adult females.
- *Disperse* (F15) – The tendency for dispersal to occur within a sampling unit.
  - TagMove (V27) – The frequency of movement of tagged animals in and out of sampling units.
  - RadioTag (V28) – Frequency of change of home range in and out of sampling units.
- *Mortal* (F16) – A measure of the level of moose mortality.
  - Accident (V29) – The number of reported accidental mortalities.
  - WintMort (V30) – The number of observed winter kills within sampling units.
  - %CafMort (V31) – The percentage of all observed and reported mortalities that were calves from the previous breeding season.
- *Natality* (F17) – A measure of the birth rate within the sampling unit.
  - Apreg (V32) – The percentage adult pregnancy rate.
  - Twins (V33) – The percentage of twins.
  - Ypreg (V34) – The percentage yearling pregnancy rate.
- *POPChang* (F18) – The change in population density resulting from the combined effects of the environmental and population mechanisms. This latent variable has no indicators, and is a composite of four latent variables, Disperse, Hunting, Mortal, and Natality.
- *Pop99* (F19) – Moose population density in 1999.
  - Census99 (V7) – Moose population counts with each sampling unit using distance sampling methods (Buckland *et al.*, 1993).

Construction of the data simulation

The model developed here is considered longitudinal because population density is measured in two waves, Pop98 (F4) and Pop99 (F19). Table 3.3 shows the chronology of data collection on the variables described above. Conceivably, snow conditions measured during the 1997–1998 winter and spring of 1998 would have affected energy availability in 1998. Ovulation rates in 1998, overwintering mortality during the 1998–1999 winter and natality in 1999 would have been impacted on by energy availability. Susceptibility to disease and hunting in 1998 affected overwintering

Table 3.3. *The chronology of data collection*

Snow	Quantity of Forage Quality of Forage Condition Grazing Census98	Genetic Ovulation Hunting Social Dispersal Disease	Mortality	Natality	Census99
Spring	**Summer**	**Fall**	**Winter**	**Spring**	**Summer**
	1998			1999	

Human development and long-term forage could be measured at any time assuming no changes occur during the course of the time sequence above.

mortality. Social organization and dispersals during 1998 also would have affected the 1999 population census.

Program input for the EQS simulation begins with a Title (Appendix 3.1). Next a Specification segment lists the sample size of randomly generated data sets produced by the model ($N = 300$), number of indicator variables ($P = 35$), methods of estimation (maximum likelihood), and the type of matrix (covariance) used by EQS. The Equations segment describes indicators (V) as linear functions of a latent variable (F) with a coefficient that is the loading and residual (E). Equations also specify the dependence relationships (paths) among endogenous latent variables and may include disturbance terms (D) for residuals. Specification of variances and covariances for the independent variables (F, D, and E) appear in the VAR and COV segments, respectively. A Contraints segment set to equal the values of the paths creating the composite variable Energy (F12) completes the model. Assigned parameter values define the model that implies population variances/covariances. Assuming multivariate normality, the simulation extracts a random sample from this population, estimating parameters with a star (*) and assuming as fixed the ones without a star. Input values for each model parameter were chosen *a priori* to create a working model of Houston's (1968) system of interrelationships.

The Simulation segment was used to generate two data sets, henceforth referred to as Data1 and Data2. A seed number was provided to initiate a random numbers generation process. Next the number of replicate data sets to be generated, their names, and how they were to be saved was specified. The final statement in the Simulation segment, Population=Model, specifies that the model input values were to be treated as the statistical

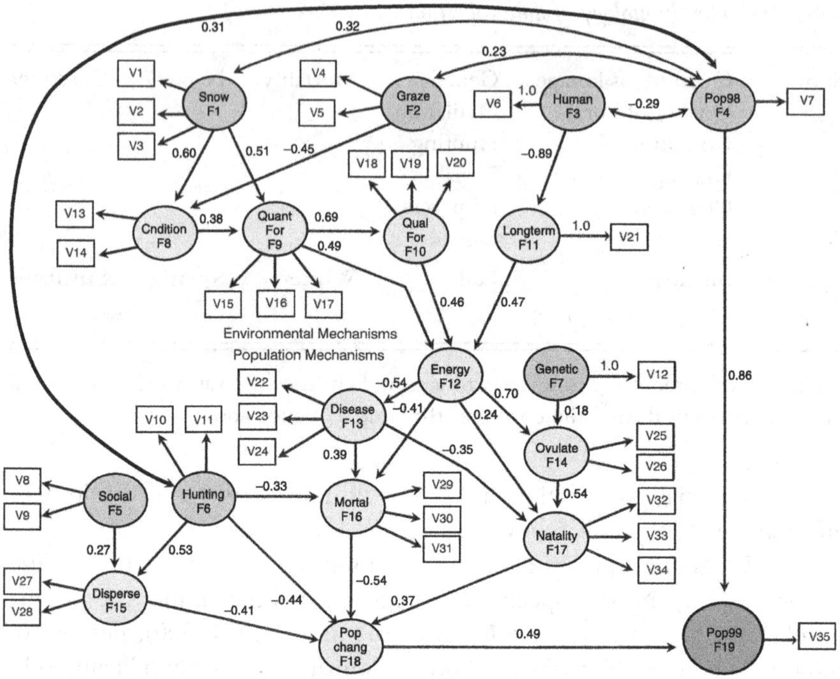

Figure 3.3. Standardized path coefficients for the population model presented in Figure 3.2. Because EQS does not produce output of the standardized values of the population model, results here are estimates based on a simulation of 100 000 cases.

population, and that the generated data sets represent statistically random samples of the population.

The Output segment requests a program output file and specifies its contents. The program was completed by the End statement. The user may consult Bentler & Wu (1995) for more information on data set simulation.

Input values for the simulated model parameters are presented in Figure 3.3. These values represent a standardized form of the population values from which sample data sets 1 and 2 were drawn. Note that no relationship between Social (F5) and Mortal (F16) was modeled to incorporate the linkage between social organization and winter mortality theorized by Houston (1968) (see Figure 3.1).

Although it is generally advisable to measure latent constructs with multiple indicators, there arise situations in which doing so is impracticable. Here, the latent variables Human (F3), Longterm (F11), Genetic (F7), Pop98 (F4), and Pop99 (F19) were measured with single indicators. Latent variables

Human, Longterm, and Genetic were constructed with equations that link the single indicator to the latent variable with a value of 1 and with no error term. This linking has the effect of setting the error variance to zero, and thus makes the indicator identical to its latent variable. Variability was incorporated into the exogenous Human and Genetic variables by including a star (*) in the equations defining values of the loadings (F variables) (Appendix 3.1; Figure 3.3). Variability was provided, in the case of the endogenous variable Longterm, by inserting a star (*) into the disturbance term.

The latent variables Pop98 and Pop99 were created somewhat differently from the other single-indicator latent variables (Appendix 3.1; Figure 3.3). I created loading and small error terms both of which were free to vary because estimates of population density also provided estimates of measurement error (Buckland *et al.*, 1993).

I also fixed the paths leading from Pop98 to Pop99 and from PopChang to Pop99 to 1.0. This had the effect of translating unit changes in the two predictors of Pop99 into fixed unit changes of one moose.

The latent variables Energy and PopChang have no indicators (Appendix 3.1; Figure 3.3). These two latent variables were composite variables (Hayduk, 1987) created by other latent variables with paths leading towards Energy and PopChang. Neither latent variable had a disturbance term. Each was a linear combination of the latent constructs that created it. Energy was the sum of forage quality, forage quantity, and long-term habitat alteration. PopChang was the net change in moose resulting from dispersal, hunting, mortality, and natality – all paths leading to PopChang (F18) from population mechanisms were fixed to 1.0 or −1.0, depending on the direction of their influence. Population density in 1999 was thus dependent on population density in 1998 and the change in population density from 1998 to 1999 resulting from the environmental and population mechanisms postulated by Houston's model.

Data analysis

Raw data from the EQS simulation data sets 1 and 2 were analyzed with PRELIS 2 (Jöreskog & Sörbom, 1996a), a component of LISREL 8 software (Jöreskog & Sörbom, 1996b). PRELIS provided preliminary diagnostic and summary statistics including tests for univariate and multivariate normality of each indicator variable. As expected, indicator variables were normal. PRELIS also created output data files consisting of the covariance matrices generated by two data sets. These matrices were used for analysis with LISREL 8.

Following methods outlined in earlier chapters, I proceeded with the analysis in three stages. First a measurement model was run on Data1 to examine the fit between indicator variables and the latent constructs. Next analyses of the full model including both the measurement and the structural submodel were run, again by using Data1. Finally, a second analysis of the full model was performed by using Data2.

Measurement model

The measurement model assessed the construction of latent variables with multiple indicators as conceived in Table 3.1. The model was run in submodel 3 of LISREL, where all latent variables are treated as exogenous ksi $\boldsymbol{\xi}$ variables and all possible correlations among $\boldsymbol{\xi}$ variables are included in the model. Here the investigator must look primarily at the fit of the model, *t*-values, and modification indices for $\boldsymbol{\Lambda}_x$ and $\boldsymbol{\Theta}_\delta$ to determine how well indicators define the latent constructs. If all goes well, the model should fit the data well. Significant *t*-values suggest that indicators are strongly determined by the latent variable. Low modification indices (<4.0) for loadings suggest that indicators do not load well on latent constructs other than the ones specified in the model. No large modification indices should appear for covariances between error terms. Such a result would indicate the presence of relationships among indicator variables that are not taken into account by the theoretical model.

A good fit was obtained for the measurement model ($\chi^2 = 392.05$ df(429), $P = 0.90$, goodness-of-fit index (GFI) $= 0.93$, root-mean-square residual (RMR) $= 0.14$). All methods of assessment of fit indicate that the model explained the data well. The model-implied variance–covariance matrix, $\boldsymbol{\Sigma}$, was very similar to the data generated variance–covariance matrix. Deviations of $\boldsymbol{\Sigma}$ from the data variance–covariance matrix were within bounds expected from random sampling variation. There were, therefore, no grounds for rejecting the measurement model.

Inspection of *t*-values revealed that each indicator variable had a significant *t*-value (i.e., $\geq$2.0) for the loading on its respective latent variable. Thus, in each case, the latent variable accounted for a significant amount of variance of the indicator. Inspection of modification indices indicated 41 instances in which model fit would be significantly improved if covariances were estimated (i.e., modification indices >4.0). The largest modification index occurred for estimation of the indicator Biomass (V15) on the latent variable Human (F3). Model χ^2 would be reduced by approximately 9.34 units with a reduction of 1df if the additional loading were

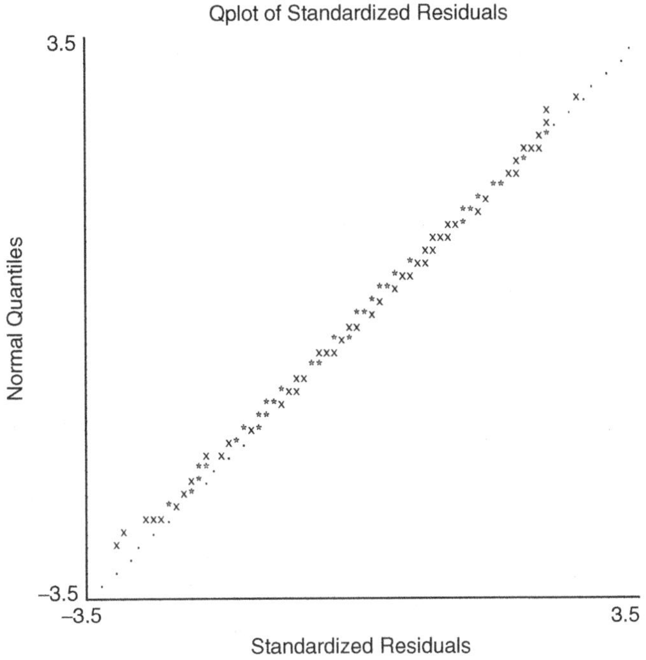

Figure 3.4. Quantile–quantile (Q–Q) plot of normalized residuals. If residuals are normally distributed, then observations should occur on or near the line that emerges from the origin at 45°.

estimated. If I could justify on the basis of my biological insight a reason to free the model constraint, it would also be justifiable to relax the model constraint and estimate this additional parameter. However, I had no theoretical grounds with which to suggest that biomass was a component of human disturbance. The Normal Quantile plot of Standardized Residuals (Figure 3.4) helps to explain why it is necessary to be conservative about freeing model constraints on the basis of information provided by modification indices. The Normal Quantile plot indicates that the residuals of estimates appear normally distributed. The unexplained variances of the indicators do not, in other words, fall outside the range one would expect on the basis of normal probability theory. Because of the large number of modification indices, a few residuals at the tails of the distribution were slightly significant. Considering the large number of modification indices estimated in this model ($N = 1059$), we should expect some Type I errors, approximately 5 out of every 100, and thus some significant modification indices that were due to random sampling variation. Here approximately 3.9% of modification indices were significant, somewhat less than 5.0%, probably because the sample size was large. More details on the inspection and utilization of modification indices are discussed in the next section.

The full model

Data1

For the purpose of exposition, I constructed an initial structural submodel that was similar to the simulated population model with two exceptions. First, a path from Social (F5) to Mortal (F16) was included to estimate an effect of social organization on mortality. However, as described earlier, no such path was incorporated into the simulation model. Second, the initial structural submodel has no path from Hunting (F6) to Mortal, although such a dependence relationship was included in the simulation. Thus the first theoretical model is somewhat flawed.

Using Data1, the initial full model had reasonably good fit ($\chi^2 = 573.51$ df(545), $P = 0.19$, GFI $= 0.90$, RMR $= 0.37$). All t-values were significant, with the exception of the path linking Social (F5) to Mortal (F16) ($t = -0.58$, not significant (NS)). Consequently, this path was omitted from the model and data were reanalyzed. Results for this model are presented in Figure 3.5. Model fit was good ($\chi^2 = 573.57$ df(546), $P = 0.20$, GFI $=$ 0.90, RMR $= 0.34$). Differences in χ^2 (0.06) resulting from the estimation of one fewer parameter were not significant. Inspection of the modification indices for $\mathbf{\Gamma}$ variables (paths linking exogenous latent variables) (Table 3.4) suggest that the model would be significantly improved by freeing two model parameters, the path from Hunting (F6) to Mortal and the path from Snow (F1) to Disease (F13). There are theoretical grounds for freeing both paths. Hunting pressure might reduce the rate of mortality by relaxing competition for suitable habitat and food. Also, heavy snowfall might reduce exposure to disease pathogens.

Using Data1, a reanalysis of the data was performed on a nested model that freed the path indicated by the largest modification index. The new model that estimated the additional path from Hunting (F6) to Mortal (F16) also fit the data well ($\chi^2 = 550.55$ df(545), $P = 0.43$, GFI $= 0.91$, RMR $= 0.36$) (Figure 3.6). The change in χ^2 (23.02) resulting from the estimation of the additional parameter was highly significant ($P < 0.001$), with a loss of a single degree of freedom. The modification index for the path from Snow (F1) to Disease (F13) remained high at 14.39.

An additional analysis was performed on Data1 in which the additional path from Snow (F1) to Disease (F13) was freed. Model fit improved to ($\chi^2 = 536.74$ df(544), $P = 0.33$, GFI $= 0.91$, RMR $= 0.36$).

Table 3.4. *Modification indices for gamma variables*

	Snow	Graze	Human	Pop98	Social	Hunting	Genetic
Cndition	—	—	1.54	0.12	0.18	0.50	0.63
QuantFor	—	0.10	3.05	0.99	0.00	0.17	0.00
QualFor	0.40	1.16	0.45	0.65	0.04	0.11	0.05
Longterm	0.26	0.44	—	0.03	1.97	0.23	1.87
Energy	5.97	0.03	1.50	1.58	0.20	0.81	0.12
Disease	14.59	1.91	1.55	1.62	0.21	0.01	0.46
Ovulate	0.64	0.67	0.57	0.44	0.16	0.19	—
Disperse	0.14	0.27	2.15	0.29	—	—	0.76
Mortal	3.42	0.03	0.52	2.22	0.33	26.71	0.00
Natality	0.60	0.70	0.00	0.43	0.45	5.11	0.08
Popchang	7.20	2.95	0.58	0.65	1.51	4.79	0.16
Pop99	7.20	2.95	0.58	0.65	1.51	4.79	0.16

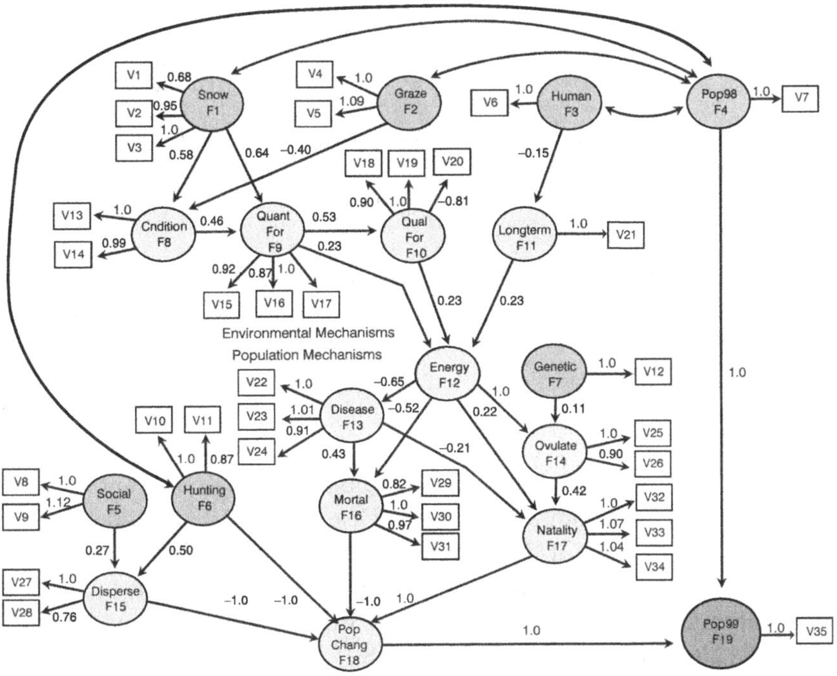

Figure 3.5. The second analysis of the full model using Data1. This model excludes the initially theorized path from Social (F5) to Mortal (F16). Path coefficients are estimated using the maximum likelihood estimation procedure.

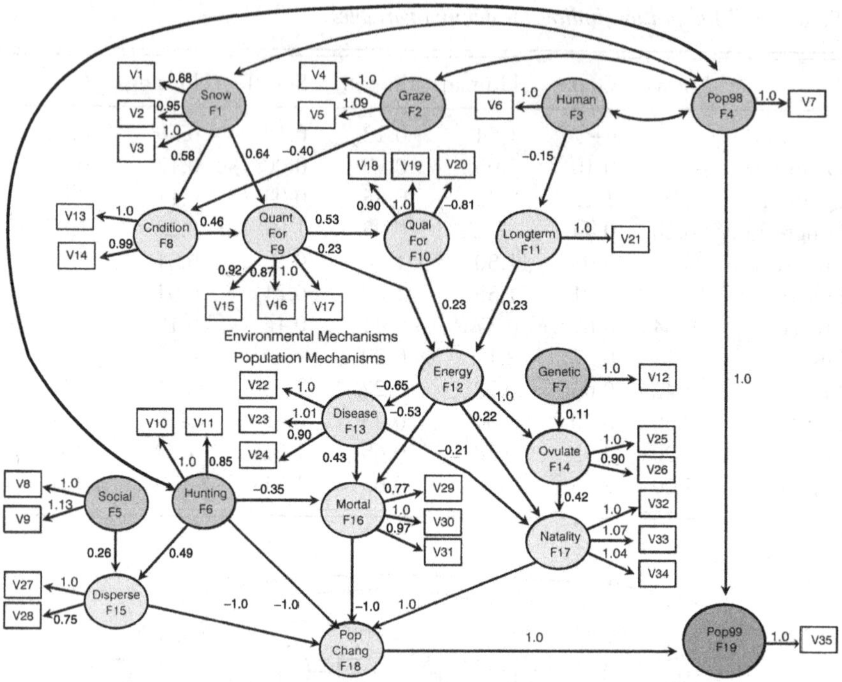

Figure 3.6. Nested full model (Data1) that frees an additional path from Hunting (F6) to Mortal (F16). Path coefficients are estimated using the maximum likelihood estimation procedure.

The change in $\chi^2(13.81)$ resulting from the estimation of the additional parameter was highly significant ($P < 0.001$), with a loss of one additional degree of freedom (Figure 3.7).

Data2

The initial analysis of the full model using Data2 was performed on a structural submodel that included the additional paths from Hunting (F6) to Mortal (F16), and from Snow (F1) to Disease (F13). The model fit the data well ($\chi^2 = 554.67$ df(544), $P = 0.37$, GFI $= 0.90$, RMR $= 0.28$). Inspection of t-values indicated that, while the path from Hunt to Mortal was significant ($t = -6.75$), the t-value for the path from Snow to Disease was not significant ($t = -0.53$) (Figure 3.8).

The final analysis using Data2 omitted the path from Snow (F1) to Disease (F16). This model fit the data equally well ($\chi^2 = 554.60$ df(545), $P = 0.38$, GFI $= 0.90$, RMR $= 0.28$) (Figure 3.9), with one additional

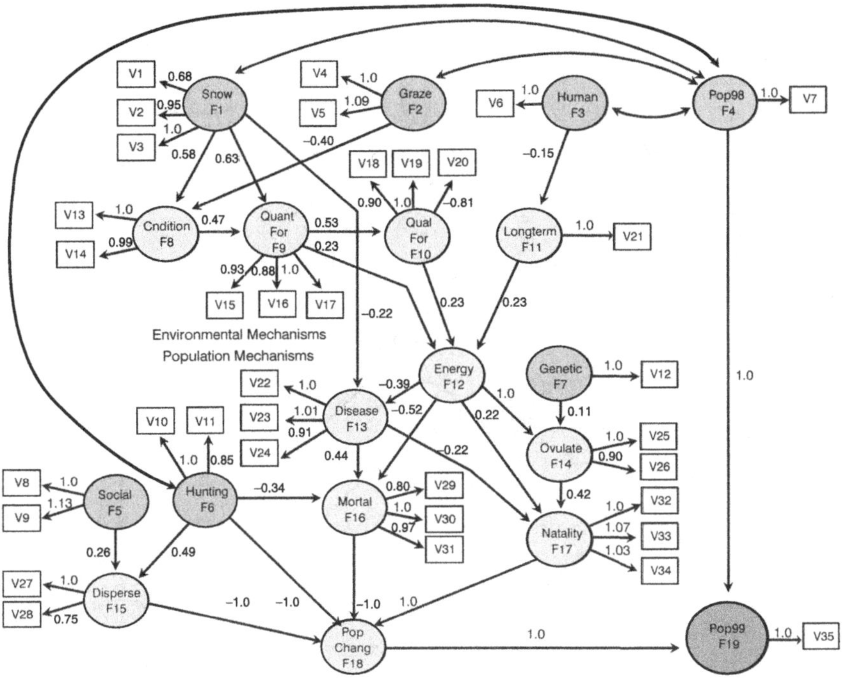

Figure 3.7. Nested full model (Data1) that frees an additional path from Snow (F1) to Disease (F13). Path coefficients estimated using the maximum likelihood estimation procedure.

degree of freedom as compared with the previous model. The change of χ^2 (0.07) with 1 df was not significant.

The standardized solution provided by LISREL 8 converts values of model paths from covariance to correlation metrics (Figure 3.9). In Data2, Pop98 (F4) accounted for 81% of the variance in Pop99 (F19), while the environmental and population mechanisms through PopChang (F18) accounted for the remaining variance of Pop99 (F19).

The effects of snowfall (Snow, F1) on quantity of forage (QuantFor, F9), and Hunting (F6) on population change (PopChang, F18) are complicated by the presence of both direct and indirect effects (Figure 3.9). In addition to the direct path, Snow affects QuantFor through plant condition (Cndition, F8), which in turn affects Quantfor. The standardized total effect (0.70) is the sum of the direct effect (0.38) plus the product corresponding to the indirect effects ($0.68 \times 0.48 = 0.32$). Similarly, Hunting affects PopChang directly (−0.51) and indirectly through dispersal (Disperse, F15)

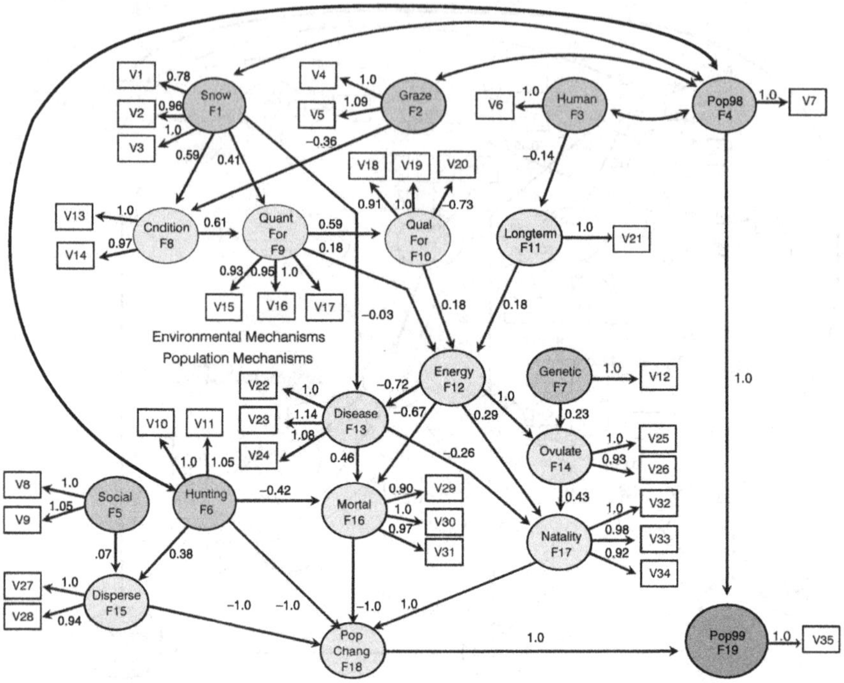

Figure 3.8. The final full model using Data2. Path coefficients are estimated using the maximum likelihood estimation procedure.

(0.56 × −0.35), and through mortality (Mortal, F16) (−0.38 × −0.57). Indirect effects are nearly offsetting; thus total effects (0.49) differ little as compared with the direct effect of Hunting on PopChang.

Discussion

The example presented here demonstrates the utility of SEM methods for testing hypotheses regarding complex systems. No other methodology would allow researchers to test such a conceptual model against data, to confirm or disconfirm its structure, to control for measurement error and collinearity, and to quantify the direct and indirect links between model variables.

Issues of the practicality of designing and testing complex models warrant consideration. The scale of such a study in terms of data requirements would in many cases not be feasible. Such would probably be the case here with the Shiras moose. Studies employing SEM methods are multivariate and require large samples. However, because of the favorable

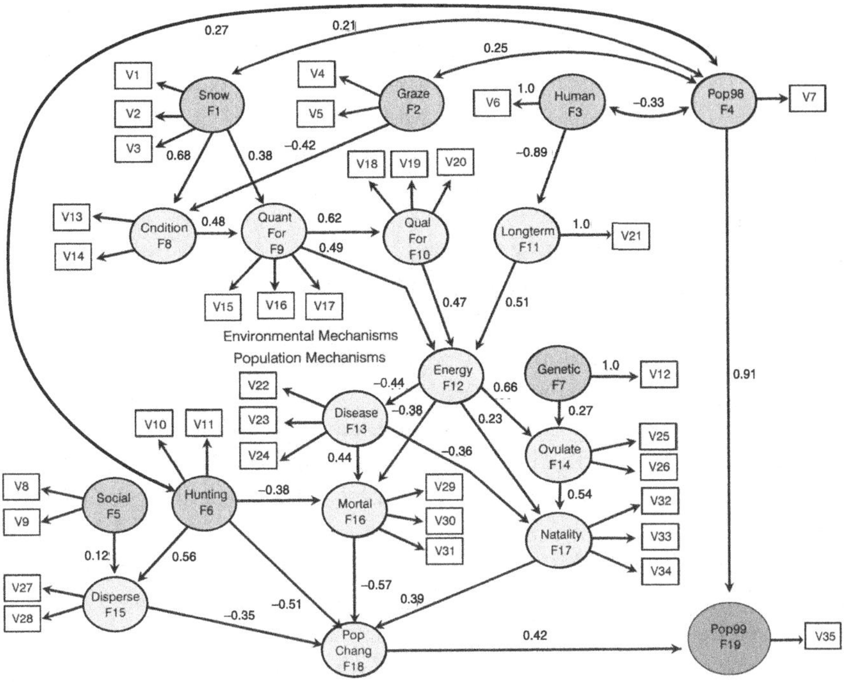

Figure 3.9. Standardized solution of the final full model (Data2).

treatment of collinearity and measurement error by SEM methods, smaller samples are required than for other types of multivariate methods (Boomsma, 1982; Raykov & Widaman, 1995; Pugesek & Grace, 1998). The sample size used here for illustrative purposes, $N = 300$, was larger than necessary. However, given the number of variables in this study, in practice, a sample of at least $N = 200$ would be advisable. Data in the simulation were well behaved, by design, in terms of normality, linearity, and fit of the measurement model. Nonnormality and nonlinearity can be dealt with by transformation of data, creating interactions among latent variables, robust estimation methods, and by specialized analyses designed for nonlinear relations discussed elsewhere in this volume (see Tomer & Pugesek, Chapter 5). As a consequence, these issues are less problematic than is the issue of measurement instruments. Although this topic has been dealt with elsewhere in this volume (see Tomer & Pugesek, Chapter 5) it is sufficiently important to reiterate the importance of designing studies with good measurement instruments. The importance of a well-defined and unambiguous measurement model increases as the size and complexity of the overall model increases.

The Shiras moose analysis demonstrates the high degree of flexibility for constructing models that reflect ecological systems afforded by SEM methods. In addition to the capacity to link latent variables in an interconnecting nexus of dependence relationships, three other features were used to accomplish the model construction.

First, the model was devised as longitudinal to investigate population change. The objectives of the analysis were to confirm or disconfirm a theory on what caused population change, and to quantify the system of interrelationships that explained that change.

The second feature used was the construction of composite variables, Energy (F12) and PopChang (F18). These two concepts were created from, as opposed to measured by, observable indicators and latent variables. Population change (PopChang) was, for example, created as a linear combination of all factors that influenced population density, dispersal, hunting, mortality, and natality. The sums of these variables defined population change.

Finally, the capability to "fix" parameters at nonzero values was used in several instances. The error terms for Census98 (V7) and Census99 (V35) were fixed because population censuses had a known level of measurement error. Estimates of measurement error are often available from population censusing methods, such as distance sampling (Buckland *et al.*, 1993) and mark–recapture analysis (Burnham *et al.*, 1987). This information can be utilized to improve SEM estimates. Paths to and from PopChang (F18), and from Pop98 (F4) to Pop99 (F19) were fixed to ± 1.0 to create a scale in which change occurred in units of one moose. This scale prevented the undesirable quantity of fractional moose. In the case of paths leading to PopChang, it was also necessary that hunting, mortality, etc., be measured in units of one moose. With paths fixed to ± 1.0, it was necessary to freely estimate the variances of dispersal, hunting, mortality, and natality. It was for this reason that the scale for latent variables was set by fixing the $\mathbf{\Lambda}$ of one indicator to 1.0 instead of fixing the diagonal of $\mathbf{\Phi}$ to unity (as in the measurement model) and also the diagonal of $\mathbf{\Psi}$ to 1.0. A common scale was also established for the three variables that created the latent variable Energy (F12). This common scale was accomplished by setting equality constraints on the paths, in this case β-values, leading from QuantFor (F9), QualFor (F10), and Longterm (F11) to Energy.

The measurement and full model analyses illustrate the procedures and some of the pitfalls of implementing SEM analysis. While over-fitting is clearly a possibility, strict adherence to scientific method reduces the chances of model over-fitting. Researchers should make a clear distinction between exploratory and confirmatory analyses in their presentations of research results, as well as in their approach to investigation. Reproducibility of results, the hallmark of sound scientific procedure, clearly strengthens the outcome. Output from SEM programs such as fit statistics, Q–Q plots, residuals, *t*-values, and modification indices provide a wealth of information with which to evaluate research results. This information can be instrumental in modifying and improving theoretical models as well as in suggesting future directions for new data collection and research.

Here the initial full model was confirmatory because an *a priori* theoretical model was tested. Diagnostic information was used to modify the model in an exploratory use of SEM. The revised model was then tested, confirmatorily, on a new data set. The path estimates of the LISREL standardized solution compares closely with the population model (Figures 3.3 and 3.9). Note that it is not possible to directly compare covariance results. The values obtained from the LISREL analysis are dependent on the scale of measurement created by fixing indicators of latent variables to 1. For example, fixing Snowmelt (V1) to 1 instead of SnowFall (V3) would rescale the latent variable Snow (F1) to Snowmelt. This rescaling would in turn cause the value of the paths from Snow to Cndition (F8) and QuantFor (F9) to change. Expression of results in covariance terms is analogous to measuring the distance between two points in metric units and in units of feet. Regardless of the scale of measurement, the distance remains the same, and only the scale changes. Results in covariance terms are useful because the effects of one latent variable on another are expressed in terms of units of known scale (e.g., meters of snow, single moose, etc.). The standardized solution converts all paths to a common metric and is useful for comparing relative importance among paths (Bollen, 1989). It is also useful here in comparing the LISREL solution to the population parameter values of the simulation model.

Once an acceptable theoretical model is fit, it is now possible to develop expert systems that will forecast future change. SEM programs provide, in addition to all path coefficients, estimates of the standard errors of each coefficient. This allows researchers to develop completely stochastic expert systems that can be used to analyze expected future change with change in various model components. The simulation capabilities of SEM programs, such as EQS and LISREL, can be used for this purpose. The approach is a powerful one. Typical simulation studies incorporate values generated piecemeal from analyses such as simple regression without properly accounting for interactions among variables. Although the intent of simulation is to study complex interactions among variables, their input values may be flawed from the outset. The SEM analysis provides input values based on simultaneous analysis of all variables, the partitioning of direct and indirect variable effects, and, in addition, promises the benefits of improved accuracy of estimates described elsewhere in this volume (see Pugesek, Chapter 2).

Appendix 3.1.

```
/TITLE SHIRAS MOOSE MODEL
/SPECIFICATION
  CAS=300; VAR=35; ME=ML; ANALYSIS=COV;
/EQUATIONS
V1=.7*F1+E1;
V2=.9*F1+E2;
V3=.9*F1+E3;
V4=.85*F2+E4;
V5=.85*F2+E5;
V6=1*F3;
V7=.95*F4+E7;
V8=.85*F5+E8;
V9=.80*F5+E9;
V10=.8*F6+E10;
V11=.8*f6+E11;
V12=1*F7;
V13=.8F8+E13;
V14=.8*F8+E14;
V15=.8*F9+E15;
V16=.8*F9+E16;
V17=.9F9+E17;
V18=.75*F10+E18;
V19=.8F10+E19;
V20=-.7*F10*+E20;
V21=1F11;
V22=.85F13+E22;
V23=.85*F13+E23;
V24=.75*F13+E24;
V25=.8F14+E25;
V26=.75*F14+E26;
V27=.65F15+E27;
V28=.55*F15+E28;
V29=.7*F16+E29;
```

```
V30=.8F16+E30;
V31=.8*F16+E31;
V32=.8F17+E32;
V33=.8*F17+E33;
V34=.8*F17+E34;
V35=.95*F19+E35;
F8=.65*F1-.4*F2+D8;
F9=.65*F1+.45*F8+D9;
F10=.65*F9+D10;
F11=-.15*F3+D11;
F12=.39*F9+.39*F10+.39*F11;
F13=-.45*F12+D13;
F14=.55F12+.20*F7+D14;
F15=.25*F5+.5*F6+D15;
F16=-.40*F6-.35*F12+.40*F13+D16;
F17=.14*F12-.25*F13+.40*F14+D17;
F18=-F15-F16-F6+F17;
F19=F18+F4;

/VAR
F1=2;
F2=3;
F3=106;
F4=16;
F5=1;
F6=1;
F7=1;
D8=1.0*;
D9=1.2*;
D10=1.5*;
D11=.6*;
D13=1*;
D16=.6*;
D14=.6*;
D15=.55*;
D17=.08*;
E1=1*;
E2=.4*;
E3=.4*;
E4=.8325*;
```

```
E5=.8325*;
E7=1.0*;
E8=.275*;
E9=.36*;
E10=.36*;
E11=.36*;
E13=.7*;
E14=.7*;
E15=1.08*;
E16=1.08*;
E17=.6*;
E18=1.3*;
E19=1.08*;
E20=1.53*;
E22=.55*;
E23=.55*;
E24=.88*;
E25=1.08*;
E26=1.32*;
E27=.34*;
E28=.50*;
E29=1*;
E30=.72*;
E31=.72*;
E32=.43*;
E33=.43*;
E34=.23*;
E35=1.0*;

/COV
F1,F2=0*;
F2,F3=3*;
F1,F3=0*;
F1,F5=0*;
F2,F5=0*;
F1,F6=0*;
F2,F6=0*;
F3,F6=0*;
F1,F7=0*;
F2,F7=0*;
```

```
F3,F7=0*;;
F4,F1=1.8;
F4,F2=1.6;
F4,F3=-12;
F4,F6=1.25;
F4,F7=0*;

/CONSTRAINTS
(F12,F9) = (F12,F10) = (F12,F11);

/SIMULATION
SEED=157567890;
REPLICATIONS=2;
DATAa='com';
SAVE=SEPARATE;
POPULATION=MODEL;
/OUTPUT
  LISTING; PARAMETER ESTIMATES; STANDARD ERRORS;
/END
```

References

Bentler, P. M. & Wu, E. J. C. (1995). *EQS for Windows: User's Guide.* Encino, CA: Multivariate Software.

Bollen, K. A. (1989). *Structural Equations with Latent Variables.* New York: Wiley.

Boomsma, A. (1982). The robustness of LISREL against small sample sizes in factor analysis models. In K. G. Jöreskog & H. Wold (eds.), *Systems under Indirect Observation: Causality, Structure, Prediction,* Part 1, pp. 149–173. Amsterdam: North-Holland.

Buckland, S. T., Anderson, D. R., Burnham, K. P. & Lake, J. L. (1993). *Distance Sampling: Estimating Abundance of Biological Populations.* London: Chapman & Hall.

Burnham, K. P., Anderson, D. R., White, G. C., Brownie, C. & Pollock, K. H. (1987). *Design and Analysis Methods for Fish Survival Experiments Based on Release–Recapture.* American Fisheries Society Monograph no. 5, Bethesda, MD.

Hayduk, L. A. (1987). *Structural Equation Modeling with LISREL: Essentials and Advances.* London: Johns Hopkins University Press.

Houston, D. B. (1968). *The Shiras Moose in Jackson Hole, Wyoming.* Grand Teton Natural History Association Technical Bulletin no. 1, Jackson, WY.

Jöreskog, L. & Sörbom, D. (1996a). *PRELIS 2: User's Reference Guide.* Chicago: Scientific Software International.

Jöreskog, L. & Sörbom, D. (1996b). *LISREL 8: User's Reference Guide.* Chicago: Scientific Software International.

Pugesek, B. H. & Grace, J. B. (1998). On the utility of path modelling for ecological and evolutionary studies. *Functional Ecology*, **12,** 843–856.

Raykov, T. & Widaman, K. F. (1995). Issues in applied structural equation modeling research. *Structural Equation Modeling*, **2,** 289–318.

4 A short history of structural equation models

Adrian Tomer

Abstract

The goal of this chapter is to provide a historical account of structural equation modeling (SEM). To do so I followed the development throughout the twentieth century of the main ideas in the four disciplines that contributed to the development of SEM: biometrics, econometrics, psychometrics, and sociometrics. Special attention is paid to the development of path analysis by Wright and its sociological applications. Also presented are the development of estimation procedures and the formulation of identification issues in econometrics and the beginning of exploratory and confirmatory factor analysis in psychology. The last part of this account presents the synthesis of these ideas in the LISREL model as well as extensions of the original model.

Introduction

A historical account of the development of structural equation modeling (SEM) is no easy endeavor. One reason for the difficulty is the complexity of the area – the fact that SEM is a huge statistical, methodological, and philosophical jigsaw. Any account should follow the streams that eventually confluenced in the remarkable synthesis that constitutes SEM. A major goal of the present account is to focus on the main streams, while ignoring the less essential streamlets. In writing it I had in mind the statistically educated reader who is familiar with, for example, the normal equations of multiple regression and with the classical least squares methods but not necessarily with other estimation methods. In writing this history of SEM, I considered it necessary, therefore, to explain statistical concepts that might be new to the reader. A historical, i.e., developmental exposition of SEM, can contribute to an intuitive understanding of the method by making explicit the (sometimes convoluted) developmental patterns. To accomplish this goal I tried to avoid mathematical formulations as much as possible without giving up on a presentation of the concepts behind the formulations. This short history is not strictly chronological. Instead, it follows the disciplinary boundaries. SEM is the common place where biometrics, econometrics, psychometrics, and sociometrics meet one another. The account will present in some detail the development of the main ideas within each of these four disciplines up to the crystallization of the LISREL model in the early 1970s. Further expansions of the model for an additional decade, to the early 1980s, are reviewed, by necessity, rather sketchily. More recent developments (that will be presented in other chapters) are not included here.

Wright's invention of path analysis

Sewall Wright published his first article that included an application of path coefficients to the problem of "size factors" in 1918. The idea in this paper was that growth factors of different broadness (general size, size of skull, of leg, etc.) may affect the size of the various bones (skull, fibia, etc.) and induce variability among them. The importance of these factors is measured by the proportion of variation determined by a factor (cause) in the size of a body part or bone (effect). Such a representation was made possible by the assumption that the growth factors are independent of each other. Wright showed that it is possible, making some plausible assumptions, to determine a numerical value for the importance of each factor on the basis of the coefficients of correlation obtained for pairs of measurements of bones. Path coefficients, corresponding to the relative importance of causal factors, were formally introduced by Wright in subsequent publications (Wright, 1920, 1921) and were applied by him to numerous problems in which the functional or causal relations between variables could be specified. An excellent presentation accompanied by many examples is given in Wright (1934). The essence of the method as presented by Wright as early as 1921 is "the combination of knowledge of the degrees of correlation among the variables in a system with such knowledge as may be possessed of the causal relations. In cases in which the causal relations are uncertain the method can be used to find the logical consequences of any particular hypothesis in regard to them" (Wright, 1921, p. 557).

The specification of a complete causal diagram is, therefore, essential to the method. Tied to such a specification, path coefficients can be defined as numbers by which variables have to be multiplied to express one variable as a linear function of other variables. When variables are given in a standardized form, these are standardized coefficients. Compound path coefficients can be defined for paths that lead from one variable to another, passing through intermediate variables. One has to multiply the simple path

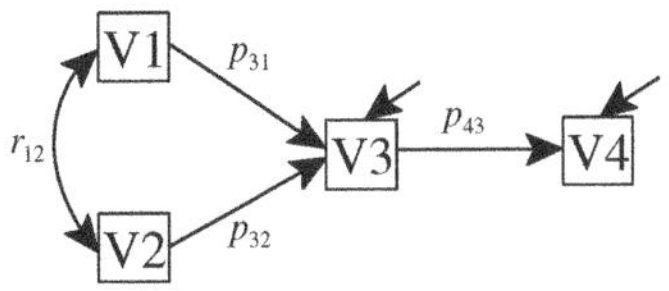

Figure 4.1. A simple path diagram illustrating relationships among four observed variables (V1–V4).

coefficients corresponding to the "legs" in the compound path. Even a complete path diagram will not include the causes for interconnections among the independent variables. Those causes are beyond the researcher's theory or intent to explain and only the intercorrelations among the independent variables are represented. In this case one should include the correlation between two such variables in the computation of a compound path that passes through the two variables. Correlations between two variables in the diagram can then be expressed in terms of sums of (compound) path coefficients for "appropriate" paths that connect the two variables – those paths that allow the causal influence to flow from one point to another in the diagram. For example in Figure 4.1 the correlation between variables V1 and V4 can be expressed using the sum of the products $r_{12}\ p_{31}\ p_{43}$ and $r_{12}\ p_{32}\ p_{43}$, where r indicates correlation and p standardized path coefficient.

Wright applied path analysis to population genetics problems to predict a correlation when the path coefficients can be specified on the basis of the theory (e.g., Wright, 1921, 1934). Another type of application – inverse path analysis – involved the estimation of coefficients based on correlations using the expressions for these correlations in terms of path coefficients.

In addition to standardized path coefficients, Wright defined "concrete path coefficients" (Wright, 1934, 1960) that can be obtained from the standardized coefficients by multiplying by the standard deviation of the target variable and dividing by the standard deviation of the origin variable. For appropriate systems these concrete path coefficients are identical with partial regression coefficients obtained by an ordinary least squares minimization procedure. To see this, consider in Figure 4.1 the regression of V3 on V1 and V2. In this case the equations defined by Wright for the correlations between V3 and V1 and between V3 and V2 are essentially the normal equations obtained using the classical least squares minimization procedure for standardized regression coefficients. To obtain the path coefficients for the whole diagram, we can conduct two regressions of variables on other variables that affect them directly; therefore to regress V3 on V1 and V2 and, also, V4 on V3.

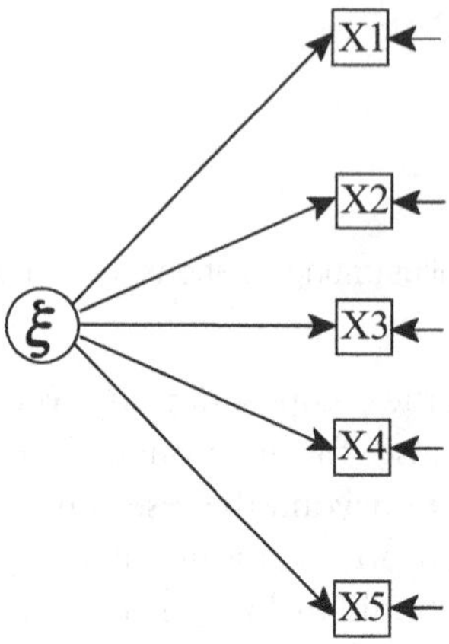

Figure 4.2. One general factor ξ explaining correlations among five observed variables (X1–X5).

The identity of path coefficients and regression coefficients in some cases may be misleading, however, and prevent us from appreciating fully the subtlety and versatility of path analysis. Wright's 1918 article on size factors can serve as an illustration. Let us revisit this example using Wright's (1932, 1934) modification of the method. In this case five observed variables (length and breadth of skull and lengths of humorus, femur, and tibia, see Figure 4.2) generate $(5 \times 4)/2 = 10$ correlations, which can be expressed in terms of path coefficients. As there are only five path coefficients, we have a case of overdetermination. One way to solve the problem is to consider residuals based on a tentative solution to predict back the correlations. Minimizing a sum of least squares will select "the best solution". Wright, however, was no slave to statistics. A model based on just one general factor was not plausible. One should expect some additional more specific factors such as a factor for leg (determining measures of femur and tibia). Therefore, positive residuals are likely. Wright proposed, therefore, to estimate the path coefficients based on reducing to a minimum only negative residuals. The positive residuals that remain indicate the necessity to use other more specific factors (Wright, 1932, 1934). In this application, path analysis clearly cannot be reduced to regression analysis. Moreover, the example illustrates some of the problems of estimation of path coefficients at a time when the theory of estimation for simultaneous equations was not yet formulated (it took until 1940, when Lawley, showed how the maximum likelihood method could be applied to factor analysis and thus econometricians could define their methods of estimation in the 1940s).

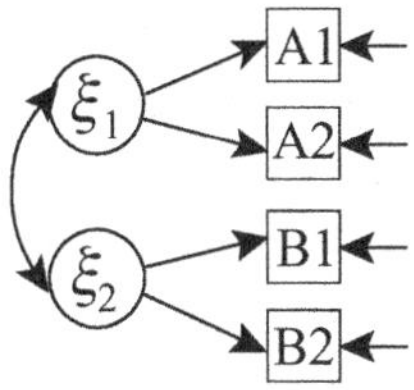

Figure 4.3. A path diagram with two unobserved (latent) variables, A and B, each with two measurements.

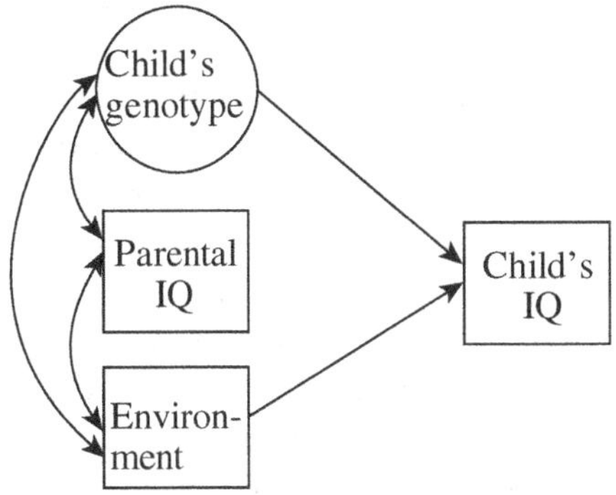

Figure 4.4. Child's IQ as a function of child's genotype, parental IQ and environment.

Another illustrative example is Wright's (1934) derivation of Spearman's correction for attenuation of correlation due to measurement error. Wright presents a diagram in which two unmeasured variables (later called latent variables) for the true scores of two traits, A and B, respectively, affect each two (parallel) measurements of those traits. Assuming that the path coefficients for the parallel measures are equal, the disattenuated correlation between the true values can be obtained by using the correlations among the parallel tests corresponding to the same latent variables as well as the cross-correlations among measurements (see Figure 4.3).

As a final example of the versatility of the method let us consider Wright's treatment of the importance of heredity to human intelligence (Figure 4.4). In this case a child's IQ is related in the path diagram to a measure of home environment and to the child's genotype (heredity). Heredity is represented as a circle to emphasize that this variable is not measured directly. The other three observed variables (represented by rectangles) generate three correlations. There are therefore more parameters to determine (three correlations and two path coefficients) than there are equations – a case of clear indeterminacy. To solve this problem Wright used information from a sample of foster children. In this case the correlations between genotype and parental IQ and between genotype and home environment can be fixed to zero. By assuming some invariances across the two populations (own versus foster children), Wright was able to estimate all the parameters and to obtain the value of path coefficients, including the path from genotype to child's IQ (0.9). A more conservative estimate is then obtained by him using a more complex model that goes back one generation to parental genotypes and home environment and allows one to take into account sources of variability that were confounded with heredity in the first analysis (Wright, 1934, p. 187).

Wright's main insight was perhaps the realization that, while in many cases the causal structure of a set of factors may be clear, the quantitative aspects of it are not. We can use a diagram to show how events are (plausibly) interconnected but such a representation would be only a "qualitative scheme" in need of quantification (Wright, 1934, p. 177). The discovery of an esthetically satisfying formula showing the decomposition of a correlation between any two variables in this scheme into contributions related to the paths in the scheme gave him the means to accomplish this. The examples above give perhaps a flavor of Wright's ingenuity and creative thinking in his application of the new method. In many respects they anticipated later developments of models that include latent variables and/or simultaneous analyses conducted in two or more groups by specifying constraints across groups (e.g., Jöreskog & Sörbom, 1989). The full importance of the method was realized by psychologists, sociologists, and econometricians much later.

Path models in sociology

Herbert Simon published the article "Spurious correlation: a causal interpretation" in a 1954 issue of the *Journal of the American Statistical Association*. The article (later reprinted in Simon's 1957 book, *Models of Man*) influenced sociological thinking on issues of causality. Simon argued convincingly that, in spite of the well-accepted adage that correlation does not mean causation, under certain assumptions correlation might indicate causation. In the case of two variables we need to assume a particular causal order and lack of correlation of disturbances to "deduce" causality. In the case of three variables the situation is more complicated. We can check whether a particular causal model is consistent with a particular pattern of correlations. If, for example, variable A is not connected directly to variable C but is connected to C via an intervening variable B, then the partial correlation between A and C controlling for B should be zero. In retrospect, given the work of Wright that preceded this article, there was nothing revolutionary in this idea and parts of the analysis might seem more or less trivial. Nevertheless, the article influenced Blalock (1962, 1964), who expanded Simon's work by considering more complex models and by examining partial correlations (or, equivalently, partial regression coefficients). Null partial correlations indicate "spurious zero-order correlations and justify the researcher in not including in her model a corresponding path of possible influence".

Blalock's awareness of Wright's work in the early 1960s is indicated by his reference to Wright's as a "related approach" (Blalock, 1964, pp. 191–2) in an Appendix to his *Causal Inferences in Nonexperimental Research* (1964). In fact, the Simon–Blalock method might seem inferior to the method of path analysis proposed by Wright and was in fact so characterized by some (e.g., Boudon, 1965; Asher, 1976). A further step was an "inverse use of path coefficients", which involved the problem of estimating the coefficients given correlational data (Duncan, 1966), a problem that was treated, as we have seen, by Wright. In this case one can estimate a recursive model that involves all the paths (a "just-identified model") and then "trim" it by eliminating paths with nonsignificant and small coefficients (see Duncan, 1966, p. 62; Heise, 1969, p. 59). This is a procedure, therefore, that goes from an unconstrained to a more constrained and over-identified model. For the latter, one has more equations than unknowns. It would seem then logical to estimate the coefficients by applying a least squares procedure to *all* equations and a procedure of this nature was advocated by Boudon (1968). However, as proven by Goldberger (1970), estimates obtained using ordinary least squares (OLS) for the regression equations have smaller variability and are therefore preferable.

The late 1960s saw increasingly more sociological applications of path models. In particular, sociologists considered longitudinal or panel data. Thus Duncan (1969), Heise (1969) and Bohrnstedt (1969) considered the path analysis formulation of two-wave, two-variable models. Duncan (1969) emphasized that a general two-wave two-variable model is under-identified and, therefore, there is a need for "*a priori* substantive assumptions" to transform it into a just-identified or over-identified model. Heise (1970) used a simulated true two-wave model with four variables measured with error to evaluate the effects of estimating the model under the assumption of perfect measurements. The development of the general LISREL model will facilitate the analysis of models like this and, in fact, much more complicated ones (for a very comprehensive analysis of longitudinal models, see Jöreskog & Sörbom, 1977). For other advanced topics, such as the estimation and identification of nonrecursive models, sociologists, starting with Blalock (1964), found that important results have already been established in the voluminous econometric literature.

Sociologists' interests in path analysis was related also to the importance of theoretical concepts in sociology and to the need to measure them, perhaps using multiple indicators in this process. This is the subject matter of the next section.

Theoretical constructs and indicators in sociology

Blalock's books *Causal Inferences in Nonexperimental Research* (1964) and *Theory Construction* (1969) were influenced by the philosopher of science F. S. C. Northrop, in particular by his *The Logic of the Sciences and the Humanities* (1952) in which he introduced the concept of "epistemic correlation" – as "a relation joining an unobserved component of anything designated by a concept by postulation to its directly inspected component denoted by a concept by intuition" (*ibid*., p. 119). Consistent with this concept, Blalock (1969) addressed the issue of measurement of theoretical constructs (such as anomie, solidarity, etc.) via the concept of indicators – variables that can be measured and that provide measures for the theoretical variables. In a simple case, an indicator, say suicide rate, can be connected to a single theoretical concept, say anomie. More frequently, however, the connections are more complex, with other constructs (such as "altruism") affecting the indicator. Blalock argued that the specification of the relationships ("the epistemic correlations" defined by Northrop) between theoretical concepts and operational indicators should be explicitly made in an "auxiliary theory". Blalock provided a simple example (based on the work of the sociologist Homans) that included three theoretical concepts. True suicide rate is determined by Individualism, which itself is determined by Protestantism. The auxiliary theory should specify how indicators for the constructs are determined by those constructs and possibly by other factors. For example "measured suicide rate" is determined by the true rate but also by other factors that introduce error. Also it is possible that this measured suicide rate and another indicator – the measured percentage of Protestants – are affected (distorted) by other common factors. In this example Blalock also entertained the possibility that one construct – Individualism – might not have any indicators, perhaps because no psychological test was yet available.

It was clear to Blalock that there is no direct testing of the theory itself that does not assume the adequacy of the auxiliary theory (which is the equivalent of the measurement model in the LISREL model, see p. 23). Evidently, a mispecification of the auxiliary theory (measurement model) might result in rejecting a true model (as specified by the basic theory). Even worse, opportunistic respecifications may lead to the acceptance of a false model. This point was further driven home by the use of simulated true models and the generation and analysis of covariance matrices according to a specified simplified or false model (e.g., Costner & Schoenberg (1973) demonstrated the dangers of respecification of measurement models on the basis of residual patterns at the level of the whole model and suggested, instead, examining simpler submodels).

Returning to the situation of sociological methodology in the 1960s, it is clear that methodological (errors in variables, multiple indicators, causality, testing theories) and philosophical considerations (measurement of theoretical concepts) intertwined to create a great interest in models that allow a systematic approach to these issues. This is well expressed in the content of the volumes of *Sociological Methodology* of the late 1960s and early 1970s. For example Edgar Borgatta, the editor of the 1969 volume of *Sociological Methodology* devotes about half of the content to path analysis. The popularity of path analysis in sociology in the late 1960s started diffusing also to the neighboring field of psychology. Duncan's (1969) article mentioned above was published in *Psychological Bulletin*. Werts & Linn published their "Path analysis: psychological examples" also in *Psychological Bulletin* in 1970. While sociologists rediscovered Wright and analyzed a variety of path models, they produced little in terms of systematization of the field. It was in econometrics that more systematic advances had been achieved. We move therefore to examine the situation in this area.

Models and structures in econometric thought – the identification problem

Econometricians started formulating and exploring statistical consequences of systems of simultaneous equations in the fourth decade of the twentieth century. By the mid to late 1940s, the terminology related to these systems, as well as important results in the areas of identification and estimation, were already established. Many of those were published in a series of monographs invited by the Cowles Commission in an effort to systematize the field. More advances were made later, particularly in the development of new methods of estimation that use a full information approach. This section, as well as the next presents briefly some of the results that, retrospectively, seem to be important in the development of SEM.

Let us consider a very simple system of equations (adopted from Goldberger, 1964) that describe relationships among variables in a model of demand and supply.

$$Q = aP + bZ + u \quad \text{(demand)}, \tag{4.1}$$

$$P = cQ + v \quad \text{(supply)}, \tag{4.2}$$

where Q stands for quantity, P for price and Z for personal income. The letters u and v indicate "errors in the equation" – latent variables that the researcher has to specify to account for the fact that the explicit variables are not sufficient to completely define the relationship. Typically, these disturbances are considered to have a mean of zero and to be independent of the exogenous variables in the equation. Z, the personal income variable, can be considered as independent (exogenous) variables and predetermined. P and Q are dependent (endogenous) variables.

The concepts of structure and model were clearly defined by Koopmans (1949), who built on previous work by Frisch (1934), Haavelmo (1943), himself (Koopmans, 1937) and others. A structure is defined by providing specific values for the parameters of the equations (a, b, and c) and the complete specification of the distribution of the disturbances. A model, on the other hand, includes specification of the form of the relationship – the equations and, most importantly, the form of the distribution of the disturbances (a point emphasized by Haavelmo, 1943). It is possible to have more than one structure that will satisfy the same model. In this case the system of equations is not identified. In the example above it is easy to show that equation (4.1) (demand) is not identified (see, e.g., Koopmans, 1949) so that there are an infinite number of values for parameters a, b satisfying the equation. Econometricians were therefore forced to consider the problem of identifiability of linear models, i.e., the formulation of conditions under which a unique structure exists. The reader may recall that Wright had encountered this problem in some of the path models he considered. The modern definition of identification and of related necessary and sufficient conditions ensuring the uniqueness of the structure (as a set of values for model parameters) was given by Koopmans and his colleagues, who built upon previous work by Frisch, Haavelmo, Mann & Wald, and Marschak (for a historical exposé, see Duo, 1993).

Koopmans and colleagues (Koopmans, 1949; Koopmans *et al.*, 1950) also formulated clearly and proved the rank condition (a necessary and sufficient condition) and the order condition (a necessary condition). The order criterion specifies how many variables should be excluded (i.e., their coefficient should be fixed to zero) from an equation to make this equation identifiable. For G equations included in the system this number is $G - 1$. To illustrate this, let us assume that, in the example above, the supply equation includes another exogenous variable that is not included in the first equation. This will make both equations identifiable since the number of exogenous variables excluded is equal to the number of endogenous included (2) minus 1. The rank condition "is that we can form at least one nonvanishing determinant of order $G - 1$ out of those coefficients, properly arranged, with which the variables excluded from that structural equation appear in the $G - 1$ other structural equations" (Koopmans, 1949, p. 113). Not only is the formulation of the rank condition more difficult, its application is also more difficult than the application of the order condition (which, however, is only a necessary condition).

It is important to mention that the rank and order condition are formulated for systems in which disturbances are allowed to correlate. If we force correlations to be zero, the conditions are not correct any more. In this last case it is enough to require the system of equations to be recursive. In a recursive system, loosely speaking, the endogenous variables, y, can be arranged in an order in which each variable can be expressed as a linear function of other endogenous variables that precede it in this order and of exogenous variables X (technically, the $\mathbf{b}$ matrix in a model of type $y = \mathbf{a}x + \mathbf{b}y + e$ is triangular).

The Cowles Commission generated additional results on the topic of identification. In particular Wald (1950) provided a generalized analysis of "identification of economic relations" that proved valuable later (see The LISREL model, p. 105). The rank and order conditions were further generalized by Fisher (1959).

Estimation of parameters in econometric models

Recursive models of the type mentioned above are not only identifiable but can also be easily estimated, using, e.g., the classical (ordinary) least squares method, separately for each equation. For this reason some econometricians, in particular Wold (see Wold & Jureen, 1953) have advocated the use of recursive models in the expression of economic relationships. Nevertheless, the structural form presented in the section above remained dominant in the econometric literature (for some of the reasons, see, e.g., Goldberger, 1964) and there was a need to find methods to estimate models expressed in this form. It is easy to show that an OLS method will not fare well. The reason is that, generally speaking, the interdependence of equations makes the disturbances and the variables on the right-hand side of the equation dependent, thus violating one of the assumptions of OLS without which these estimators are no longer unbiased. This point is well known today but it took some time to be widely recognized. Frisch (1934), examining a particular equilibrium model, found that OLS will not provide consistent estimates for the parameters. At the beginning of the 1940s, Haavelmo (1943) thought it was still important to explain this point in detail and to warn against the practice adopted by "many economists" of applying the least squares method separately to the equations of a system. Instead, he recommended the use of a maximum likelihood procedure taking into account the "joint probability distribution of the observable variables" (Haavelmo, 1943, p. 7). Haavelmo's paper played an important role in the "fall of OLS" in structural estimation (see Epstein, 1989). The development of MLE for systems of equations was accomplished in the 1940s and generated several methods that can be applied either at the level of a single equation or at the level of the whole system of equations. Thus Anderson & Rubin (1949, 1956) used the MLE at the level of the equations. The method was called LI/LGRV, LI standing for limited information and LGRV for least generalized residual variance The corresponding method at the level of the system was dubbed FI/LGRV, FI for full information (we will use also the abbreviation FIML). The fundamental statistical work was accomplished by Mann & Wald (1943), who studied properties of large sample MLE estimators under assumptions of multivariate normality of disturbances (error terms) of the equations. Somewhat later, Koopmans *et al.* (1950) generalized Mann & Wald's results to systems of equations that also include exogenous variables. The final work was published in 1950 and forms a large part of Monograph No. 10 of the Cowles Commission. According to Epstein (1989) a crude form of this work already existed in 1943. The next section explains the logic of MLE.

Estimation by the full information maximum likelihood method (FIML)

Let us consider a parent population that is completely determined but for the value of a parameter θ. Let us also consider a sample of N observations. We want to estimate a numerical value for θ based on the given observations. A maximum likelihood estimator does this by (loosely speaking) maximizing the probability of obtaining data like the ones that were, in fact, obtained. Technically, the researcher formulates a likelihood function as the joint probability function taken as a function of unknown parameters for the obtained variables. Then he or she tries to determine a maximum for this function (usually, in fact, for its logarithm). The theory of MLE was developed in the 1920s and 1930s by Sir Ronald Fisher and later by Rao, Cramér and others (e.g., Cramér, 1946). ML estimators have some very desirable asymptotic properties (e.g., features they increasingly approach when the sample size grows indefinitely). In particular, under general conditions, the MLE estimators display asymptotic normality and have minimum variance. Specifically, as the sample size increases, the sampling variances and covariances (multiplied by the number of observations) will approach the negative inverse of the expected values of the second derivatives of the log likelihood function. The expected values are collected in the "information matrix". Theses values was established by the Cramér–Rao inequality (Rao, 1945; Cramér, 1946) to be minimum variance bound in the sense that no estimator can have lower variability. Knowing the asymptotic distribution and having estimates for standard deviations makes possible the use of statistical tests in large samples.

The application to systems of equations presents various problems of identification, specification of the log likelihood function and computation. Koopmans *et al.* (1950) produced the form of the log likelihood under the assumption of multivariate normality of the disturbance terms and examined the properties of the function under several assumptions. Asymptotic sampling variances and covariances of the MLE estimates for a set of identifiable parameters are shown to be estimated by calculating the inverse matrix of second-order derivatives of the likelihood function (as a function of those parameters) at the point where the function reaches a maximum (see above). The computational part of Koopmans *et al.*'s chapter (which benefited from the input of John von Neumann) considered the use of several algorithms for determination of the maximum. While the work accomplished was impressive, it was also highly mathematical and required heavy calculations. As late as 1972 Johnston, in his well-known text *Econometric Methods*, wrote about FIML estimators: "We will not present the details here as the resultant equations are computationally expensive and the least used of the simultaneous estimation methods" (*ibid.*, p. 400). There was a need to develop cheaper methods of estimation.

Other methods of estimation of econometrics models

Other methods were added to the estimation "arsenal' in the 1950s and 60s. Two-stage least squares analysis (2SLS) was developed independently by Theil (1953) and Bassman (1957) and can be considered a single-equation method. The general idea is to regress y variables that appear on the right-hand sides of an equation system of type $\boldsymbol{y} = \boldsymbol{y}\mathbf{a} + \boldsymbol{x}\mathbf{b} + \boldsymbol{e}$ on $\boldsymbol{x}$ variables. Then, in a second stage, one uses estimates for $\boldsymbol{y}$ from the first step as variables, along with $\boldsymbol{x}$ variables, in regressing $\boldsymbol{y}$ variables that appear on the left-hand side of the equations of the system.

The more complex three-stage least squares (3SLS) was developed by Zellner & Theil (1962) as a simpler alternative to FIML. In this case, one uses a generalization of general least squares to a system of equations, making use of estimates of variances and covariances of residuals provided by a 2SLS. We return now to the maximum likelihood method and its use to estimate parameters in econometric models.

Model building and tests of significance

While sociologists tended to create relatively unconstrained models and to trim them later, econometricians tended to work in the opposite direction. If one starts with a relatively constrained model, the results of the tests can then be used to decide where there is a need to "complicate" the model by relaxing some assumptions (e.g., by allowing some coefficients to be estimated instead of being fixed to zero). This direction was called by Gilbert (describing British econometrics) simple to general (Gilbert, 1989). Statistical tests can be used to test the assumption that one or more coefficients are zero. Likelihood ratio tests are based on the computation of two log likelihoods. One is computed assuming the model tested is true. Another is computed for an alternative model. The difference has a χ^2 distribution with a number of degrees of freedom equal to the difference in number of parameters estimated in the two models. The principle can be applied also at the level of one equation. For the equation of type $\boldsymbol{ax} + \boldsymbol{by} = e$ estimated using MLE, Anderson & Rubin (1949) proposed a likelihood ratio statistic testing the hypothesis that the coefficients in vector $\boldsymbol{a}$ are zero. In the 1970s the strategy of starting with a relatively unrestricted model and imposing more restrictions (general to simple, see Gilbert, 1989) has become more common, at least among British econometricians, after Anderson's (1971) demonstration of the power of this sequential strategy. One potential problem is the existence of multiple, alternative (more restricted) models, each necessitating the computation of a log likelihood. A solution recommended by Sargan (1980) was the use of a test developed by Wald (1950) that requires the computation of a statistic based on parameters that the researcher may consider fixing. The statistic is calculated for the unrestricted model only and has a χ^2 distribution. For one parameter, the statistic reduces to (the square of) a z statistic testing the assumption that the estimated parameter is fixed, say, to zero (see, e.g., Bollen, 1989, p. 294).

Latent variables in econometrics

Econometricians recognized from the beginning the existence of latent variables or unobservables in econometrics. Two important types were errors of measurement ("errors in variables") and errors in the equation. As Griliches (1974) mentioned these distinctions were made by Frisch as early as 1934. However, there was a tendency in applied work to treat errors in variables as negligible and to obtain some simplification in the process. In addition, there was a tendency in the 1930s and 40s to limit the use of unobservables in the formulation of models to the minimum necessary on philosophical/methodological grounds. The change from narrow operationalism and empiricism to more liberal attitudes *vis-à-vis* theoretical terms (e.g., Carnap, 1950) was probably instrumental in effecting a change in "attitude" towards latent variables. Econometricians realized that latent variables may in fact allow a better or more efficient estimation of some of the models. This was shown by Zellner in a paper circulated in 1967 and published in 1970 for a model that included two equations with two observables (such as observed consumption and observed income) expressed as functions of a theoretical variable (such as "permanent income"). To make the model identifiable, a third equation connected the theoretical variable to other observables (such as age, education, etc.) in a deterministic way (without a stochastic component). Zellner showed that the parameters of the model can be estimated using a full-information procedure that takes advantage of all the restrictions implicitly imposed by the presence of the theoretical variable in the equations of the system. Zellner's approach was further generalized by Goldberger (1972), who considered the case in which the theoretical variable is not completely determined by the observed variables. As we will see, Goldberger played a special role in the realization of a synthesis of econometric models and factor analytical models (in which the latent variables are factors).

Charles Spearman and the beginning of exploratory and confirmatory factor analysis

The British psychologist Charles Spearman is considered to be the discoverer of factor analysis, a technique he used in an effort to elucidate the nature of human intelligence. In his first paper in 1904 (and even more so in the 1923 and 1927 books) we already find the idea that different intellectual activities have in common one function or factor, whereas other remaining factors are specific to each one of these activities. Spearman arrived at this conclusion on the basis of the fact that, for any pair of abilities that are positively correlated, this correlation can be explained by postulating that these abilities depend on a common factor *g*. When he applied this reasoning to more than one pair, he found that the common factors so derived correlate among themselves almost perfectly. Somewhat later he formulated the tetrad equation according to which the product of the correlations among two pairs of variables (first variable correlated with the second, and third correlated with the fourth) equals a product based on the same variables, which, however, are associated differently (say, first with third and second with fourth). The tetrad difference (i.e., the difference between the two products) should be zero, except for sampling error.

Spearman was not sure about the nature of this general *g* factor on which every mental activity loads, although he proposed several hypotheses about its possible nature. We can see Spearman's work as a first example of a confirmatory type of factor analysis preceding much of the later factor analytic work that was largely exploratory. Another element of Spearman's work is notable. While Spearman was interested very much in the "nature of intelligence", his theory about intelligence was vague to a considerable degree. As mentioned above, the nature of *g* was not clear. Nor would the theory, of course, allow him to specify *a priori* the loadings for the specific mental tasks (or tests), which had therefore to be determined empirically, on the basis of correlations.

Spearman also investigated the idea that one factor might not suffice and that there might be a need to postulate "group factors", representing special abilities, to account for correlations. The technique of investigation of those models was again based on an examination of the tetrad differences. Although Spearman presented the end result as mostly "negative", he did recognize the existence of four group factors or special abilities – logical, mechanical, psychological, and arithmetical. To those he added a fifth special ability: the ability to appreciate music.

After Spearman, factor analysis developed mainly in an exploratory direction, the goal being to determine (particularly in the field of intelligence) the number and the character of the factors. The fundamental model of factor analysis (e.g., Thurstone, 1931, 1935) expresses variables as weighted sums of factors. From this formulation follows the expression of the population variance–covariance matrix in terms of weights or loadings, factor variances–covariances, and unique variances:

$$\mathbf{\Sigma} = \mathbf{\Lambda}\mathbf{\Phi}\mathbf{\Lambda}^{\mathrm{T}} + \mathbf{\Psi}, \tag{4.3}$$

where $\mathbf{\Sigma}$ stands for the variance–covariance matrix, $\mathbf{\Lambda}$ for the matrix of loadings, $\mathbf{\Phi}$ for the factor covariance matrix and $\mathbf{\Psi}$ for the matrix of unique variances or residuals. By replacing $\mathbf{\Sigma}$ with $\mathbf{S}$, the sample variance–covariance matrix, the problem of factor analysis can be formulated as one of finding loadings and factor variances–covariances that will reproduce $\mathbf{S}$ as well as possible. Assuming that we have already a good estimate for the unique variances collected in $\mathbf{\Psi}$ (or, alternatively, for the their counterparts – the communalities that are the diagonal elements of $\mathbf{\Lambda}\,\mathbf{\Phi}\mathbf{\Lambda}^{\mathrm{T}}$) – and assuming further uncorrelated factors ($\mathbf{\Phi} = \mathbf{I}$) the problem of finding $\mathbf{\Lambda}$ is reduced to the problem of finding the eigenvectors and eigenvalues of $\mathbf{\Sigma} - \mathbf{\Psi}$. Usually, the solution will involve an iterative procedure designed to improve the approximation for $\mathbf{\Psi}$ based on communalities given by the "last solution". This is the essence of the principal axes method introduced by Hotelling (1933) and Thurstone (1933). The sum of the squares of the loadings of the tests on the first factor (first axis) is maximized so that the first factor accounts for maximum variability in the variables. Then the second factor (axis) accounts for maximum variance of the residuals, etc. The calculational burden, particularly in cases of large matrices, was huge in the 1930s and 40s, before the advent and dissemination of computers. Alternative procedures, such as the centroid method (e.g., Burt, 1940; Thurstone 1935, 1947) were used to make hand calculations less onerous.

In the case of multiple factors, as indicated by Thurstone (1935), Thomson (1936) and others, factor analysis suffers from an indeterminacy problem. A solution found using the method delineated above can be transformed into another solution by multiplying the matrix of factor loadings by an appropriate (orthogonal) matrix, i.e., by executing a rotation of factors considered as "axes". One way to "remove" the indeterminacy is to impose a constraint, in addition to orthogonality. A commonly used constraint is: $\mathbf{\Lambda}^{\mathrm{T}}\mathbf{\Psi}^{-1}\mathbf{\Lambda} =$ diagonal (Lawley, 1940; Lawley & Maxwell, 1963), which is helpful in a MLE procedure (see p. 105). This type of constraint does not require any theoretical assumptions. It can indeed be shown that, given a pattern of loadings, it is possible by the use of orthogonal transformations to obtain another pattern of loadings that will satisfy the constraint $\mathbf{\Lambda}^{\mathrm{T}}\mathbf{\Psi}^{-1}\mathbf{\Lambda} =$ diagonal. Of course, a maximum likelihood solution can also be rotated to increase interpretability.

Another way to remove the indeterminacy is to select out of an infinite number of good solutions one that is the "simplest". The criterion of a simple structure was formulated by Thurstone (1935, 1947). According to the simple structure criterion, it is preferable that each variable load on as few factors as possible. Technically he required at least one zero loading for each variable (more elaborate definitions are also given by Thurstone in his 1947 book – see p. 107). Thurstone hoped that a simple structure, if found, would point to an "underlying order in the abilities" (Thurstone, 1938, p. 90) and will, therefore, prove robust across studies. The concept of simple structure took factor analysis in an "explanatory" or confirmatory direction. A distinction between descriptive and inferential factor analysis was made by Hartley in his dissertation (1952; cited in Henrysson, 1957). According to this view, factors may have in some cases "surplus meaning", which makes them explanatory hypothetical constructs rather than intervening variables (cf. MacCorquodale & Meehl, 1948). Consequently, and consistent with Carnap's (1950) discussion of degrees of confirmation of scientific theories, inferential solutions in factor analysis should be accepted on the basis of prior results and further generalized (confirmed) across populations of individuals and tests.

The interest in using factor analysis to provide evidence for a theory has increased even for some who had earlier taken a position that emphasized the descriptive character of the method. Burt, for example, who compared factors in his 1940 book to meridians and parallels defined on the surface of the globe to facilitate location, was, in the late 1940s, interested in providing evidence for his hierarchical theory of cognitive processes. According to this theory, there was one general factor and a hierarchy of group factors (Burt, 1951). The theory influenced the creation or choice of the items to be factor analyzed, whether or not a rotation was attempted and what rotation procedure was used. As Burt (1951, p. 45) indicated "the method of factor analysis employed must be one that is automatically capable of confirming or disproving the hypothesis in question". Still there was no rigorous method to test whether or not the results of the analysis fit the theory. Burt (1940) was indeed aware that factors, considered as sources of variability in the population, can generate hypotheses about average means in the population that can be tested using an analysis of variance. In two papers from 1947 and 1952 he even proposed specific tests of significance. A more comprehensive and rigorous approach to testing factors from an analysis of variance perspective was developed by Creasy (1957) and, even more so, by Bock (1960).

Another force pushing in a confirmatory direction was statistical–methodological: the indeterminacy of the factor analytical solution could be resolved by adding *a priori* constraints according to a theory or to previous results. In fact one of the oldest types of factor analysis, the multiple group method (e.g., Burt, 1950), had a confirmatory character. In the case of multiple group factor analysis, factors are defined as weighted sums of variables. The researcher can specify, for example, that, with four variables, one factor is the sum of the first two and a second factor is the sum of the other two variables. Multiple group analysis can be considered confirmatory, since assumptions are made about the weights, typically by fixing some of them to zero and by specifying weights for the remaining variables (for a clear exposition of the method, see Gorsuch, 1983; the multiple group method has seen more recently a resurgence motivated by the idea of finding a simple substitute to the more complex factors determined in a confirmatory factor analysis, see Mulaik, 1989). Indeed, Guttman in a 1952 paper in *Psychometrika* emphasized the distinction between an "*a posteriori* theory" and an "*a priori* theory" (*ibid.*, p. 215). In the first case, which corresponds to exploratory factor analysis, the researcher formulates hypotheses after inspection of the data, either before rotation (but after obtaining a preliminary factor solution) or after. In the second case, corresponding to confirmatory factor analysis (according to Bentler (1986), Tucker was the first to use the terminology exploratory–confirmatory in an article published in 1955 in *Psychometrika*), the researcher chooses the weights of the variables (typically but not necessarily 1s and 0s) in the definition of the group factors according to a theory he or she believes is true. To cite Guttman (1952, p. 215): "The results of a factor analysis would seem to be more trustworthy if the weight matrix $\mathbf{X}$ is chosen according to an *a priori* psychological theory, which is then *tested* by the data".

Rather than considering factors as weighted sums of variables, the fundamental equation of factor analysis expresses variables as weighted sums of factors (see above). Therefore a natural way to proceed in a confirmatory direction would be to analyze cases in which some or all of these loadings are specified in advance. The case that assumes *a priori* loadings to be zero was analyzed by Howe (1955), Anderson & Rubin (1956) and, subsequently, by Lawley (1958). All of them used the MLE that was previously applied by Lawley (1940) to exploratory factor analysis. The case in which the whole matrix of loadings is considered to be known, and the problem is to estimate the parameters of the $\mathbf{\Phi}$ and $\mathbf{\Psi}$ matrices, was treated by Bock & Bargmann (1966), also using MLE. In the same vein, Jöreskog (1966) formulated the confirmatory problem in factor analysis as one in which the simple structure is given *a priori* (rather than approximated, perhaps by rotation, *a posteriori*), by specifying appropriate zero loadings. MLE can then be used to estimate the parameters and to test the simple structure hypothesis.

In 1969 was published Jöreskog's article on confirmatory factor analysis, which gives the most general treatment to it by using MLE and appropriate numerical methods. In the next sections I review in some detail the use of MLE by Lawley and Jöreskog in exploratory and confirmatory factor analysis.

Maximum likelihood estimation of factor loadings in exploratory factor analysis

Lawley used MLE for estimation of factor loadings in a 1940 paper and further developed the topic in two additional papers in 1942 and 1943. A somewhat more recent and accessible presentation was given by Lawley & Maxwell (1963). The log of the likelihood function is expressible in this case as a function of loadings of the variables on factors and of residual variances. The problem of maximization of this function is then treated by calculating the analytical expressions for the partial derivatives with respect to loadings and residuals and equating them to zero. This procedures generates two basic equations in matrix form. The first equation has the form:

residuals variances = function (sample variance–covariance, loadings).

The second equation has the form:

loadings = function (loadings, residual variances, sample variances–covariances).

These equations can then be solved by a process of iterations. We start with approximate values for loadings and residuals on factors $1, 2 \ldots k$ (iteration 1). From here and by substituting these approximations in the second equation for loadings (on the right side) we obtain better approximations for loadings on the left side (iteration 2). Also, by substituting the newly obtained solutions for loadings in the first equation, we obtain better approximations for residual variances.

The process can be continued to obtain increasingly better approximations in successive iterations. Lawley indicated that, while no general theorem of convergence was provided, practice showed that the procedure will usually converge, although sometimes very slowly (Howe, 1955; Lawley & Maxwell, 1963). In fact the procedure will sometimes break down (Howe, 1955; Jöreskog, 1967), particularly when the initial approximations are not good enough. For this reason Jöreskog (1967), after important correspondence on this issue with Lawley, developed a new computational method based on an application of the numerical algorithm of Davidon–Fletcher–Powell (Fletcher & Powell, 1963). Instead of equating to zero the first partial derivatives (as in Lawley's treatment) those derivatives can be used in the numerical method to approximate the minimum by using a clever adaptation of the method of the steepest descent (for a short description of the Fletcher and Powell algorithm, see the next section). One potential problem was that, even using a very efficient algorithm, the simultaneous minimization of a function of, say, hundreds of variables presented serious problems for computers in the late 1960s. Therefore Jöreskog decided to use the idea (suggested to him by a correspondence with Lawley, see Jöreskog, 1967, p. 445) of minimizing a conditional function of the residual variances, instead of minimizing the unconditional fit function, which depends on loadings, as well. This was economical since, with tens of variables and multiple factors, the number of loadings (as independent variables that affect the value of the function) may be easily in the hundreds.

Maximum likelihood estimation of factor loadings in confirmatory factor analysis and SEM in general

The estimation of factor loadings in confirmatory factor analysis (CFA) started by Howe (1955), Anderson & Rubin (1956) and Lawley (1958) suffered from the problems presented for the use of MLE in an exploratory situation. These problems were, in fact, exacerbated by the need to invert nondiagonal matrices (see Lawley & Maxwell, 1963, p. 75). The computational problems were probably a result of the fact that, until the mid 1960s not much attention was paid to the confirmatory approach (see Jöreskog, 1966, p. 165). In addition to the need to find efficient numerical methods of estimation, there was a need to generalize CFA to cover cases where loadings are fixed to values different from zero and those in which the factors can be intercorrelated. This was accomplished by Jöreskog, in part by extending his work on exploratory factor analysis (Jöreskog, 1966, 1967). The idea of minimizing a conditional function was not applicable to a confirmatory situation in which constraints are imposed on the whole factor space (typically by equating many of the loadings to zero, see p. 103). Different numerical algorithms were available to solve the problem. Thus the Newton–Raphson method makes use of second-order derivatives to improve the approximation provided by a simple steepest descent procedure. Bock & Bargmann (1966) used Newton–Raphson for the estimation of specific confirmatory models after obtaining expressions for second-order derivatives. Approximate analytical expressions for these derivatives, in the general case, were obtained by Lawley (1967). There was a drawback though. The computation of the inverse matrix of second derivatives (an estimate of the information matrix) at every iteration for matrices of hundreds of elements was prohibitive for the computational power available in the late 1960s. The Davidon–Fletcher–Powell algorithm solved this problem elegantly by approximating the information matrix **E** in successive iterations "using the information built up about the function" (Jöreskog, 1969, p. 187). In Jöreskog's application, the method of steepest descent was used in the beginning for five iterations followed by computations of **E** in successive iterations designed to approach the minimum, once the steepest descent iterations came pretty close to it.

Although the method was first presented for CFA, it can be extended to analysis of more general covariance structures, as shown by Jöreskog (1970). Another type of extension is to a simultaneous CFA in several populations. The latter topic had a distinguished history starting with Pearson and continuing with Lawley (1943–44) and Meredith (1964). The problem was to specify under what conditions one could expect a factor pattern to be invariant over several populations and how to determine this common factor pattern. The last problem can be investigated using a general CFA model and imposing constraints sequentially. For example, if a model that imposes equality of loadings is acceptable, the researcher can impose, in addition, the constraint that the unique variances are equal, etc.

Likelihood ratio tests

ML estimators allow also for the construction of significance tests, as was shown in 1928 by Neyman and Pearson. The idea is to consider a null hypothesis that indicates that several parameters have specified values versus an alternative hypothesis that does not require any specific value for parameters. We consider the likelihood ratio created by maximizing the likelihood function under the constraints of the null hypothesis versus an unconstrained maximization. Lawley used the method in exploratory factor analysis via MLE to test the hypothesis that specifies the number of factors to be a specific number. Following Lawley & Maxwell (1963), we can specify the log likelihood under the assumption of a true null hypothesis (e.g., that the number of factors $= k$) by substituting the estimated matrix of population variances–covariance in the formula for log likelihood. Let us call this quantity L_0. Similarly, assuming that the parameters can have any value, we will maximize the log likelihood if we assume that the covariance matrix in the population is as found in the sample. Substituting the sample matrix–covariance matrix in the log likelihood formula we obtain L_1. Then, in large samples, $2(L_1 - L_0)$ is distributed approximately as χ^2, with a number of degrees of freedom equal to the number of variances and covariance-effective number of parameters estimated. The same procedure can be applied to test for a particular model that specifies some loadings to be fixed in CFA (Lawley & Maxwell, 1963). In this case, the number of degrees of freedom for χ^2 is given by subtracting from the number of variances–covariances the number of loadings estimated and the number of residuals for variables (assuming that the factors are not correlated).

Identification problems in confirmatory factor analysis

The reader is already aware of the indeterminacy problem in exploratory factor analysis (EFA). The problem of identification, however, goes beyond the problem of indeterminacy. To see this, consider a simple situation in which a solution was found that would reproduce well the correlation between two variables using two factors that are also allowed to correlate. The solution cannot be unique, since we can reproduce the same correlation by changing both the loadings and the factor intercorrelation! There is, therefore, an infinite number of solutions. A similar situation will happen with, say, two factors subsuming four variables, unless further restrictions are placed on the $\mathbf{\Lambda}$ matrix.

In EFA the investigation of identification will typically consider a hypothetical solution and will investigate the conditions that this solution should obey in order to be unique. Much of this investigation was related to Thurstone's concept of a simple structure. Thurstone tried to formulate a set of rules that will ensure the uniqueness of the solution. For example he required (rule 1) that each row (corresponding to a variable or test) in the oblique factor matrix have at least a zero and (rule 2) that for each column (factor) there should be a number of zeros equal at least to the number of factors (Thurstone, 1947, p. 335). Moreover (still rule 2), the tests corresponding to these zeros were required to be linearly independent, a requirement that clearly cannot be satisfied, since it contradicts the first part of the rule – the existence of a minimum number of zeros (cf. Anderson & Rubin, 1956, p. 122). Other conditions, such as that (rule 4) for any two columns (factors) there should be a large proportion of tests with zero loadings on these factors, were not mathematically precise. Reiersøl (1950) modified Thurstone's rules and formulated accurate, necessary and sufficient conditions of unique identification of the simple structure, under the assumption that the residual matrix, $\mathbf{\Psi}$, is identified. Closely related to these investigations are those (CFA investigations) in which elements of the $\mathbf{\Lambda}$ matrix are specified *a priori* (usually 0s and 1s) and the question is whether or not the estimated loadings for the estimated elements are unique under rotation. Rules for uniqueness were provided independently by Howe (1955) and by Anderson & Rubin (1956) and were also used by Jöreskog (1966, 1969, 1979), who defined a unique solution to exist "if all linear transformations that leave the fixed parameters unchanged also leave the free parameters unchanged" (*ibid.*, 1969, p. 185). Uniqueness, in this sense, ensures only that an alternative solution cannot be obtained by rotation (e.g., multiplication on the right of the weight matrix by a nonsingular matrix). It does not ensure that there is no other solution (that cannot be obtained by rotation), as it was thought by many for some time (see Bollen & Jöreskog, 1985, on this point).

Anderson & Rubin (1956) in their remarkable "Statistical inference in factor analysis" (where they also discuss estimation in cases of *a priori* specification of loadings) were able to formulate a sufficient condition for local identification. Later, a simple sufficient two-indicator rule for identification was given by Wiley (1973; see also Bollen, 1989, pp. 244–6) requiring each variable to load on one factor and to have a residual that does not correlate to any other residual. Generally speaking, the study of identification of CFA in the 1960s and 70s was less satisfactory than the parallel study in econometrics (cf. Davis, 1993). However, the fact that there were no good, practical rules to establish identification status (other than to prove identification algebraically by showing that parameters can be expressed as functions of variances–covariances) was not a serious obstacle for the formulation of a general model (see The LISREL model, below).

General comment on exploratory and confirmatory factor analysis

The developments in EFA and CFA for about half of the twentieth century were not influenced much by path analysis and by developments in econometrics. Nevertheless, as mentioned by Bentler (1986), there were a few exceptions. Thurstone, for example, in his 1947 *Multiple Factor Analysis* (pp. 417–418) presented two second-order factor models using path diagrams of the type introduced by Wright. In addition, the development of CFA, in which the researcher typically specifies some loadings to be fixed to zero, brought the field close to econometric models that also specify many possible paths to be zero. It is not by happenstance that Anderson & Rubin, whose 1956 paper included important innovations in the field of factor analysis (estimation of models with loadings specified to be zero and discussion of identification issues), started their collaboration on the topic of factor analysis while being associates of the Cowles Commission for Research in Economics. It took one more decade, however, for the need for a synthesis to be clearly recognized.

The need for a synthesis

The need for the formulation of a model that will accommodate the needs of psychologists, sociologists, and econometricians had increasingly become clear by the late 1960s. The awareness regarding communalities between the different disciplines increased as the interdisciplinary interaction intensified. In particular, the econometrician Goldberger played a major role in the great synthesis, both directly by his own contributions, and indirectly by encouraging the interdisciplinary interaction and the creation of a synthesis. In his "Econometrics and psychometrics: a survey of communalities" Goldberger (1971) pointed to the "common elements" in the two disciplines. For example, zero factor loadings in CFA have their parallel in the over-identified econometric models, achieved frequently by fixing path coefficients to zero. Similarly, estimation of loadings in factor analysis using MLE has its parallel in the use of MLE to estimate path coefficients. In both cases, the "trick" is to estimate simultaneously the structural coefficients *and* the error variances (or communalities) left after the coefficients have "done their work".

In addition to this increased awareness of the existence of common elements and common methodological and statistical needs, the advances in the development of numerical methods and in computer power in the 1960s made feasible the creation of computer programs for the implementation of the statistical model. LISREL, both as a mathematical model and as a computer application, was in the making.

The LISREL model: Keesling, Jöreskog, and Wiley

The LISREL model was formulated in essentially equivalent ways by Keesling, Jöreskog, and Wiley in the early 1970s and is called sometimes, for this reason, the JKW model (Bentler, 1980).

Keesling's formulation

James Ward Keesling presented his formulation of the LISREL model in his 1972 dissertation that he submitted when a student in the Department of Education at the University of Chicago. Four chapters represent his original contributions and they were written, according to Keesling's own remarks, under the influence of David Wiley and of the Swedish statistician Herman Wold. In Chapter 7 of his thesis, Keesling presented "a covariance structure model for causal flow analysis". The model is formulated succinctly, using one equation that presents observed variables as functions of other observed variables as well as of unmeasured sources of variation in the endogenous variables. The formulation is in econometric tradition. Variances and covariances of the measured variables (exogenous and endogenous) are expressed as functions of the parameters of the model. This expression was further modified by Keesling to allow for latent variables in two additional roles: as true values of measured variables (thus allowing for the modeling of errors of measurement) and as the source of common variation (common factor) between two parallel measures. In this last formulation the model is close to its most general formulation and is equivalent to the models presented by Jöreskog and Wiley. The use of one matrix equation that incorporates complex matrices makes the structure of relationships less transparent than it is in Jöreskog's or Wiley's formulation. Keesling's treatment is comprehensive, including also a discussion of estimation of the model using maximum likelihood estimation and numerical methods that can be used for minimization. The treatment also includes a discussion of local identification based on the above-mentioned theorem proved by Wald (1950) (see also below).

Jöreskog and Wiley's formulations

Karl Jöreskog presented his LISREL model in a general form at the conference organized and chaired by Arthur Goldberger in November 1970 at Madison, Wisconsin. The conference papers were published in 1973 in the volume *Structural Equation Models in the Social Sciences*, edited by Arthur Goldberger and Otis Duncan. Jöreskog formulated his general model using three equations. The first equation presents the relationship between latent variables, using the distinction between exogenous or independent variables and endogenous or dependent variables. A third vector of variables introduces disturbances or "errors in equations". Two additional equations express observed variables (deviations from means) as sums of latent variables and errors of measurement. The variance–covariance matrix is then expressible as a function of six matrices (of coefficients, variances–covariances of the latent variables and error variances). The model so formulated comes very close to the usual formulation of the LISREL model but it assumes one observed variable for each latent variable. Jöreskog pointed in the chapter, to the necessity to extend the model to accommodate multiple indicators per latent variable and indicated briefly that this goal could be achieved by including two more matrices of loadings that represent the weights that the latent variables receive in relation to a particular indicator. An important part of Jöreskog's presentation deals with the estimation problem that requires the minimization of the maximum likelihood fit function. Standard errors for the parameters are obtained by calculating the inverse of the information matrix at the point of minimization. The method advocated for minimization is the Davidon–Fletcher–Powell method (Fletcher & Powell, 1963) implemented, for the general model, by a LISREL computer program. The algorithm uses an approximation of the inverse of the information matrix that becomes better in successive iterations (see above). Once estimated, a model can be tested using a likelihood ratio test (see above). The value for χ^2 is easily computable from the value of the fitting function at the minimization (in terms of the fitting function given by Jöreskog & Sörbom (1981) in the manual for LISREL 5, this value is particularly simple – $(N - 1) \times F_{\min}$.

Wiley specified a general model in his discussion at the conference of "the identification problem for structural equation models with unmeasured variables". Essentially, the model is similar to the one specified by Keesling and to the model presented by Jöreskog. In his discussion of identification, Wiley, as Keesling did in his dissertation, suggested using a result obtained by Wald (1950) to check for the existence of local identification. Wald considered a case in which a set of (identified) parameters is a function of another set of parameters that one has to determine the identification status of the first set. In our case the first set of (identified) parameters are variances–covariances implied by the model. The second is the set of the parameters themselves. We compute the matrix of first derivatives of the vector of variances–covariances with respect to parameters and then we substitute "reasonable values" for the parameters to obtain numerical values for the first derivatives. The final step is to evaluate the rank of the resulting matrix. According to Wald, for local identification, it is sufficient and necessary that the rank be equal to the number of parameters. Alternatively, as pointed out by Wiley (1973, p. 82), we can rely on the information matrix (calculated by taking the probability limits of the second-order derivatives of the fitting function). This matrix is calculated anyway, at the point of minimization, by the LISREL program (or by an equivalent program) to derive standard deviations of the estimated parameters. Moreover, its rank equals the rank of the matrix of first derivatives, so that we can use it to decide whether the model is locally identified (for a discussion of local identification, see Bollen, 1989).

The new general model brought under one umbrella a large variety of apparently very diverse models (let us say, for example, a second-order factor analytical model – see Jöreskog, 1970 – and an econometric model). In addition, it allowed the combination of those models into even more complex models. The work done, mainly by Jöreskog (for an evaluation of Jöreskog's importance, see also Bentler, 1980, 1986), on implementation of MLE and numerical methods to the estimation of those models and the construction of appropriate computer programs, provided researchers in a variety of disciplinary areas with a new and powerful tool. Since the formulation of the model in the early 1970s, alternative formulations have been offered. Perhaps the best known is the formulation provided by Bentler & Weeks (1979, 1980), who used only two types of variables – independent and dependent – and only three matrices of parameters (fewer matrices mean, however, more complex matrices). The Bentler–Weeks model does not require some of the assumptions of orthogonality of the error terms that the LISREL model requires (e.g., between errors in x and errors in y variables). The following sections are devoted to several important expansions of the general model.

The extension of the LISREL model to include means

In the specification of the LISREL model, variables are defined as deviations from their respective means. Sometimes, however, the researcher might be interested in the means themselves. For example, two populations may differ in the structure of a set of variables (e.g., in the values of the path coefficients or loadings) but also in the mean levels. Sörbom (1974) extended the classical LISREL model to include means. The extension involves the introduction of intercept terms in each of the three equations of the model (see above) and by introducing a fourth parameter for the means of the endogenous variables. The expanded model implies the means for the observed matrices, in addition to covariances. Correspondingly, the fitting function includes an expression for differences between implied and obtained means. The additional four vectors cannot be determined simultaneously within one group because of the fact that means of the observed variables are determined by both intercepts and means of latent variables, and one can compensate for the other. They become identifiable though when more than one group are used and when the relative differences in means, rather than absolute values, are important. For example, Sörbom (1976) examined changes in true scores by constructing a model in which a latent variable with two indicators is measured at two time points. To make the model identified he further considered two populations, such as a treatment and a control group. While we cannot identify change within each one of the two groups, we can identify differences across the groups in amount of change.

Generalized least squares

MLE, as we have seen, was the main method of estimation of SEM. In addition, least square approaches were developed in the early 1970s. Those approaches are based on a comparison between the sample variances–covariances and the population variances–covariances, those that are implied by the (assumably true) model. As in a regression situation, the idea is to determine values for the free parameters that will make the population variances–covariances as close as possible to their sample counterparts. There are, however, problems. In classical regression models one commonly assumes that error variances in y are constant across values of the independent variables (homoskedasticity). Another assumption of zero autocorrelations imposes zero correlations among error variances. Those assumptions are not realistic in most SEM applications. For example, the scale of variables is almost always arbitrary to some extent. A change in scale will affect unweighted least squares that minimize (sum of) squared differences between the two types of covariances. Similarly, sample covariances are likely to be correlated for models that imply relationships among variables. To understand this, consider a population of couples and four variables measured: height and weight per unit of height (index of slimness) for the male and for the female partner. Intuitively, it is clear that a sample which, by chance, shows a higher than average covariance between the two properties in males, is likely to show a higher than average covariance between the two properties in females as well! The unweighted least squares approach was therefore replaced by a generalized least squares approach, previously developed by Aitken (1934) in the context of linear regression. Here the squares are weighted using as a weight matrix, the inverse of the matrix, of variances–covariances of the error terms.

The idea of applying a generalized least squares approach to SEM was first developed for unrestricted factor analysis models by Jöreskog &

Goldberger (1972). Using the analogy with Aitken's generalized least squares (GLS) approach, the weight matrix should be a matrix of variances–covariances of the residuals obtained for the observed variables inversed. Each residual is a difference between a sample and a population covariance, so that this weight matrix is obtained by reversing a matrix with elements expressing sampling variability of sample variances and covariances. Fortunately, under some assumptions that are always fulfilled in the case of multivariate normal distribution, this matrix can be expressed by a product of the population variance–covariance matrix by itself (and multiplied by $2/N - 1$). The last matrix can be now be estimated using the sample matrix **S**. Jöreskog & Goldberger demonstrated that the GLS estimator so obtained is scale free and it has asymptotically the same nice properties that the MLE has – e.g., minimum variance.

Along the same lines, Anderson (1973) analyzed GLS estimators for covariance structures that can be expressed as a linear function of parameters. Generally, however, covariance structures are nonlinear functions of parameters. It was left to Browne (1974) to show that, under general conditions, GLS estimators can be applied to linear and nonlinear covariance structures. He demonstrated that they are consistent, asymptotically normally distributed and asymptotically efficient. In fact, Browne showed that a GLS estimator obtained by using the covariance matrix implied by the MLE estimator will be very close to the MLE estimators in large samples. In this sense MLE estimators can be considered as members of the class of generalized least squares estimators. As in the case of MLE, the value of the fit function at the point of minimization multiplied by $N - 1$ produces a statistic that has, at limit (e.g., for very large samples), a χ^2 distribution with a number of degrees of freedom equal to the number of variances–covariances minus the number of parameters estimated.

The extension of the LISREL model to include dichotomous and ordered variables

Dichotomous variables may be considered to be expressions of continuous latent variables that "underlie" the observed variables. For example, a subject can correctly answer an item if his or her "response strength" exceeds a certain threshold (Bock & Lieberman, 1970, p. 181). This approach (which goes back to the calculation of tetrachoric correlation coefficients by Pearson (1901)) was first taken by Bock & Lieberman (1970), Christoffersson (1975) and Muthén (1978). Subsequently, Muthén generalized the model to deal with simultaneous factor analysis of dichotomous variables in multiple groups (Muthén & Christoffersson, 1981) and eventually presented a general SEM that can handle dichotomous, ordered and continuous variables (Muthén, 1984). The model can deal with situations that assume normality of the latent endogenous variables but not necessarily of the exogenous variables ($\boldsymbol{x}$). In the case of dichotomous and ordered variables polychoric correlations are estimated according to Olsson's (1979) two-step method.

Distribution free estimators

The MLE and the GLS method of estimation (as presented above) require rather stringent assumptions. For example, the MLE method is commonly based on the maximization of a log L function, which assumes a Wishart distribution of sample variances and covariances. This distribution, in turn, assumes multivariate normality of the observed variables. The GLS requires, strictly speaking, less stringent assumptions (only the assumption of null fourth cumulants, an assumption that is satisfied by distributions that show no excessive kurtosis), but in practice the difference is not large. Strong violations of these assumptions may translate into wrong coefficients, wrong standard deviations and wrong tests. It would be nice to have a method of estimation that requires less demanding assumptions. This was Browne's motivation in developing the theory for "asymptotically distribution free best GLS estimators" in his 1982 and 1984 publications. His remarkable insight/result was that the use of the matrix of variances–covariances of sample covariances as the weight matrix will make the resultant estimators "distribution free".

In Browne's treatment of the problem, variables are considered to be continuous but not necessarily displaying a multivariate normal distribution and not even satisfying the somewhat weaker condition required by the standard GLS (see previous section). Browne considered minimization of GLS functions that use a weight matrix of a certain type. Minimization generates parameters that are consistent and have at limit a multivariate normal distribution. Moreover, selecting the weight matrix to be a covariance matrix of the residuals (see above) multiplied by $(N-1)^{1/2}$ will produce estimators with asymptotic variances that are minimal for estimators of this kind (so-called "best generalized least square", or BGLS, estimator). In addition, substituting the parameters in the function that was minimized produces a statistic that, as $N \rightarrow$ infinity, has a χ^2 distribution with $\frac{1}{2}p(p+1)$ minus q degrees of freedom. Here, $k = \frac{1}{2}p(p+1)$ is the number of variances–covariances among p variables and q is the number of estimated parameters.

The weight matrix can be estimated using the matrix of fourth moments (obtained by multiplying deviations from the mean for four variables and dividing by N) and the matrix of variances–covariances. Therefore the use of distribution free methods require as an input matrix a matrix of order $k \times k$, where $k = \frac{1}{2}p(p+1)$ and this matrix needs to be further inversed. With 50 variables this means inverting a matrix of more than a thousand elements. To deal with this problem, Browne (1984) developed a theory of estimation for elliptical distributions that include, but are not limited to, normal distributions. Similarly, Bentler & Dijkstra (1985) suggested other ways to reduce the computational burden.

Indices of fit

In many applications of SEM, tests of significance, although usually available, are not sufficient. One reason for this is the status of the hypothesis to be confirmed/disconfirmed. This is a null hypothesis. Confirmation may be a result of a small sample. Disconfirmation of a very good but not perfect model may be a result of using a very large sample. Keesling, in his 1972 dissertation, cautions about the interpretation of the likelihood ratio test, since "the number of degrees of freedom on which the estimated dispersion matrix, **S**, is based is a multiplier in the formula for the likelihood ratio Chi-Square statistic" (*ibid.*, p. 101–2). Some indices, such as the nonnormed fit index (Bentler & Bonett, 1980), were adapted from EFA. Thus Tucker & Lewis (1973) provided an index to evaluate an exploratory model with k factors, estimated via MLE, by calculating the increment in fit *vis-à-vis* a null model that assumes that variables are independent (zero factors). Bentler & Bonett generalized this index to any kind of covariance structure model and proposed also a "normed fit index", with the property that it lies in the interval (0, 1). Typically, the null model will be a highly restrictive model, although in some cases it still can contain free parameters (Bentler & Bonett, 1980, p. 600).

A somewhat different approach is reflected by LISREL's indices of fit introduced first in LISREL V (Jöreskog & Sörbom, 1981). Those include the goodness-of-fit index (GFI), the adjusted goodness-of-fit index (AGFI) and the root mean square residual (RMSR) and are based on a comparison between the sample variance–covariances and the ones fitted or implied by the model. The magnitude of the χ^2 statistic itself, divided by the number of corresponding number of degrees of freedom, may serve also as an indication of fit, and Jöreskog & Sörbom (1981) recommended it. Other indices of fit incorporate an adjustment for model parsimony (James *et al.*, 1982).

A one-factor model explaining covariances between, say, 10 variables, may produce a better parsimonious index than a two-factor model explaining the same, although the last model will reproduce the sample variances–covariances with more fidelity. The AGFI also includes an adjustment for parsimony.

Most indices of fit mentioned have unknown sampling distributions and their behavior (e.g., dependability on sample size) was studied in the 1980s in simulation studies. A large number of additional indices were developed during the 1980s and 90s and many were incorporated in SEM computer programs such as LISREL and EQS.

In addition to indices of fit, more specific indices may help in respecifying the model to fit better. In the LISREL program these are modification indices (MI) developed by Sörbom (see Jöreskog & Sörbom, 1981, 1989; Sörbom, 1989) that indicate the reduction in χ^2 to be expected by freeing a particular path that was fixed in the original model. A generalized version of the MI is the Lagrange multiplier (included in the EQS program) that will test for the release of a set of path coefficients (Buse, 1982; Bentler, 1986). The Wald test mentioned above is used to test for the inverse operation of fixing paths that were originally allowed to be estimated (see, e.g., Bollen, 1989).

Conclusion

The history of SEM reflects in many respects developments in statistical theory, methodology, and philosophy of science in the twentieth century. While, over the years, SEM has had its virulent critics (and still does), the past two decades have seen a tremendous increase in implementation of SEM made increasingly easier by the existence of very efficient computer programs that use a variety of estimation methods and provide additional information needed to assess fit and respecify models, if needed. The rich development of SEM continues today with the publication of a journal devoted to SEM issues (*Structural Equation Modeling*) and of numerous technical and methodological journal publications and books on SEM topics.

References

Aitken, A. C. (1934). On least-squares and linear combinations of observations. *Proceedings of the Royal Society of Edinburgh*, **55**, 42–48.

Anderson, T. W. (1971). *The Statistical Analysis of Time Series.* New York: Wiley.

Anderson, T. W. (1973). Asymptotically efficient estimation of covariance matrices with linear structure. *Annals of Statistics*, **12**, 1–45.

Anderson, T. W. & Rubin, H. (1949). Estimation of the parameters of a single stochastic difference equation in a complete system. *Annals of Mathematical Statistics*, **20**, 46–63. Included in Cowles Commision Paper, New series, no. 36.

Anderson, T. W. & Rubin, H (1956). Statistical inference in factor analysis. In J. Neyman (ed.), *Proceedings of the Third Berkeley Symposium on Mathematical Statistics and Probability*, vol. V, pp. 111–150. Berkeley, CA: University of California Press.

Asher, H. B. (1976). *Causal Modeling*, Beverly Hills, CA: Sage.

Bassman, R. L. (1957). A generalized classical method of linear estimation of coefficients in a structural equation. *Econometrica*, **25**, 77–83.

Bentler, P. M. (1980). Multivariate analysis with latent variables: causal modeling. *Annual Review of Psychology*, **31**, 419–456.

Bentler, P. M. (1986). Structural modeling and *Psychometrika*: an historical perspective on growth and achievements. *Psychometrika*, **51**, 35–51.

Bentler, P. M. & Bonett, D. G. (1980). Significance tests and goodness of fit in the analysis of covariance structures. *Psychological Bulletin*, **88**, 588–606.

Bentler, P. M. & Dijkstra, T. (1985). Efficient estimation via linearization in structural models. In P. R. Krishnaiah (ed.), *Multivariate Analysis*, VI, pp. 9–42. Amsterdam: North-Holland.

Bentler, P. M. & Weeks, D. G. (1979). Interrelations among models for the analysis of moment structures. *Multivariate Behavioral Research*, **14**, 169–185.

Bentler, P. M. & Weeks, D. G. (1980). Linear structural equations with latent variables. *Psychometrika*, **45**, 289–308.

Blalock, H. M. (1962). Four-variable causal models and partial correlations. *American Journal of Sociology*, **68**, 182–194.

Blalock, H. M. (1964). *Causal Inferences in Nonexperimental Research*. Chapel Hill, NC: University of North Carolina.

Blalock, H. M. (1969). *Theory Construction*. Englewood Cliffs, NJ: Prentice-Hall, Inc.

Bock, R. D. (1960). Components of variance analysis as a structural and a discriminal analysis for psychological tests. *British Journal of Statistical Psychology*, **13**, 151–163.

Bock, R. D. & Bargmann, R. E. (1966). Analysis of covariance structures. *Psychometrika*, **31**, 507–534.

Bock, R. D. & Lieberman, M. (1970). Fitting a response model for *n* dichotomously scored items. *Psychometrika*, **35**, 179–197.

Bohrnstedt, G. W. (1969). Observations on the measurement of change. In E. F. Borgatta (ed.), *Sociological Methodology*, pp. 113–133. San Francisco: Jossey Bass.

Bollen, K. A. (1989). *Structural Equations with Latent Variables*. New York: Wiley.

Bollen, K. A. & Jöreskog, K. G. (1985). Uniqueness does not imply identification: a note on confirmatory factor analysis. *Sociological Methods and Research*, **14**, 155–163.

Boudon, R. (1965). A method of linear causal analysis: dependence analysis. *American Sociological Review*, **30**, 365–374.

Boudon, R. (1968). A new look at correlation analysis. In H. M. Blalock & A. B. Blalock (eds.), *Methodology in Social Research*, pp. 199–235. New York: McGraw Hill.

Browne, M. W. (1974). Generalized least square estimators in the analysis of covariance structures. *South African Statistical Journal*, **8**, 1–24.

Browne, M. W. (1982). Covariance structures. In D. M. Hawkins (ed.), *Topics in Multivariate Analysis*, pp. 72–141. Cambridge: Cambridge University Press.

Browne, M. W. (1984). Asymptotic distribution free methods in analysis of covariance structures. *British Journal of Mathematical and Statistical Psychology*, **37**, 62–83.

Burt, C. (1940). *The Factors of Mind.* London: University of London Press.

Burt, C. (1947). Factor analysis and analysis of variance. *British Journal of Psychological Statistics Sect.*, **1**, 3–26.

Burt, C. (1950). Group factor analysis. *British Journal of Psychological Statistics Sect.*, **3**, 40–76.

Burt, C. (1951). The factorial study of the mind. In *Essays in Psychology dedicated to David Katz*, pp. 18–45. Uppsala: Almqvist & Wiksell.

Burt, C. (1952). Tests of significance in factor studies. *British Journal of Psychological Statistics, Sect.* **5**, 109–133.

Buse, A. (1982). The likelihood ratio, Wald and Lagrange multiplier tests: an expository note. *American Statistician*, **36**, 153–157.

Carnap, R. (1950). *Logical Foundations of Probability.* Chicago: University of Chicago Press.

Christoffersson, A. (1975). Factor analysis of dichotomized variables. *Psychometrika*, **40**, 5–32.

Costner, H. L. & Schoenberg, R. (1973). Diagnosing indicator ills in multiple indicators models. In A. S. Goldberger & O. D. Duncan (eds.), *Structural Equationns in Social Sciences*, pp. 167–199. New York: Seminar Press.

Cramér, H. (1946). *Mathematical Methods of Statistics.* Princeton, NJ: Princeton University Press.

Creasy, M. A. (1957). Analysis of variance as an alternative to factor analysis. *Journal of the Royal Statistical Society, Series B*, **19**, 318–325.

Davis, W. R. (1993). The FC1 rule of identification for confirmatory factor analysis. *Sociological Methods and Research*, **21**, 403–437.

Duncan, O. D. (1966). Path analysis: sociological examples. *American Journal of Sociology*, **72**, 1–16.

Duncan, O. D. (1969). Some linear models for two-wave, two-variable panel analysis. *Psychological Bulletin*, **72**, 177–182.

Duo, Q. (1993). *The Formation of Econometrics.* Oxford: Clarendon Press.

Epstein, R. (1989). The fall of OLS in structural estimation. In N. de Marchi & C. Gilbert (eds.), *History and Methodology of Econometrics*, pp. 94–107. Oxford: Clarendon Press.

Fisher, F. M. (1959). Generalization of the rank and order conditions for identifiability. *Econometrica*, **27**, 431–447.

Fletcher, R. & Powell, M. J. D. (1963). A rapidly convergent decent method for minimization. *Computer Journal*, **6**, 163–168.

Frisch, R. (1934). *Statistical Confluence Analysis by Means of Complete Regression Systems.* Economics Institute Publication, no. 5. Oslo: Oslo University.

Gilbert, C. (1989). LSE and the British approach in time series econometrics. In N. de Marchi & C. Gilbert (eds.), *History and Methodology of Econometrics*, pp. 108–128. Oxford: Clarendon Press.

Goldberger, A. S. (1964). *Econometric Theory.* New York: Wiley.

Goldberger, A. S. (1970). On Boudon's method of linear causal analysis. *American Sociological Review*, **35**, 97–101.

Goldberger, A. S. (1971). Econometrics and psychometrics: a survey of commonalities, *Psychometrika*, **36**, 83–107.

Goldberger, A. S. (1972). Maximum likelihood estimation of regressions containing unobservable independent variables. *International Economic Review*, **13**, 1–15.

Goldberger, A. S. & Duncan, O. D. (eds.). (1973). *Structural Equation Models in the Social Sciences.* New York: Seminar Press.

Gorsuch, R. L. (1983). *Factor Analysis.* Hillsdale, NJ: Lawrence Erlbaum Associates.

Griliches, Z. (1974). Errors in variables and other unobservables. In D. J. Aigner & A. S. Goldberger (eds.), *Latent Variables in Socio-economic Models*, pp. 1–37. Amsterdam: North-Holland.

Guttman, L. (1952). Multiple group methods for common-factor analysis: their basis, computation, and interpretation. *Psychometrika*, **17**, 209–222.

Haavelmo, T. (1943). The statistical implications of a system of simultaneous equations. *Econometrica*, **11**, 1–12.

Hartley, R. E. (1952). On the logical foundations of factor analysis. thesis no. 1638, University of Chicago.

Heise, D. R. (1969). Problems in path analysis and path inference. In E. F. Borgatta (ed.), *Sociological Methodology 1969*, pp. 38–73. San Francisco: Jossey Bass.

Heise, D. R. (1970). Causal inference from panel data. In E.F. Borgatta & G. W. Bohrnstedt (eds.) *Sociological Methodology*, pp. 3–27. San Francisco: Jossey Bass.

Henrysson, S. (1957). *Applicability of Factor Analysis in the Behavioral Sciences.* Stockholm: Almqvist & Wiksell.

Hotelling, H. (1933). Analysis of a complex of statistical variables into principal components. *Journal of Educational Psychology*, **24**, 417–441, 498–520.

Howe, W. G. (1955). *Some Contributions to Factor Analysis*. Report no. ORNL-1919. Oak Ridge, TN: Oak Ridge National Laboratory.

James, L. R., Mulaik, S. A. & Brett, J. M. (1982). *Causal Analysis*. Beverly Hills, CA: Sage.

Johnston, J. (1972). *Econometric Methods*. New York: McGraw Hill.

Jöreskog, K. G. (1966). Testing a simple structure hypothesis in factor analysis. *Psychometrika*, **31**, 165–178.

Jöreskog, K. G. (1967). Some contributions to maximum likelihood factor analysis. *Psychometrika*, **32**, 443–482.

Jöreskog, K. G. (1969). A general approach to confirmatory maximum likelihood factor analysis. *Psychometrika*, **34**, 183–202.

Jöreskog, K. G. (1970). A general method for analysis of covariance structures, *Biometrika*, **57**, 239–251.

Jöreskog, K. G. (1973). A general method for estimating a linear structural equation system. In A. S. Goldberger & O. D. Duncan (eds.), *Structural Equation Models in the Social Sciences*. New York: Seminar Press.

Jöreskog, K. G. (1979). Statistical estimation of structural models in longitudinal developmental investigations. In J. R. Nesselroade & P. B. Baltes (eds.), *Longitudinal Research in the Study of Behavior and Development*, pp. 303–351. New York: Academic Press.

Jöreskog, K. G. & Goldberger, A. S. (1972). Factor analysis by generalized least squares. *Psychometrika*, **37**, 243–260.

Jöreskog, K. G. & Sörbom, D. (1977). Statistical models and methods for analysis of longitudinal data. In D. J. Aigner & A. S. Goldberger (eds.), *Latent Variables in Socio-economic Models*, pp. 285–325. Amsterdam: North-Holland.

Jöreskog, K. G. & Sörbom, D. (eds.) (1979). *Advances in Factor Analysis and Structural Equation Models*. Cambridge, MA: Abt Books.

Jöreskog, K. G. & Sörbom, D. (1981). *LISREL V: Analysis of Linear Structural Relationships by Maximum Likelihood and Least Squares Methods*. Chicago: National Educational Resources. Distributed by International Educational Services, Chicago.

Jöreskog, K. G. & Sörbom, D. (1989). *LISREL 7 – A Guide to the Program and Applications*, 2nd edition. Chicago: SPSS Publications.

Keesling, J. W. (1972). Maximum Likelihood Approaches to Causal Flow Analysis. Ph.D. dissertation, Department of Education, University of Chicago.

Koopmans, T. C. (1937). *Linear Regression Analysis in Economic Time Series*. Netherlands Economic Institute.

Koopmans, T. C. (1949). Identification problems in economic model construction. *Econometrica*, **17**, 125–143.

Koopmans, T. C., Rubin, H. & Leipnik, R. B. (1950). Measuring the equation systems of dynamic economics. In T. C. Koopmans (ed.) *Statistical Inference in*

Dynamic Economic Models, Cowles Commission for Research in Economics, Monograph no. 10, pp. 53–237. New York: Wiley.

Lawley, D. N. (1940). The estimation of factor loadings by the method of maximum likelihood. *Proceedings of the Royal Society of Edinburgh*, **60**, 64–82.

Lawley, D. N. (1942). Further investigations in factor estimation. *Proceedings of the Royal Society of Edinburgh*, **61**, 175–185.

Lawley, D. N. (1943). The application of the maximum likelihood method to factor analysis. *British Journal of Psychology*, **33**, 172–175.

Lawley, D. N. (1943–44). A note on Karl Pearson's selection formulae. *Proceedings of the Royal Society of Edinburgh, Section A*, **62**, 28–30.

Lawley, D. N. (1958). Estimation in factor analysis under various initial assumptions. *British Journal of Statistical Psychology*, **11**, 1–12.

Lawley D. N. (1967). Estimation in factor analysis under various initial assumptions. *British Journal of Statistical Psychology*, **11**, 1–12.

Lawley, D. N. & Maxwell, A. E. (1963). *Factor Analysis as a Statistical Method.* New York: Butterworths.

MacCorquodale, K. & Meehl, P. E. (1948). On a distinction between hypothetical constructs and intervening variables. *Psychological Review*, **55**, 95–107.

Mann, H. B. & Wald, E. (1943). On the statistical treatment of linear stochastic difference equations. *Econometrica*, **11**, 173–220.

Meredith, W. (1964). Notes on factorial invariance. *Psychometrika*, **29**, 177–185.

Mulaik, S. A. (1989). Confirmatory factor analysis. In J. R. Nesselroade & R. B. Cattell (eds.), *Handbook of Multivariate Experimental Psychology*, pp. 259–288. New York: Academic Press.

Muthén, B. (1978). Contributions to factor analysis of dichotomous variables. *Psychometrika*, **43**, 551–560.

Muthén, B. (1984). A general structure equation model with dichotomous, ordered categorical, and continuous latent variable indicators. *Psychometrika*, **49**, 115–132.

Muthén, B. & Christoffersson, A. (1981). Simultaneous factor analysis of several groups of dichotomous variables. *Psychometrika*, **46**, 407–419.

Neyman, J. & Pearson, E. S. (1928). On the use and interpretation of certain test criteria for the purposes of statistical inference. *Biometrika*, **20A**, Part I, 175–240, and Part II, 263–294.

Northrop, F. S. C. (1952). *The Logic of the Sciences and the Humanities.* New York: Macmillan.

Olsson, U. (1979). Maximum likelihood estimation of the polychoric correlation coefficient. *Psychometrika*, **44**, 443–460.

Pearson, K. (1901). Mathematical contributions to the theory of evolution. VII: On the correlation of characters not quantitatively measurable. *Philosophical Transactions of the Royal Society of London, Series A*, **195**, 1–47.

Rao, C. R. (1945). Information and accuracy attainable in the estimation of statistical parameters. *Bulletin of the Calcutta Mathematical Society*, **37**, 81.

Reiersøl, O. (1950). On the identifiability of parameters in Thurstone's multiple factor analysis. *Psychometrika*, **15**, 121–149.

Sargan, J. D. (1980). Some tests of dynamic specification. *Econometrica*, **48**, 879–897.

Simon, H. A. (1954). Spurious correlation: a causal interpretation. *Journal of the American Statistical Association*, **49**, 467–479.

Simon, H. A. (1957). *Models of Man*. New York: Wiley.

Sörbom, D. (1974). A general method for studying differences in factor means and factor structures between groups. *British Journal of Mathematical and Statistical Psychology*, **27**, 229–239.

Sörbom, D. (1976). A statistical model for the measurement of change in true scores. In D. N. M. de Gruijter & J. L. Th. Van der Kamp (eds.), *Advances in Psychological and Educational Measurement*, pp. 159–169. New York: Wiley.

Sörbom, D. (1989). Model modification. *Psychometrika*, **54**, 371–384.

Spearman, C. (1904). "General intelligence", objectively determined and measured. *American Journal of Psychology*, **15**, 201–292.

Spearman, C. (1923). *The Nature of "Intelligence" and the Principles of Cognition*. London: Macmillan.

Spearman, C. (1927). *The Abilities of Man*. New York: J. J. Little and Ives Company.

Theil, H. (1953). *Estimation and Simultaneous Correlation in Complete Equation Systems*. The Hague: Centraal Planbureau.

Thomson, G. H. (1936). Some points of mathematical technique in the factorial analysis of ability. *Journal of Educational Psychology*, **27**, 37–54.

Thurstone, L. L. (1931). Multiple factor analysis. *Psychological Review*, **XXXVIII**, 406–427.

Thurstone, L. L. (1933). *The Theory of Multiple Factors*. Ann Arbor, MI.: Edwards Bros.

Thurstone, L. L. (1935). *The Vectors of the Mind*. Chicago: Chicago University Press.

Thurstone, L. L. (1938). *Primary Mental Abilities*. Chicago: Chicago University Press.

Thurstone, L. L. (1947). *Multiple Factor Analysis*. Chicago: Chicago University Press.

Tucker, L. R. & Lewis, C. (1973). A reliability coefficient for maximum likelihood factor analysis. *Psychometrika*, **38**, 1–10.

Wald, A. (1950). Remarks on the estimation of unknown parameters in incomplete systems of equations. In T. C. Koopmans (ed.), *Statistical Inference in Dynamic Economic Models*. Cowles Commission for Research in Economics, Monograph no. 10, pp. 305–310. New York: Wiley.

Werts, C. E. & Linn, R. L. (1970). Path Analysis: Psychological Examples. *Psychological Bulletin*, **74**, 193–212.

Wiley, D. E. (1973). The identification problem for structural equation models with unmeasured variables. In A. S. Goldberger & O. D. Duncan (eds.), *Structural Equation Models in the Social Sciences*. New York: Seminar Press A.S.

Wold, H. O. A. & Jureen, L. (1953). *Demand Analysis*. New York: Wiley.

Wright, S. (1918). On the nature of size factors. *Genetics*, **3**, 367–374.

Wright, S. (1920). The relative importance of heredity and environment in determining the piebald pattern of guines pigs. *Proceedings of the National Academy of Sciences, USA*, **6**, 320–332.

Wright, S. (1921). Correlation and causation. *Journal of Agricultural Research*, **XX**, 557–215.

Wright, S. (1932). General, group and special size factors. *Genetics*, **17,** 603–619.

Wright, S. (1934). The method of path coefficients. *Annals of Mathematical Statistics*, **V**, 161–215.

Wright, S. (1960). Path coefficients and path regressions: alternative or complementary concepts? *Biometrics*, **16**, 189–202.

Zellner, A. (1970). Estimation of regression relationships containing unobservable independent variables. *International Economic Review*, **11**, 441–454.

Zellner, A. & Theil, H. (1962). Three stage least squares; simultaneous estimation of simultaneous equations. *Econometrica*, **30**, 54–78.

5 Guidelines for the implementation and publication of structural equation models

Adrian Tomer and Bruce H. Pugesek

Abstract

Although distinctions are commonly made between exploratory and confirmatory models, structural equation modeling (SEM) is not an exploratory statistical method per se. The successful implementation of a structural equation model requires considerable *a priori* knowledge of the subject matter under investigation. The researcher will usually have a theoretical model in mind to test, and measurement instruments, including latent variables, devised to measure and relate constructs within the model. Research with SEM is a process in which theory is devised, data are collected and analyzed, and models are tested, modified, and confirmed with new data in an iterative fashion. In this context, SEM is rightly viewed as a confirmatory method. As a consequence, a number of epistemological and technical issues require consideration over and above the pure mathematics of the SEM model. In this chapter, we provide background on the philosophical aspects of the study of dependence relationships with SEM, the formulation of latent constructs, model justification, model identification, sample size and power, estimation methods, evaluation of model fit, model modification, interpretation of results, and publication of results. Our objective is to provide a guide that researchers can use to successfully devise and report the results of an SEM study.

Cause and correlation

Biologists have been slow to embrace and employ structural equation modeling (SEM). At the same time, its use has blossomed in other areas of inquiry such as econometrics, sociology, and psychology. Ironically, the foundation for SEM was laid by the eminent evolutionary geneticist Sewall Wright, with his seminal work on path analysis (Wright, 1921). SEM is, after all, an extension of path analysis in which latent variables are used in place of single indicator variables.

Shipley (2000) suggested that the predominant philosophical view of the era in which path analysis was introduced caused organismal biologists to move away from path modeling, and that this philosophical view has been passed on from one generation of biologists to the next until the present day. Wright's work on path analysis was immediately criticized on philosophical grounds by Niles (1922), who wrote the following:

> "Causation" has been popularly used to express the condition of association, when applied to natural phenomena. There is no philosophical basis for giving it a wider meaning than partial or absolute association.

Fisher (1925) also viewed causation as separate from correlation, and developed new experimental designs with randomization and experimental control as the basis for studying causation. His experimental designs complete with inferential tests of significance found favor for their ease of use and their clear-cut application of experimental design to the problem of studying causation. Wright's path analysis, which depended on the assumption that the direction of pathways linking variables was correct, was largely ignored.

Indeed, today's graduate biology programs teach largely Fisher's approach to the study of causation and the philosophy that correlation does not imply causation. Few training biologists learn path analysis. Current thinking among ecologists is that the research attack on natural systems should combine nonexperimental field data with controlled randomized experiments to derive our understanding of natural phenomena (e.g., James & McCulloch, 1990).

Jöreskog (1970) solved many of the problems of path analysis. The use of latent variables in path models enables the researcher to estimate theoretical models that account for error variance. No longer must the researcher assume that modeled direct and indirect pathways are correct. Latent variables also reduce problems of measurement error and bias/distortion to path coefficients inherent in earlier regression models such as path analysis (Pugesek & Tomer, 1995).

Still, logical or plausible assumptions about the direction of some pathways must be made. These can often be tested (see below). However, in many circumstances, the researcher's knowledge can be put to use to make logical conclusions about the directionality of pathways. For example, one can assume that greater habitat quality may result in higher reproductive success, and that higher reproductive success is not likely to improve habitat quality. Thus the pathway linking these two variables in an SEM model would run from habitat quality to reproductive success.

Advances in the field of artificial intelligence, for example, d-separation and its relationship to partial correlations (Shipley, 2000), software programs such as TETRAD (see below) and SEM have modified our view of correlational data and its ability to assert causal relationships. These techniques have changed the landscape with regard to the traditions that have been passed down among biologists. Scientists now make the assertion that "correlation does not imply causation", although true for two variables, is not necessarily true of three or more variables (Spirtes *et al.*, 1993). We suggest that the reader reviews Pearl (1988) for insights on SEM and the study of causation. Also Shipley (2000) provides a discourse on the evolution of thinking among biometricians as well as the strengths and limitations of the various methods that are employed in the study of causation.

Latent variables

Another reason that biologists have been slow to adopt SEM may be that few are accustomed to thinking in terms of measurement with latent variables. The latent variable is key to SEM. Latent variables with multiple indicators make it possible to model error variance and they assist in the process of formulating over-identified models and thus allow both the estimation of models as a set of simultaneous linear equations and the estimation of model fit. Good latent variables are also critical for creating measurement instruments with the highest construct validity and a commensurate reduction of measurement error. Biologists are, however, largely unfamiliar with factor analytical methods and even less so with the notion that factors can themselves become variables with which to measure dependence relationships.

Two major approaches, principal components analysis (PCA) and principal factor analysis (PFA), are considered as data reduction methods (Rummel, 1970; Everritt, 1984) in which patterns of covariation among a set of variables are assembled into factors. PCA is overwhelmingly used by biologists. The distinction between PCA and PFA lies in the ways in which factors are defined and assumptions about the nature of residuals. In PCA, factors are derived in a manner that accounts for the maximum amount of explained variation of the observed variables. In PFA, factors are derived in a manner that maximizes common variance (i.e., variance explained in the observed variables by the factors) (Gorsuch, 1983). PCA assumes that the residuals of observed variables are small, while this is not the case for PFA. PFA assumes that there is a considerable amount of uniqueness in each variable and takes into consideration only that portion of variability that is correlated with other variables.

Confirmatory factor analysis (CFA) used in SEM bears a closer resemblance to PFA in the manner in which factors are estimated. CFA differs from both PCA and PFA in that factors are derived from the researcher's knowledge and theory and not from the actual data. In CFA the researcher specifies which variables comprise factors in the model and then employs a methodology to assess the accuracy of the theorized structure. In PCA and PFA all variables are allowed to load on all factors and the number of factors is determined empirically. Both methods employ predetermined rotational algorithms in order to maximize interpretability. CFA does not employ rotational algorithms. Researchers who wish to explore CFA in more detail should consult Bollen (1989) and Reyment & Jöreskog (1993).

The justification of the proposed model

Models need, of course, to be justified. The type of justification varies widely. Some models are based on existing theory in the sense that, given the theory, most people will agree that the specification of the model follows. In this case we speak about a good match between the theory and the model, one in which the mathematical–statistical constraints of the model clearly emulate the ones specified in natural language by the theory (on this point, see Hayduk, 1996). It is not uncommon, however, that the relationship between the model and the theory is weaker. For example, the researcher might decide to present a relationship between two variables as a one-directional causal relationship, although the theory might suggest a reciprocal relationship. It is also common that, in the desire to specify a model that is identified, researchers simplify a theory or focus on specific parts of it, ignoring other aspects. Moreover, theories are frequently vague, allowing a variety of interpretations and, therefore, a variety of models (e.g., Scheines *et al.*, 1998). In all cases, it is important to present the relationship between the model and the theory in an explicit way that reflects the state of affairs prior to the estimation and evaluation of the model.

It is also possible that the model is, in a sense, the theory. This may happen when a researcher presents a new theoretical formulation in a form that is very close to a SEM formulation. Justification in these cases can be provided using empirical data and/or existing theories to argue that the relationships specified by the model are consistent with existing data and theories and therefore are plausible. While some authors are opposed to the building of integrative models that "reflect the literature" (e.g., Hayduk, 1996, p. 3) and advise commitment to more narrowly defined perspectives, there is a place for syntheses that do not sacrifice too much parsimony and falsifiability. It is also possible that empirical results would "suggest" a particular confirmatory model. The paradigmatic example here is that of an exploratory factor analysis (EFA) that suggests a certain structure or measurement model. Indeed results, such as those obtained by Gerbing & Hamilton (1996), show that EFA can be a very useful heuristic device in the construction of measurement models. The latter can then be further analyzed using CFA and, eventually, cross-validated using new data.

In addition to the proposed model, researchers should consider alternative models and, in particular, equivalent models (see Interpretation of results and Model modification, below).

Specification of the model identification issues

Except for very complex models, the specification of the model (or at least the most important parts of the specification) can be presented graphically. Very complex models might include so many paths that a graphical presentation becomes impractical. Authors may then decide to use a simplified version (e.g., by presenting groups of variables) and by providing a full description of the model in an appendix. If there are concerns regarding identification, those need to be addressed. In many cases it will be clear that the model is identified because it clearly satisfies a sufficient condition for identification. In other cases the identification status might be less clear and the researcher might have only good reasons to believe that the model is locally identified at the point of estimation. A somewhat doubtful identification status should be explicitly acknowledged.

Sample size and number of indicators per latent variable

There are good reasons to prefer larger sample sizes, certainly $N > 100$ and, if possible, $N > 200$ or more (Boomsma, 1982; Gerbing & Anderson, 1993). A larger N will increase power (see Considerations on power, below). Also the estimation of the parameters is based on asymptotic theory and is likely to be distorted in small samples. More is better also in relation to the number of indicators, as shown by Marsh *et al.* (1998) using CFA simulation studies. In particular, the combination of small samples and few indicators per factor (<4) should be avoided, if at all possible.

The selection of the estimation method

The case of a true model

Maximum likelihood estimation (MLE) continues to be the most frequently used method of estimation. When variables display multivariate normality and the theoretical model is a true model, the choice of MLE or generalized least squares (GLS) is likely to be inconsequential for the estimated values of the parameters or the characteristics (chi-square (χ^2), indices of fit, etc.) of the model. When deviations from multivariate normality are severe (e.g., skewness of 2.0 or more and kurtosis of 7.0 or more; see West *et al.*, 1995 Curran *et al.*, 1996), and the sample size is not very large, it seems that corrected MLE or GLS statistics, such as the Satorra–Bentler χ^2 (Satorra, 1990) and robust standard error estimates reported by the EQS program are a good choice (e.g., Hu & Bentler, 1995; Curran *et al.*, 1996; Finch *et al.*, 1997). Such statistics avoid the problem of overestimation of the χ^2 statistic and underestimation of the standard errors that typically occur under conditions of nonnormality. The only case in which the asymptotic distribution-free (ADF) method of estimation (Browne, 1984) appears to have an advantage is one in which the sample size is 1000 or more, the model is relatively simple, and the data show severe deviations from multivariate normality (Chou & Bentler, 1995).

The case of a simplified model

In many, perhaps most cases, the theoretical model is a simplified approximation of the true model in the sense of including relevant variables and paths while also excluding other variables and paths. The researcher would like in this case to obtain realistic fit estimates (that would indicate the need for improving the model) and estimates for the included paths that are not too remote from the true values. Ollson *et al.* (1999), using CFA models, showed that the GLS method provides indices that are inflated (indicating very good "empirical fit" in spite of not so good "theoretical fit") precisely because of the use of distorted values for parameter estimates and misleading modification indices. Assuming that their results can be generalized to other types of model, MLE appears to be the method of preference when the estimated model is an approximation.

Evaluation of fit

Indices of fit were typically evaluated in simulation studies in which models were specified as true (so that sample covariance matrices were generated from the population matrix determined by the model) or as misspecified to some extent (from "small" to "severe").

True models

Good indices of fit should not be influenced too much by sample size or nonnormality. As indicated above, Hoyle & Panter (1995) presented a helpful comparison of several types of indices of fit with general recommendations. Chi-square values for the models should be reported as well as an index of absolute fit such as the goodness-of-fit index (Jöreskog & Sörbom, 1981). Among type 2 indices (Hu & Bentler, 1995), the incremental fit index (IFI) appears to behave well in smaller samples and the nonnormed fit index (Bentler & Bonett, 1980) appears to behave well in sample sizes larger than 150. The comparative fit index (CFI; Bentler, 1989), a type 3 index based on noncentral χ^2 distribution can also be used, particularly when the sample is small and the distribution of variables nonnormal (West *et al.*, 1995).

As indicated by MacCallum *et al.* (1996), it is an important advantage to use indexes of fit with known distributions for which one can specify confidence intervals (CIs). Such an index is the root-mean-square error of approximation (RMSEA) proposed by Steiger & Lind (1980) (see also Browne & Cudek, 1993). Indexes with known distribution allow also a type of hypothesis testing in which the researcher retains the proposed model by rejecting a null hypothesis that specifies a value for the index that would indicate poor fit. This type of hypothesis testing is discussed further under Considerations of power, below.

Misspecified models

A good index of fit would be affected by misspecification of the model but relatively little by sample size, method of estimation, etc. From this point of view, Fan & Wang (1998), and Fan *et al.* (1999), using slightly and moderately misspecified models, found the McDonald's Centrality Index (Mc; McDonald, 1989), the RMSEA and the GFI to be "good performers", in the sense of being sensitive to the degree of misspecification introduced in the model. However, when the indicators have low reliability (low loadings), indices such as GFI and RMSEA are likely to be too "optimistic", failing to reject misspecified models (Bandalos, 1997; Shevlin & Miles, 1998). In such cases, special caution is required (Bandalos, 1997).

Cut-off criteria

For incremental fit indices such as the CFI, a cut-off point of 0.90 is usually adopted as a "rule of thumb" (e.g., Bentler & Bonett, 1980), whereas 0.05 is used for RMSEA (Steiger, 1989; Browne & Cudeck, 1993). The goal is to adopt criteria that would result in not too many seriously misspecified models being retained (a Type II error) without rejecting too many models that may be only correct or only slightly incorrect (a Type I error). This is tricky, particularly because the corresponding probabilities are likely to change with model, parameter values, variable distributions, definition of seriousness of misspecification, etc. Recently more stringent criteria were suggested. Thus Hu & Bentler (1999), basing themselves on simulation work, suggested that combinations of two indices may help to prevent acceptance of too many misspecified models. Assuming MLE, they proposed using a cut-off value of about 0.08 for the SRMR (standardized root mean residual), together with 0.95 for most of the indexes investigated including the CFI and the TLI (Tucker–Lewis index) or NNFI (nonnormed fit index). For Mc a value of 0.90 is recommended and, for RMSEA, 0.06. Using these rules in smaller samples sizes, of 250 or less, tended to over-reject true models when the RMSEA, TLI or Mc were used under conditions of nonnormality.

Indexes for what purpose?

The selection of an index should be based, in part, on the reason for its use. Particularly important is the distinction between prediction versus development of a very accurate causal model (e.g., Kaplan, 1991). If the latter is the case, an index such as RMSEA that estimates the error of approximation seems appropriate. If the goal is prediction, however, an estimate of the "overall error" (Browne & Cudeck, 1993), such as the cross-validation index (CVI; see Browne & Cudeck, 1993) is more pertinent (on this point, see Bandalos, 1997).

Considerations of power

Issues of power in SEM arise when we want to say with some confidence (let us say with 80% probability) that a particular model is correct, or approximately correct, when, in fact, this is the case. The particular model may be, for example, a model that has for one of the parameters a value different from the value specified (e.g., zero) in the model under investigation.

Alternatively, the particular model may be the one investigated when we want to make a confident judgment that this model is approximately correct by rejecting the hypothesis that it is not. In the first case we want to know what is the probability of rejecting our original model given that this model is false. Satorra (1989), who developed with Sarris (Satorra & Saris, 1985) the original approach to the problem of power, suggested using LISREL modification indices, or EQS Lagrange multipliers, together with tables of the noncentral χ^2 distribution, to assess power. Reporting the power of the model may be particularly important when an alternative model that has a nonzero parameter for one of the paths is plausible but allowing the corresponding path coefficient to be estimated does not result in a significantly better model. It is possible that there was not enough power to find out that the alternative model is significantly better.

While small samples make rejection of the hypothesis that the model fits more unlikely (on the basis of the usual χ^2 test), they make it more likely that one can obtain results suggesting that the model is a poor approximation, when in fact the model is pretty good. We are often interested in providing evidence that the model is not a poor approximation but rather a good one (this is the second case mentioned above). An index that is appropriate here is Steiger's RMSEA (see Steiger, 1989; Browne & Cudeck, 1993) that supposedly is less than 0.5 for very good models (close fit), less than 0.1 for models that do not fit poorly, and larger than 0.1 for poor models. Using this index, MacCallum *et al.*, (1996) suggested a test of not-close fit, in which the null hypothesis is that the RMSEA is equal to, or larger than, 0.05 in the population (or, perhaps, larger than 0.1 for a not-poor fit) and the researcher is interested in rejecting this hypothesis when it is not true, therefore concluding in the favor of "close fit". Computer programs such as LISREL, Amos or CALIS provide CI for RMSEA so that the researcher may reject the hypothesis of not-close fit if the upper bound of a (say) 90% CI is less than 0.05. The question of power is the following: what is the probability for the upper bound of RMSEA to be less than 0.05 when, in the population, the RMSEA is very small, let us say 0.01? If this probability is relatively low (let us say much less than 80%), an RMSEA for the model in excess of 0.1 does not constitute a good reason to reject the model. Therefore the power should be reported in those cases in which a poor RMSEA was obtained. Tables of power can be used (see e.g., Loehlin, 1998, p. 264, based on MacCallum *et al.*, 1996) as well as computer programs. A short SAS program for computing power was provided by MacCallum *et al.* (1996).

The interpretation of the results – the issue of equivalent models

Equivalent models (Stelzl, 1986; Lee & Hershberger, 1990; Hershberger, 1994) are defined formally as models that imply identical covariance matrices and have identical χ^2 values. They should be viewed as the equivalent of alternative interpretations or confounds in experimental settings (see MacCallum *et al*, 1993; MacCallum, 1995), and their dismissal as implausible is crucial in establishing the validity of the interpretation. Recent study (Williams *et al.*, 1996) of those models has shown that parameter values of paths common to equivalent values vary dramatically among models, emphasizing the need to be able to eliminate models. In spite of their importance, they are rarely reported by researchers (see MacCallum *et al.*, 1993), perhaps because of lack of awareness and/or because the equivalent models are likely to weaken or undermine their conclusions (MacCallum *et al.*, 1993). The possibility of equivalent models should be considered at the stage of formulation of the model, when the researcher might be able to eliminate them by reformulating the model, by adopting a longitudinal approach, etc. (see MacCallum *et al.*, 1993).

The formulation of simple rules to generate equivalent models (Stelzl, 1986; Lee & Hershberger, 1990; for extensions to multigroup models, see also Raykov, 1997) made an examination of such models a relatively easy task.

In addition to equivalent models one should consider other models that may be plausible theoretically and might fit the data well.

Model modification

The researcher is typically inclined to modify the original model when the fit is not satisfactory and a specification error is suspected. Several tools are used to respecify the model, the most popular being modification indexes (MI; Sörbom, 1989) and Lagrange multipliers (LM; Bentler, 1986). Those are tests for improvement of fit contingent on removal of one restriction (MI) or multiple restrictions (LM) at a time. Additional strategies were developed for model modification and specification search. Thus Marcoulides *et al.* (1998) used a "Tabu search" to select the best model out of a large number of possible models obtained by releasing and/or fixing parameters. The Tabu search method requires a special FORTRAN program together with a SEM program. This is the case also with another method of model modification that examines behavior of covariances in sets of four variables (tetrads) – the vanishing of tetrad differences. The method requires the use of the TETRAD II program (Spirtes *et al.*, 1990; Scheines *et al.*, 1998). The existence of alternative methods of specification search raises the issue of the "best method". Unfortunately there is no clear consensus on this point (for a comparison of TETRAD II on the one hand and LISREL and EQS on the other hand, see e.g., Hayduk, 1996, Chapter 4). However, there is complete consensus on the importance of modifying models taking into account substantive considerations to avoid capitalizing on chance (i.e., on the peculiarities of the sample) too much. In particular, when the sample is small, opportunistic modifications may be completely misleading. In fact, as shown by MacCallum *et al.* (1992), even with samples as large as 400, modifications are likely to differ much across various samples. Complete stability was not achieved even at $N = 1200$. Power considerations may also be brought into play by considering not only the size of MI but also using (fully standardized, see Chou & Bentler, 1990) expected value of the coefficient (EPC) for a particular path were this path to be freely estimated (Kaplan, 1995). First priority should then be given to large MIs and large standardized EPCs that are theoretically plausible.

Opportunistic model modification can sometimes be prevented altogether by *a priori* specification of alternative models that may correspond to different theoretical positions (Jöreskog, 1993). In particular, stronger and weaker versions of a theory can be formulated as nested models that can afterwards be easily compared using χ^2 differences, indices of fit, etc.

Finally, there is a place for data-driven model modification when the model fits poorly, theory is vague, and/or relatively simple modifications are not very helpful. In those cases the researcher, while remembering the need for cross-validation, might decide to have recourse to a Tabu search or to a TETRAD II exploration.

Existing guidelines

Structural equation models analyses, no matter what program was used (LISREL, EQS, AMOS, CALIS, etc.), typically provide an enormous amount of information. Moreover, the processes of formulation and estimation of models, as well as the interpretation of the results and the possible modifications of the model, pose complex questions that the researcher, the reviewer of the article, and the reader have to address. The need for guidelines was increasingly felt by editors of journals facing complex decisions. The article "Reporting structural equation modeling results in psychology and aging: some proposed guidelines", by Raykov *et al.* (1991), was one of the first attempts to deal with this need. Raykov *et al.* formulated several guidelines focusing on the issues of presentation and specification of the model and of the method of estimation (e.g., MLE, GLS, etc.), its evaluation (using χ^2 statistics, indices of fit, etc.) and presentation of the solution (path coefficients, etc.), as well as on issues of model modification (using, for example, modification indexes) and model comparisons (i.e., nested models).

A few recommendations were included in the fourth edition of the *Publication Manual of the American Psychological Association* (1994, pp. 130–131) regarding the presentation of "path and LISREL (linear structural relations) tables". The APA *Manual* (p. 132) points out that "results of analyses of structural models are often presented in a figure" and that model comparisons via χ^2 values and indices of fit should be clearly summarized in a table.

The SEM methodology has recently made important advances and the computer programs implementing SEM today provide an even larger variety of choices of method of estimation, types of scaling of the solution, indices of fit, etc., than were available a decade ago. Substantial progress has been made, in particular in the areas of indices of fit, power, and selection of estimation method. These advances created the need for additional recommendations and guidelines. For example, Hoyle & Panter (1995) provided detailed recommendations for "writing about structural equation models". In particular, they devote a large space to the topic of fit criteria based on the results obtained via simulation studies on the performance of various indices. Other useful suggestions, recommendations and applicable results are spread throughout the literature and we used many of them in the present formulation.

References

American Psychological Association (1994). *Publication Manual of the American Psychological Association*, 4th edition. Washington, DC: APA.

Bandalos, D. L. (1997). Assessing sources of error in structural equation models: the effects of sample size, reliability, and model misspecification. *Structural Equation Modeling*, **4**, 177–192.

Bentler, P. (1986). *Lagrange Multiplier and Wald Tests for EQS and EQS/PC.* Los Angeles: BMDP Statistical Software.

Bentler, P. (1989). *EQS Structural Equations Program Manual.* Los Angeles: BMDP Statistical Software.

Bentler, P. M. & Bonett, D. G. (1980). Significance tests and goodness of fit in the analysis of covariance structures. *Psychological Bulletin*, **88**, 588–606.

Bollen, K. A. (1989). *Structural Equations with Latent Variables.* New York: Wiley.

Boomsma, A. (1982). Robustness of LISREL against small sample sizes in factor analysis models. In K. G. Jöreskog & H. Wold (eds.), *Systems under Indirect*

Observation: Causality Structure, Prediction Part I, pp. 149–173. Amsterdam: North Holland.

Browne, M. W. (1984). Asymptotically distribution-free methods for the analysis of covariance structures. *British Journal of Mathematics and Statistical Psychology*, **37**, 62–83.

Browne, M. W. & Cudeck, R. (1993). Alternative ways of assessing model fit. In K. A. Bollen & J. S. Long (eds.), *Testing Structural Equation Models*, pp. 136–162. Thousand Oaks:, CA: Sage.

Chou, C.-P. & Bentler, P. M. (1990). Model modification in covariance structure modeling: a comparison among likelihood ratio, Lagrange multiplier, and Wald tests. *Multivariate Behavioral Research*, **25**, 115–136.

Chou, C.-P. & Bentler, P. M. (1995). Estimates and tests in structural equation modeling. In R. Hoyle (ed.), *Structural Equation Modeling: Issues, Concepts, and Applications*, pp. 37–55. Newbury Park, CA: Sage.

Curran, P. J., West, S. G. & Finch, J. F. (1996). The robustness of test statistics to nonnormality and specification error in confirmatory factor analysis. *Psychological Methods*, **1**, 16–29.

Everitt, B. S. (1984). *An Introduction to Latent Variable Models.* London: Chapman & Hall.

Fan, X. & Wang, L. (1998). Effects of potential confounding factors on fit indices and parameter estimates for true and misspecified SEM models. *Educational and Psychological Measurement*, **58**, 701–735.

Fan, X., Thompson, B. & Wang, L. (1999). Effects of sample size, estimation methods, and model specification on structural equation modeling fit indexes. *Structural Equation Modeling*, **6**, 56–83.

Finch, J. F., West, S. G. & MacKinnon, D. P. (1997). Effects of sample size and nonnormality on the estimation of mediated effects in latent variable models. *Structural Equation Modeling*, **4**, 87–107.

Fisher, R. A. (1925). *Statistical Methods for Research Workers.* Edinburgh: Oliver & Boyd.

Gerbing, D. W. & Anderson, J. C. (1993). Monte Carlo evaluations of goodness-of-fit indices for structural equation models. In K. A. Bollen & J. S. Long (eds.), *Testing Structural Equation Models*, pp. 40–65. Newbury Park, CA: Sage.

Gerbing, D. W. & Hamilton, J. G. (1996). Viability of exploratory factor analysis as a precursor to confirmatory factor analysis. *Structural Equation Modeling*, **3**, 62–72.

Gorsuch, R. L. (1983). *Factor Analysis.* Hillsdale, NJ: Lawrence Erlbaum Associates.

Hayduk, L. A. (1996). *LISREL Issues, Debates, and Strategies*. Baltimore, MD: Johns Hopkins University Press.

Hershberger, S. L. (1994). The specification of equivalent models before the collection of data. In A. von Eye & C. C. Clogg (eds.), *Latent Variable Analysis: Applications for Developmental Research*, pp. 68–105. Thousand Oaks, CA: Sage.

Hoyle, R. H. & Panter, A. T. (1995). Writing about structural equation models. In R. Hoyle (ed.), *Structural Equation Modeling: Issues, Concepts, and Applications*, pp. 158–176. Newbury Park, CA: Sage.

Hu, L.-T. & Bentler, P. M. (1995). Evaluating model fit. In R. Hoyle (ed.), *Structural Equation Modeling: Issues, Concepts, and Applications*, pp. 76–99. Newbury Park, CA: Sage.

Hu, L.-T. & Bentler, P. M. (1999). Cutoff criteria for fit indexes in covariance structure analysis: conventional criteria versus new alternatives. *Structural Equation Modeling*, **6**, 1–55.

James, F. C. & McCulloch, C. E. (1990). Multivariate analysis in ecology and systematics: panacea or Pandora's box? *Annual Review of Ecology and Systematics*, **21**, 129–166.

Jöreskog, K. G. (1970). A general method for the analysis of covariance structures. *Biometrica*, **57**, 239–251.

Jöreskog, K. G. (1993). Testing structural equation models. In K. A. Bollen & J. S. Long (eds.), *Testing Structural Equation Models*, pp. 294–316. Newbury Park, CA: Sage.

Jöreskog, K. G. & Sörbom, D. (1981). *LISREL V: Analysis of Linear Structural Relationships by the Method of Maximum Likelihood.* Chicago: National Educational Resources.

Kaplan, D. (1991). On the modification and predictive validity of covariance structure models. *Quality and Quantity*, **25**, 307– 314.

Kaplan, D. (1995). Statistical power in structural equation modeling. In R. Hoyle (ed.), *Structural Equation Modeling: Issues, Concepts, and Applications*, pp. 100–117. Newbury Park, CA: Sage.

Lee, S. & Hershberger, S. (1990). A simple rule for generating equivalent models in covariance structure modeling. *Multivariate Behavioral Research*, **25**, 313–334.

Loehlin, J. C. (1998). *Latent Variable Models*, 3rd edition. Hillsdale, NJ: Lawrence Erlbaum Associates.

MacCallum, R. C. (1995). Model specification: procedures, strategies, and related issues. In R. Hoyle (ed.), *Structural Equation Modeling: Issues, Concepts, and Applications*, pp. 16–36. Newbury Park, CA: Sage.

MacCallum, R. C., Roznowski, M. & Necowitz, L. B. (1992). Model modifications in covariance structure analysis: the problem of capitalization on chance. *Psychological Bulletin*, **111**, 490–504.

MacCallum, R. C, Wegener, D. T., Uchino, B. N. & Fabrigar, L. R. (1993). The problem of equivalent models in applications of covariance structure analysis. *Psychological Bulletin*, **114**, 185–199.

MacCallum, R. C., Browne, M. W. & Sugawara, H. M. (1996). Power analysis and determination of sample size for covariance structure modeling. *Psychological Methods*, **1**, 130–149.

Marcoulides, G. A., Drezner, Z. & Schumacker, R. E. (1998). Model specification searches in structural equation modeling using Tabu search. *Structural Equation Modeling*, **5**, 365–376.

Marsh, H. W., Hau, K.-T, Balla, J. R. & Grayson, D. (1998). Is more ever too much? The number of indicators per factor in confirmatory factor analysis. *Multivariate Behavioral Research*, **33**, 181–220.

McDonald, R. P. (1989). An index of goodness-of-fit based on noncentrality. *Journal of Classification*, **6**, 97–103.

Niles, H. E. (1922). Correlation, causation and Wright's theory of 'path coefficients'. *Genetics*, **7**, 258–273.

Olsson, U., Troye, S. V. & Howell, R. D. (1999). Theoretic fit and empirical fit: the performance of maximum likelihood method versus generalized least squares estimation in structural equation models. *Multivariate Behavioral Research*, **34**, 31–58.

Pearl, J. (1988). *Probabilistic Reasoning in Intelligent Systems: Networks of Plausible Inference*. San Mateo, CA: Morgan Kaufmann.

Pugesek, B. H. & Tomer, A. (1995). Determination of selection gradients using multiple regression versus structural equation models (SEM). *Biometrical Journal*, **37**, 449–462.

Raykov, T. (1997). Equivalent structural equation models and group equality constraints. *Multivariate Behavioral Research*, **32**, 95–104.

Raykov, T., Tomer, A. & Nesselroade, J. R. (1991). Reporting structural equation modeling results in *Psychology and Aging:* some proposed guidelines. *Psychology and Aging*, **6**, 499–503.

Reyment, R. A. & Jöreskog, K. G. (1993). *Applied Factor Analysis in the Natural Sciences.* Cambridge: Cambridge University Press.

Rummel, R. J. (1970). *Applied Factor Analysis.* Evanston, IL: Northwestern University Press.

Satorra, A. (1989). Alternative test criteria in covariance structure analysis; a unified approach. *Psychometrika*, **54**, 131–151.

Satorra, A. (1990). Robustness issues in structural equation modeling: a review of recent developments. *Quality & Quantity*, **24**, 367–386.

Satorra, A. & Saris, W. E. (1985). Power of the likelihood ratio test in covariance structure analysis. *Psychometrika*, **50**, 83–90.

Scheines, R., Spirtes, P., Glymour, C., Meek, C. & Richardson, T. (1998). The TETRAD project: constraint based aids to causal model specification. *Multivariate Behavioral Research*, **33**, 65–117.

Shevlin, M. & Miles, J. N. V. (1998). Effects of sample size, model specification and factor loadings on the GFI in confirmatory factor analysis. *Personality and Individual Differences*, **25**, 85–90.

Shipley, B. (2000). *Cause and Correlation in Biology: A User's Guide to Path Analysis, Structural Equations and Causal Inference.* Cambridge: Cambridge University Press.

Sörbom, D. (1989). Model modification. *Psychometrika*, **54**, 371–383.

Spirtes, P., Scheines, R. & Glymour, C. (1990). Simulation studies of the reliability of computer-aided model specification using the TETRAD II, EQS, and LISREL programs. *Sociological Methods and Research*, **19**, 3–66.

Spirtes, P., Glymour, C. & Scheines, R. (1993). *Causation, Prediction, and Search*. New York: Springer-Verlag.

Steiger, J. H. (1989). *EzPATH: Causal Modeling*. Evanston, IL: SYSTAT Inc.

Steiger, J. H. & Lind, J. M. (1980). Statistically based tests for the number of common factors. Paper presented at the annual meeting of the Psychometric Society, Iowa City, IA.

Stelzl, I. (1986). Changing a causal hypothesis without changing the fit: some rules for generating equivalent path models. *Multivariate Behavioral Research*, **21**, 309–331.

West, S. G., Finch, J. F. & Curran, P. J. (1995). Structural equation models with nonnormal variables: problems and remedies. In R. Hoyle (ed.), *Structural Equation Modeling: Issues, Concepts, and Applications*, pp. 56–75. Newbury Park, CA: Sage.

Williams, L. J., Bozdogan, H. & Aiman-Smith, L. (1996). Inference problems with equivalent models. In G. A. Marcoulides and R. E. Schumacker (eds.), *Advanced Structural Equation Modeling*, pp. 279–314. Mahwah, NJ: Lawrence Erlbaum Associates.

Wright, S. (1921). Correlation and causation. *Journal of Agricultural Research*, **10**, 557–585.

Section 2 Applications

6 Modeling intraindividual variability and change in bio-behavioral developmental processes

Patricia. H. Hawley and Todd D. Little

Abstract

We present a basic rationale for studying intraindividual change processes in development and discuss some of the advantages and disadvantages of P-technique factor analyses for representing such change. An empirical example is presented to elucidate the application of the techniques and the interpretation of output. The data for our example come from observations of individual-level variables and social interactions of five Asian elephants (*Elephas maximus*). By referring to such a data base, we address many issues affecting observationally based data collection within intraindividual change designs. Although not commonly employed by developmentalists within the social and biological sciences, such designs can provide rich information for evaluating developmental theories.

Introduction

For developmentalists, intraindividual variability and change reflect the essence of their scientific inquires. All too often, however, researchers are reluctant to employ designs that are ideally suited to significant developmental questions: to understand the individual, one could study the individual over time. At first glance, this situation seems surprising because intraindividual differences designs and analysis techniques have been presented regularly for over 50 years (e.g., Cattell *et al.*, 1947). On the one hand, this reluctance is quite understandable given the myriad criticisms of "small-*n* research". On the other hand, hasty criticisms distract from the unquestionable utility of such approaches. The goals of this chapter are (1) to present a basic rationale for studying intraindividual change processes in development, (2) to address criticisms of such approaches and discuss some of the advantages and disadvantages of a factor analytical technique conceived to represent such change, and (3) to clarify the application of such techniques for interpreting the dimensions of change contained in observationally based data.

Small-*n* research and modes of selection

Those conducting research on small subject pools (e.g., captive animals, preschool children) often have the common experience that their work is considered less interesting or less rigorous than that of colleagues who enjoy large sample sizes. It is well known that small samples constrain statistical power and limit the generalizations that can be drawn. A single datum, however, represents more than a piece of specific information on a specific individual. It also includes other modes of classification, including a time component such as the measurement occasion (Cattell, 1952; Nesselroade & Jones, 1991). Inevitably, a data set includes various effects associated with several classification modes. Although data derived from large subject pools have the advantage of evincing trends across many individuals, inadvertent and unavoidable effects result from selection of the subjects mode. Such studies, for example, implicitly assume intraindividual stability and thus cannot adequately address how variables covary over time within individuals (i.e., intraindividual variability).

In contrast, small-*n* research contexts offer unique opportunities that behavioral researchers would do well to tout – namely, the opportunity to study variability and change in individuals and the degree of stability within a set of measures over time (Roberts & Nesselroade, 1986; Nesselroade & Jones, 1991). Idiosyncrasies of individuals can hinder the identification of nomothetic regularities when more common large-sample techniques are employed (Daly *et al.*, 1974; Machlis *et al.*, 1985). Individual differences need not be treated as error, but are worthy targets of study themselves (Allport, 1937; Nesselroade & Jones, 1991).

The study at hand

The original impetus for the present study was to explore intragroup social processes and interindividual differences in the behavior structure of five captive Asian elephants. From early on, workers in animal behavior have been concerned with identifying meaningful and useful dimensions of behavior in order to make interindividual and interspecific functional comparisons (Darwin, 1872; Morgan, 1894; Jennings, 1906). Much of this work has focused on patterns of apparent temporal associations that speak to issues of structure and function (Lorenz, 1941; Sebeok, 1965; Hinde, 1966). More recently, researchers have relied on various quantitative methodologies in order to minimize the effect of subjectivity when creating these categorizations (e.g., Senar, 1990; Figueredo, *et al.*, 1992; Golani, 1992). Such quantitative approaches generally involve sampling from a broader range of behaviors than might otherwise be considered in order to minimize subjective bias in the selection process – subjective perceptions of importance, relative frequency, or ease of measurement need not be the driving considerations of variable selection. By considering a broader range of behaviors, subtle and often complex relationships among variables can be detected. When such a micro-level approach is used, the potential problems of having too much information are countered by using analytical tools appropriate for its integration and organization.

Organizing analytical tools

Analytical tools that are appropriate for integration and organization of complex underlying relationships meld aspects of both psychometrics and measurement theory, two themes generally associated with and central to psychology. Perhaps this required interdisciplinary approach underlies an apparent reluctance to pursue quantitative descriptions of behavioral repertoires. Yet the union of psychology and approaches of animal behaviorists lends itself to the description and measurement of "natural units of behavior" (i.e., behavioral dimensions) such that the distinctiveness and coherence of these dimensions can be determined with the application of appropriate analytical tools. Whether one considers these dimensions or natural units as mere mathematical groupings or as underlying causes is largely a matter of personal philosophy (e.g., Maraun, 1996; Mulaik, 1996) and bears more on interpretation than on the derivation process.

Factor analytical approaches

Factor analytical techniques in general measure the degree to which observed variables are associated with the same underlying dimension as well as the degree to which these dimensions are distinct from other dimensions (Campbell & Fiske, 1959). Factor loadings (i.e., the regression weight predicting the variable from the factor) reflect the strength and direction of a relationship between a variable and the underlying dimension. The degree to which these factors reflect distinct behavioral dimensions is indicated by: (1) a number of high factor loadings for a cluster of behaviors (convergence), with these variables having functionally zero loadings on the other factors (divergence); and (2) low to moderate correlations among other dimensions (discrimination). These relations among variables and factors speak to the validity of a construct (e.g., Cronbach, 1951; Cronbach & Meehl, 1955). Thus such analytical approaches are a useful means to address the substantive concerns and biologically based questions of animal behavior scientists (i.e., questions of structure) as well as measurement issues derived from psychology (i.e., questions of validity).

Previous research

Quantitative based approaches have been used to successfully derive the structure of behavior across many taxa (e.g., Blurton Jones, 1972; Senar, 1990), resulting in illuminating insights into ontogenetic function. For example,"rough and tumble" behavior of children is less related to aggression than it is to play (Blurton Jones, 1972), while rough behavior in the zebra finch is related to reproduction and courtship (Figueredo *et al.*, 1992). Although these studies employed different analytical techniques, they had one quality in common: data were aggregated over many individuals measured over multiple occasions. While such an aggregation is acceptable when the goal is to establish group-level commonality (i.e., structural aspects of behavior common to all), this mixing of intraindividual (i.e., within subjects) and interindividual (i.e., between subjects) variability eclipses the uniqueness of each subject. That is to say, such an approach may overlook important aspects of subjects' individuality due to their unique personality characteristics, social status, or developmental stage.

Common and unique processes

In contrast to these works, the current study specifically explores individual differences across intraindividual structures. Despite the fact that individuals within a species might adopt dissimilar behavioral strategies to adapt to the local social conditions (e.g., Parker, 1984), individual differences in animals are seldom the target of scientific study. In contrast, within the framework of this study, a generalized "elephantine" pattern of behavior is expected, as well as variations in this pattern reflecting individuals' developmental stages and dominance ranks. For this reason, an individual-focused approach was adopted to uncover such differences.

Describing intraindividual variability

In the course of this study, three primary aspects of behavioral structure are explored. First, the structure of Asian elephants' micro-level behaviors are represented in a coherent and meaningful manner. As a result, behavioral processes are not described at the micro-level, but rather at a more informative and manageable molar-level (i.e., dimensions). Second, the similarities among individuals over this meaningful structure are assessed. That is to say, does each animal have an idiosyncratic pattern, the same pattern, or a combination of uniqueness and similarity? And third, such behavioral structures illuminate some aspects of the behavioral repertoire of captive Asian elephants by providing the temporally based context in which micro-gestures occur. The function of micro-level behaviors is suggested by what consistently co-occurs with it over time.

Our tool

To achieve these ends, we applied P-technique factor analysis. Under certain conditions (see below), P-technique is very useful in creating a quantitative description of behavior that objectively and precisely describes intra-individual variation in behavior patterns over time and empirically creates higher-order molar categories. In our view, three general situations for which P-technique factor analysis is quite appropriate are when (1) sufficiently large occasions have been assessed, (2) cross-occasion dependencies are nonsignificant (see below), and (3) broad-based micro-level behaviors have been measured. In such situations, P-technique can reduce the welter of detail from an individual to meaningful dimensions and establish coherent patterns of intraindividual change among the dimensions.

P-technique factor analysis

Introduced as a way to investigate intraindividual variability (Cattell *et al.*, 1947), P-technique, in contrast to the more oft applied factor analytical techniques, is performed on a single individual who is assessed across multiple variables on multiple occasions. The number of factors obtained from an individual's data reflects the complexity of the measured behavioral processes over time and the obtained factor pattern indicates the structure of those processes (Wessman & Ricks, 1966; Jones & Nesselroade, 1990). Thus the uniqueness of the individual is not lost, but is represented in a unique combination of common traits as manifested in the correlational nexus among multiple measures. For these reasons, P-technique was introduced as a means of representing meaningful relations among the measured variables for a single individual.

Individual differences and nomothetic generalizations

In contrast to P-technique, the more often applied R-technique factor analysis empirically determines normative trends across multiple individuals on multiple variables (see Cattell, 1952, 1963). As mentioned above, these normative trends do not reflect intraindividual variability because measurements are typically made at one occasion. In the same way that no individual may be at the group mean, P-technique patterns may not correlate highly with patterns derived by R-technique. Accordingly, nomothetic generalizations derived from R-technique could be misleading (Roberts & Nesselroade, 1986).

Although labor intensive, structures of intraindividual change derived from P-technique can be compared qualitatively and quantitatively to explore similarities and differences across individuals. In this way, idiosyncrasies that hinder identifying nomothetic regularities and stable traits can be uncovered (Roberts & Nesselroade, 1986; Hooker *et al.*, 1987) and multiple P-technique factor patterns can be compared to establish meaningful (nomothetic) regularities across a number of individuals. Simply stated, differences among solutions indicate unique, intraindividual variability, and similarities among solutions indicate common behavioral processes for which interindividual variability can be examined (Hawley, 1994).

The dynamic nature of multi-occasion data

Although P-technique factor analysis has been used in the field of psychology in, for example, the domains of mood, personality, and locus of control (for a review, see Jones & Nesselroade, 1990), it has not been applied to data derived from behavioral observation. We believe that this situation is not due to a lack of interest in questions of intraindividual variability, change, and stability. Instead, it is probably in part due to the effort required in conducting longitudinal observational studies, and, perhaps more critically, due to the fact that P-technique has been criticized as being inherently flawed (Anderson, 1963; Molenaar, 1985; Schmitz, 1987).

The criticisms of P-technique are by no means inconsequential as they penetrate the very core of the analysis: namely, the dynamic nature of multi-occasion correlations. Although measurements are taken over many occasions, traditional P-technique factor analysis does not address the time-ordered dependencies of the relationships among a set of variables (Anderson, 1963). Rather, the correlation matrix addressed by P-technique factor analysis reflects relationships among variables that are measured within the same observation occasion. Correlations reflect the degree to which variables co-occur at the same time point and do not reflect the dynamic relationships or dependencies among variables across sequential occasions. As can be seen in Figure 6.1, the time-related dependencies depicted for the two example variables would be overlooked using the P-technique paradigm. By focusing on the simultaneous relations recorded at occasions 1 through 6, one would erroneously conclude that there is no relationship between variables 1 and 2.

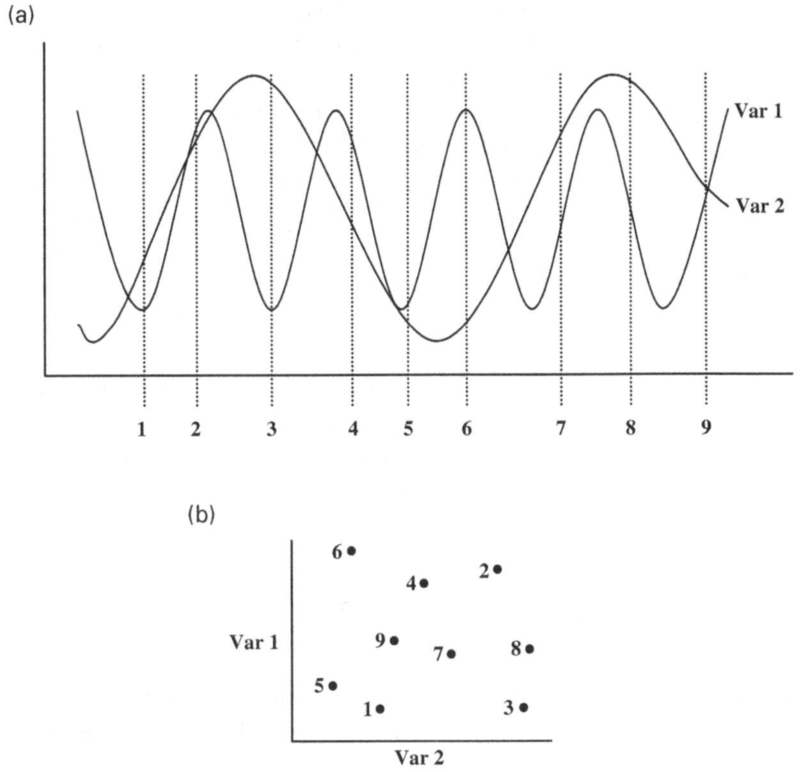

Figure 6.1. Graphic example of possible time-dependent processes affecting two measured variables (Var 1 and Var 2). (a) Observed occasions, unevenly spaced in time, (b) plotted occasions.

In addition, with time-series data, an assumption of factor analysis may be violated: namely, errors of measurement across occasions may be correlated (Anderson, 1963; Molenaar, 1985). Quite conceivably, the time-dependent process that uniquely affects a measured variable on one occasion may continue to do so across repeated occasions. As a consequence, correlations analyzed by standard P-technique may be inflated by such time-dependent lag processes affecting the observations. Schmitz (1987) has shown that apparent simultaneous relations may emerge as artifacts from asynchronous (autocorrelational and cross-lagged) relations among a set of variables. Theoretically, such spurious relations can emerge in a time-ordered series if the sampled occasions sufficiently tap into the time-series process and if the time-series process differentially affects the synchronous observed variable indicators of two or more factors. Under such conditions, a correlation between two or more variables may emerge not because of a common synchronously measured characteristic but because of the effect of the differential time-dependent process.

Under these circumstances, some authors recommend analyzing a *lagged* covariance matrix, or covariances calculated upon lagged relationships instead of simultaneous relationships among variables (Anderson, 1963; Molenaar, 1985, 1994; Molenaar *et al.*, 1992; Wood & Brown, 1994). This newer modeling paradigm has proven valuable both in theory and application. However, these criticisms of standard P-technique and the recommended use of dynamic P-technique as a means to account for the lagged information inherent in a standard time-ordered data set may discount certain conditions of applicability as well as some important merits of a standard P-technique approach. Before turning to the conditions for which a standard P-technique approach is appropriate, we first discuss the basic relationship between standard and dynamic P-technique (see also Wood & Brown, 1994).

Relationship between standard P-technique and dynamic P-technique

The relationship between standard P-technique and dynamic P-technique can be seen in Figures 6.2, 6.3, and 6.4. First, Figure 6.2 shows a

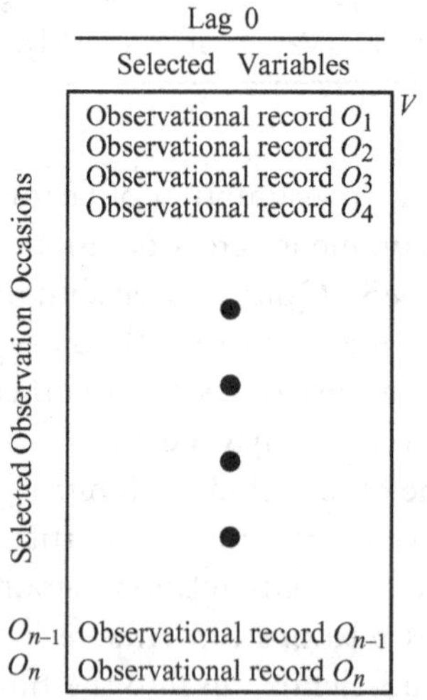

Figure 6.2. Schematic representation of a standard P-technique data structure.

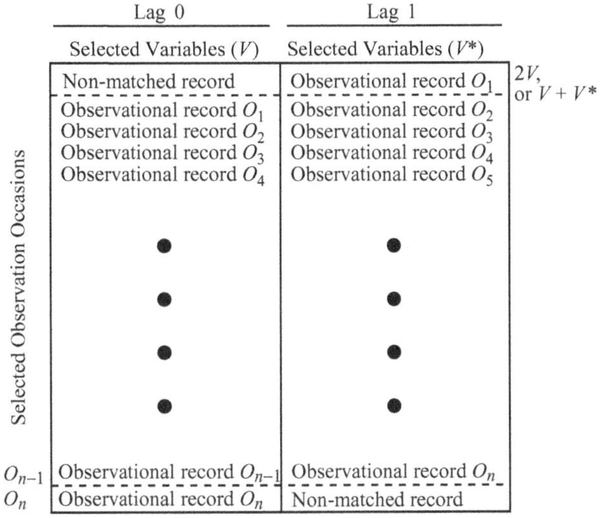

Figure 6.3. Schematic representation of a dynamic P-technique data structure. (Note, because the same variables are represented twice in a dynamic data structure, an asterisk denotes the Lag 1 variables.)

time-ordered raw data structure of a standard P-technique analysis which has the dimensions of O_n (number of observations) by V (number of variables). This data matrix is labeled Lag 0 as it represents simultaneous relationships among the variables (i.e., no time lags). In Figure 6.3, a singly lagged dynamic P-technique structure of the time-ordered data is depicted. This lagged structure is obtained simply by matching each observational record with the record following it in time. Observation occasion 1, O_1, at Lag 0 is paired with observation occasion 2, O_2, at Lag 1 and O_2 at Lag 0 is paired with O_3 at Lag 1 and so on.[1]

Figure 6.4 depicts a lagged covariance matrix generated from the lagged data structure shown in Figure 6.3 and is the input covariance matrix for a dynamic P-technique analysis. Illustrated with three variables, the structure of the lagged covariance matrix contains three distinct elements.

[1] If, however, the nonmatched records of the dynamic data structure (Figure 6.3) are treated as missing data and estimated, it is possible, depending on the method of estimation, to obtain identical covariance structures. That is to say, if all nonmatched records are replaced (yielding an O_{n+1} by $2V$ raw data matrix), or, if only one nonmatched cell is estimated (yielding an O_n by $2V$ raw data matrix), constraints on the estimation process could be placed such that the resulting covariance structures among the variables V and V^* would be the same, thereby eliminating a source of variability that could affect the factor solutions.

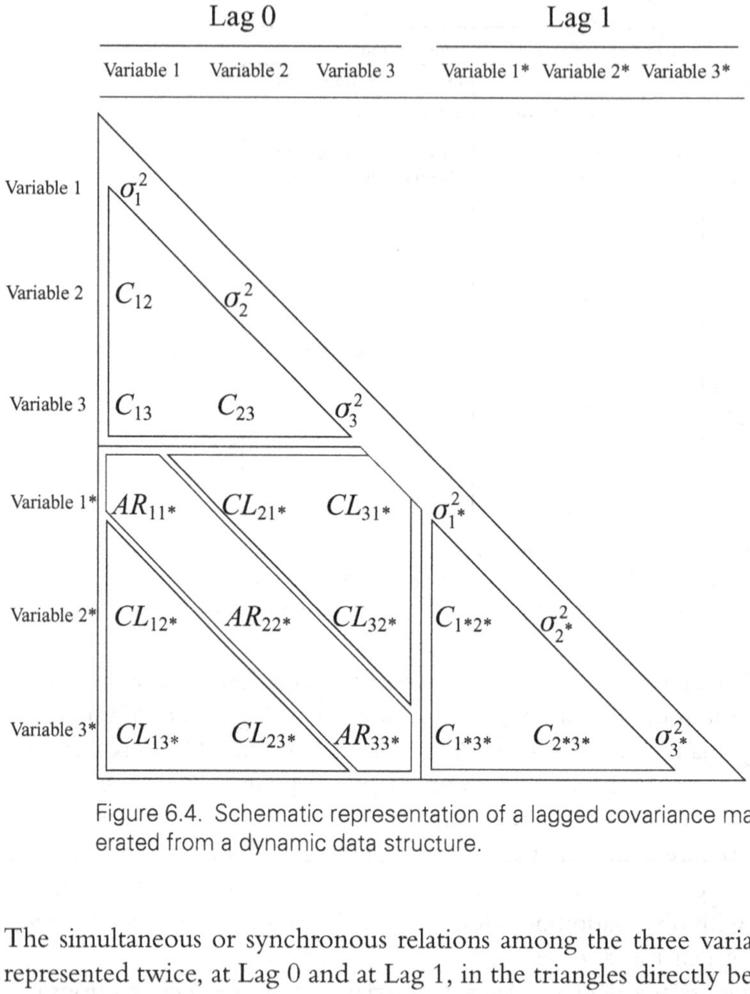

Figure 6.4. Schematic representation of a lagged covariance matrix generated from a dynamic data structure.

The simultaneous or synchronous relations among the three variables are represented twice, at Lag 0 and at Lag 1, in the triangles directly below the major diagonal. At each lag, the variances of the variables are located along the major diagonals and covariances are located off the diagonals. Depending on how nonmatched records are treated and the number of observations, the corresponding elements between these two sections would be nearly or exactly identical (see note on p. 151). The lower quadrant of the lagged covariance matrix contains the lagged information among the variables. Two sources of lag information are contained in this matrix. First, autoregressive (AR) lagged relations between each variable are represented along the diagonal of this lower quadrant (AR_{11*}, AR_{22*}, and AR_{33*}), i.e., a variable's correlation with itself between Lag 0 and Lag 1. Cross-lagged (*CL*) relations among the variables are represented in the upper and lower triangles of this quadrant (e.g., CL_{12^*}, CL_{13^*}, and CL_{23^*}), which represent, for example, covariation between variable 1 at Lag 0 (V_1) and variable 2 at Lag 1 (V_{2^*}), V_1 and V_{3^*}, V_2 and V_{3^*}, and so on. In a dynamic P-technique analysis, this lag information is explicitly modeled (see e.g., Molenaar, 1985; Molenaar *et al.*, 1992), while it is not modeled in traditional P-technique. For these reasons, dynamic P-technique can be considered a general model in which one or more lags are evaluated and standard P-technique as a specific model in which zero lags are specified. In this sense, standard P-technique can be considered a special case of dynamic P-technique (Wood & Brown, 1994).

Conditions of applicability

The possibility of factors arising as artifacts from asynchronous time-dependent processes threaten to invalidate the outcome of traditional P-technique factor analysis. For this reason, such artifactual relations should be carefully considered when applying a traditional P-technique model. Monte Carlo statistical analysis work, however, has demonstrated that the synchronous relations among a set of variables will emerge despite asynchronous time-dependent influences such that the correct number of synchronous or simultaneous dimensions (i.e., factors) can be accurately recovered. More specifically, if the strength of the asynchronous influence is equal to or less than the synchronous influence (Schmitz, 1987), the correct number of factors will emerge.

Sampling protocols also influence the applicability of standard P-technique. If the sampled occasions do not sufficiently tap into the time-series process (e.g., because the intervals between the sampled occasions are longer than the time-series process affecting the variables or randomly tap into the time-series process) and/or if the time-series process uniformly affects the observed indicators of a factor, then the likelihood of a spurious factorial dimension appearing is substantially diminished. If this is the case, then the possibility of correlated errors among the synchronously measured variables is also significantly reduced and the effects of a time-series process will probably influence the factor indicators uniformly such that no additional spurious factors will emerge. Thus one must carefully consider possible lagged relations inherent in a P-technique data structure.

Importantly, the potential influence of lagged relations prior to conducting a standard P-technique analysis can, and should be, empirically evaluated. If little or no lag information is inherent in a lagged covariance matrix, a standard P-technique analysis would provide a complete description of the intraindividual change relations among a set of variables. Although tests for lag information generally should be evaluated across as many lags as are deemed necessary or fruitful, two sources of information can be tested for evidence of lag relations at each level of a lagged structure: the autoregressive relations between each variable (i.e., a variable's correlation with itself across lags; represented on the minor diagonal of the lower quadrant in Figure 6.4) and the cross-lagged relations among the variables (i.e., a variable's correlation with the remaining variables across lags; represented in the lower and upper triangles within the lower quadrant in Figure 6.4). General equations can be employed that yield an approximate, asymptotic χ^2 value and associated degrees of freedom that can be evaluated for significance (e.g., Jöreskog & Sörbom, 1993). Furthermore, as we demonstrate below, such equations can be applied to the autoregressive correlations, the cross-lagged correlations, or both. If the outcomes of such tests are not significant (i.e., these elements of the lagged covariance matrix are essentially zero), then little or no lag information is contained among the tested elements and one can confidently evaluate and interpret a standard P-technique representation of the simultaneous relations among the variables.

It should be noted that not all lagged information impacts equally on the interpretability of a standard P-technique. For example, consistent lag information in the autoregressive correlations would influence the corresponding factors at each lag with a significant cross-lag contribution between lags. By consistent information, we mean that the set of variables indicating a factor should show significant autoregressive relations that are approximately the same magnitude and in the same direction. Here, the number of factors would be the same for both a standard and dynamic P-technique analysis. If the tests for lag information in the cross-lagged relations are significant, then important time-dependent information may be contained within the lagged structure. Such a condition may still not pose a problem for employing standard P-technique because, as mentioned above, synchronous relations can still be recovered in the presence of cross-lagged, asynchronous relations (Schmitz, 1987).

We offer one final theoretical consideration. If cross-lagged information is inherent within a lagged covariance matrix, the origin of the lag information can stem from two sources: a common source associated with the factors and/or a unique source associated with the variables. If the source of the lag information stems from the factors, an invariant factor pattern matrix for each lag can be estimated, and the nature of the asynchronous relations between two or more lags can be represented as lagged effects among the factors rather than as lagged effects among the variables. Although specific lag effects may also emerge among the variables, these effects may be trivial relative to the factor-level effects (for a discussion of these issues in relation to cross-cultural comparisons, see Little, 1997). Such possibilities can be tested using confirmatory dynamic models to determine the degree to which the lag information is adequately captured as relations among the factors relative to relations among the variables.

Under some (perhaps many) conditions, traditional P-technique factor analysis may provide a useful and sufficient representation of covariation patterns among a set of variables. We will now go one step further to suggest that in some cases dynamic relations should be evaluated further on the *obtained factors* from the standard P-technique. Because the unique variance in a single observed variable may contain significant proportions of reliable nonconstruct-related variance (i.e., variable specific) and error (i.e., unreliable) variance, the nature of the reliable construct variance of a variable may be overshadowed by the accompanying unique variance such that the lag relations are hidden. In contrast, the information contained at the construct level may evince lagged relations because this information is represented as a linear composite of many observed variables such that the proportion of reliable, common construct variance among the variables is increased relative to their unique specificities (Nunnally, 1978; see also Little *et al.*, 1999).

Advantages of P-technique factor analysis

As discussed above, not only can traditional P-technique withstand some very important empirically and theoretically based criticisms, it also has several practical advantages over the dynamic approach. Practically speaking, dynamic P-technique is a useful adjunct to a standard P-technique analysis. Representing in interpretable form the nature of a set of dynamic relations becomes more cumbersome as the number of variables increases (i.e., the number of possible dynamic relations increases multiplicatively as the number of variables increases). In our view, the results of a standard P-technique provide a description of the simultaneous or synchronous relations among variables that allow a researcher to reduce the number of dimensions in a meaningful way against which some form of dynamic analysis would be evaluated. Although a dynamic approach provides a representation of the asynchronous or lagged relations within a given data set, standard P-technique provides a means by which the synchronous or simultaneous relations can be reduced to a manageable, practical set of dimensions.

In addition, the exploratory uses of dynamic and standard P-technique differ in an important way. The nature of the identification process in dynamic P-technique requires the extracted factors to be orthogonal because there are insufficient degrees of freedom to estimate factor correlations. In contrast, P-technique can be implemented using a standard exploratory common factor analysis model (e.g., Gorsuch, 1983) and therefore allows oblique (correlated) factor solutions. To estimate oblique factors in a dynamic model, one must place restrictions on the model (e.g., by fixing certain factor loadings to zero, a known value, or to equality with other model parameters). Such model constraints are not done arbitrarily, but rather according to some theoretical rationale. As such, theory-guided constraints move one into the realm of *confirmatory* models.

Confirmatory applications

Exploratory models require only the specification of the number of presumed common factors and the collection of observed variables. No additional specifications or restrictions are made by the researcher because all observed variables are assumed to be affected by all common factors and all unique variances of the measured variables are assumed to be uncorrelated. This approach is appropriate at early stages of theory development when more specific causal relationships among a complex array of variables are unknown. In contrast, theory-guided constraints are placed on confirmatory models (often referred to as structual equation models, LISREL models, or covariance structure models). Behavioral scientists often have very clear ideas about, for example, which observed variables and common factors should be related, which unique variances are intercorrelated, which relationships are of equal magnitude, etc. Statistical tests aid in determining whether the specified model is consistent with a body of data.

Confirmatory approaches to P-technique data are not only in principle possible, but are also highly desirable. A confirmatory approach is a straightforward extension of exploratory techniques involving the same data structure and consideration for time-lag influences. Several domains of research seem to be well prepared to support theory-guided P-technique models (e.g., personality, social dominance). Nonetheless, they appear, at the time of writing, to be rare (but see, Nesselroade, 1994; Russell *et al.*, 1996).

Summary

We consider standard P-technique to be a useful and justifiable procedure to analyze data from intraindividual change designs in exploratory research. But because of the time-ordered nature of such data, various types of dynamic analysis should be conducted to determine the degree of generalizability from a standard P-technique approach. We have focused on the rationales for, as well as the uses and interpretations of, standard P-technique. Furthermore, information on various dynamic approaches to analyses can be found in Molenaar (1985, 1994), Molenaar *et al.* (1992), Schmitz (1987), and Wood & Brown (1994).

Empirical illustration of P-technique factor analysis

Subjects and procedure

Subjects

Five female Asian elephants (*Elephas maximus*) were observed from September 1992 through November 1993 from the outside of a 2300 m^2 outdoor enclosure at Tierpark, Berlin. Two were adults (aged 36 and 19 years), two were adolescents (13 years), and one was a juvenile (9 years). Only two animals (the 19-year-old adult and the juvenile) will be discussed for the sake of illustration and brevity. Both animals are known to be of highest and lowest dominance rank, respectively.

Behavioral checklist

Behaviors were recorded from a wide array of contexts in order to minimize biases in variable choice. Measures included: (1) gradations of behavior, such as relative distance from others (e.g., proximity); (2) postural variables, such as ear, head, and trunk position; (3) individual activities that do not require the presence of a social partner (e.g., object manipulation and sand tossing); and (4) dyadic interactions, such as touching and approaching. For each focal animal, dyadic interactions were coded as "active" if the focal animal was the active agent of the behavior, or "passive" if the focal animal was the recipient of the behavior.

In the original study, 75 micro-level variables were observed and recorded, 51 of which showed adequate variability across all animals to be included in the analyses. These analyses resulted in five behavioral structures (15 factors each) specific to each animal. Only a portion of these results reflecting 18 of the 51 variables (6 factors) will be presented here because of space constraints. These 18 variables can be found in Table 6.1.

Data collection

At least two of the five elephants were selected randomly each day and filmed with social partners in 5-minute segments. Video tapes were scored using a time sampling procedure in which a behavior was recorded if it occurred at least once within each of 20 15-second intervals (e.g., Rhine &

Table 6.1. *Measured variables and criteria of occurrence*

Variable	Description
Locomote	Forward movement of at least one body length at a walking pace
Approach	Focal animal approaches another, closing a distance of at least two body lengths (approx. 6 m) to a body length (approx. 3 m). Route must be direct, locomotion forward
Leave	Focal animal increases the distance to a partner from one to at least two body lengths
Sand Throw	Sand is tossed with trunk on the head, back, sides, underside, or in the air
Ear Flap	A full forward and backward fanning of the ears
Air Temperature	Recorded daily from newspaper records in degrees Celsius
Manipulate Object	Holding an object with trunk or mouth
Whip Object	Object is struck with the trunk against the body, front legs, or head
Touch Head*	Trunk of one animal touches the head, ear orifice, nostril, eye, or temple of another
Touch Mouth*	Trunk of one animal touches the mouth of another
Back Into*	Animal reduces the distance between herself and a partner in a backwards orientation such that her rump nears the other animal. The body orientation may be rump–rump, rump–head, rump–side, etc.
Body Contact*	Body–body touching. Excludes trunk contact
Hindquarters-to-Face*	The hindquarters of one animal is within 25 cm of the face of another
	Reflects relative body position only, not how they got that way (e.g., turning into, backing into, or approaching)

This is a partial list included in this study for exemplary purposes (i.e., 18 of 51 variables). Other variables included trunk positioning (on head, out, high) and tail positions (out, twisting), etc.

(*) Indicates the variable has an active and passive version (see text).

Ender, 1983). Each subject was observed for at least 248 such 5-minute occasions (range 248 to 278). Rater reliability, as reflected in percentage agreement over an observation occasion coded at two time points, ranged from 0.75 to 1.0.

Pre-analysis data treatment

In order to minimize the possibility of capitalizing on chance linear relationships, outliers were attenuated for each behavior for each focal animal separately. Values lying beyond the 99% confidence interval of the mean by behavior and by animal were accordingly adjusted to the 99% confidence level. Positively skewed variables were transformed to meet the conditions of normalcy required of regression-based analytical techniques.

Model specifications

P-technique factor analyses were performed for each of the two example elephants individually on the 18 variables using SAS's PROC FACTOR procedure. This analysis is similar to other factor analytical techniques except that the data in the matrix are drawn from a single individual and the rows in the matrix are observation occasions rather than subjects. An iterated principal axis solution and a Harris–Kaiser oblique rotation were specified. The number of factors appropriate for extraction was assessed by scree plots, eigenvalues, and rotated factor loading patterns as well as more objective indices such as the maximum likelihood χ^2 statistic, the Tucker–Lewis ρ coefficient (Tucker & Lewis, 1973), and the Akaike information criterion (AIC; Akaike, 1987). Coefficients of congruence (e.g., Korth, 1978) were used to compare factor solutions across individuals so that each factor of one rotated factor solution was compared with each factor of the other solution.

Results

Empirical tests of dynamic relations in the analyzed covariance matrices

Before conducting the standard P-technique analyses of the intraindividual covariance matrices, we tested them for evidence of dynamic autoregressive and cross-lagged information. Given the random nature of the observation schedule, we generated a lagged covariance matrix that contained only one lag. This covariance matrix was then analyzed using the LISREL structural equation modeling package (LISREL 8; Jöreskog & Sörbom, 1993). First, we specified each corresponding element of the simultaneous relations submatrices to be identical (i.e., the variances and covariances at Lag 0 were equated with their corresponding elements at Lag 1). In addition, we tested whether the cross-lagged covariances were significantly different from zero. To do so, we fixed all these covariances to be zero. The resulting χ^2 and degrees of freedom reflect the degree to which these covariances contain

Table 6.2. *Tests for dynamic (lag) relations in the intraindividual datasets*

	Adult			Juvenile		
Test	χ^2	df	p	χ^2	df	p
No cross-lagged relations	451.7	477	0.792	397.0	477	0.997
No autoregressive relations	572.4	495	0.009	500.5	495	0.422
One autoregressive relation[a]	510.3	494	0.295	—	—	—

These tests represent a nested model comparison of the lag correlations among the variables. The nested model specifies fixed zero correlations for the lag relations being tested. It is compared with the model in which all correlations are freely estimated.
[a]Because the test of no autoregressive relations was significant we evaluated the freely estimated parameters, fitted residuals, and modification indices to determine which and how many autoregressive correlations were significant. We found one autoregressive estimate that was significant for the adult and therefore it was freely estimated for this test.

cross-lagged information. As seen in Table 6.2, these cross-lag covariances did not differ from zero for both subjects. Interpretationally, this outcome indicates that asynchronous, time-dependent processes have not significantly affected the variables in the analyses.

Finally, we examined the autoregressive elements of the dynamic matrices. As with the cross-lagged elements, we specified the autoregressive covariances to be zero and maintained the previously specified restrictions. Thus this final model is quite restricted and tests (1) the equality of the two simultaneous submatrices at Lag 0 and Lag 1, (2) the evidence of nonzero information in the cross-lagged covariances, as well as (3) the autoregressive covariances. As seen in Table 6.2, all these restrictions were fully supported for the juvenile. For the adult, we found a single significant autoregressive correlation. Given these outcomes, we can conclude that (1) little or no lagged information has affected these variables, and (2) using a standard P-technique analysis is empirically justifiable.

Standard P-technique factor patterns

A six-factor pattern based on the 18 variables was found for each subject (for full results of all possible variables, see Hawley, 1994). Tables 6.3 and 6.4 show the rotated factor pattern for these subjects.

Table 6.3.

A. Rotated factor pattern for the adult

Variable	F1	F2	F3	F4	F5	F6	h^2
Locomote	0.73	0.01	−0.09	−0.10	0.02	0.06	0.76
Active Approach	0.41	0.00	−0.01	0.11	0.02	−0.03	0.61
Active Leave	0.79	0.01	0.17	0.04	−0.05	−0.08	0.55
Active Back Into	0.14	0.65	−0.02	0.09	0.10	−0.07	0.53
Active Body Contact	−0.08	0.74	0.15	−0.02	−0.03	−0.01	0.62
Active Hindquarters-to-Face	−0.08	0.19	−0.18	−0.05	−0.20	0.11	0.17
Passive Back Into	0.11	0.04	0.76	−0.07	−0.07	0.00	0.65
Passive Body Contact	−0.18	0.20	0.41	0.03	0.04	0.06	0.37
Passive Hindquarters-to-Face	0.01	0.16	0.57	−0.02	−0.09	0.09	0.47
Sand Throw	−0.11	−0.13	0.16	0.30	−0.03	−0.08	0.31
Ear Flap	−0.09	−0.13	0.16	0.43	0.39	0.02	0.52
Air Temperature	0.00	0.05	−0.08	0.82	−0.13	0.00	0.66
Manipulate Object	−0.03	−0.04	−0.09	−0.10	0.84	−0.00	0.70
Whip Object	0.02	−0.22	0.04	−0.11	0.64	0.06	0.48
Passive Touch Head	0.03	0.02	−0.04	0.02	−0.09	0.67	0.51
Active Touch Head	0.02	−0.11	−0.04	0.03	−0.11	0.57	0.60
Active Touch Mouth	−0.04	0.06	−0.02	0.01	0.02	0.71	0.50
Passive Touch Mouth	−0.03	−0.08	0.12	−0.04	0.11	0.77	0.61

B. Interfactor correlations and reliabilities

	F1	F2	F3	F4	F5	F6
Factor 1	(0.70)					
Factor 2	0.07	(0.64)				
Factor 3	−0.08	0.28	(0.68)			
Factor 4	−0.09	−0.13	−0.19	(0.53)		
Factor 5	−0.15	−0.02	0.00	0.26	(0.69)	
Factor 6	0.27	0.12	0.13	−0.16	−0.02	(0.76)

C. Summary statistics

	F1	F2	F3	F4	F5	F6
PR^2	0.11	0.09	0.09	0.08	0.10	0.12
SR^2	0.06	0.05	0.05	0.05	0.07	0.07

F, factor; h^2, community estimate; PR^2, partial r^2 statistic; SR^2, semi-partial r^2 statistic; underlined values indicate pronounced loadings (i.e., ≥ 0.3).

Table 6.4.

A. Rotated factor pattern for the juvenile

Variable	F1	F2	F3	F4	F5	F6	h^2
Locomote	0.78	−0.04	0.03	0.03	−0.06	0.01	0.71
Active Approach	0.48	−0.02	−0.11	−0.08	−0.17	−0.05	0.45
Active Leave	0.58	0.21	−0.02	−0.02	0.15	0.08	0.46
Active Back Into	−0.07	0.73	−0.09	−0.11	−0.03	−0.10	0.63
Active Body Contact	−0.10	0.30	0.22	−0.06	−0.14	0.17	0.43
Active Hindquarters-to-Face	0.08	0.79	0.01	0.11	0.03	0.03	0.56
Passive Back Into	−0.01	−0.08	0.78	0.00	−0.01	−0.04	0.60
Passive Body Contact	0.03	0.08	0.56	−0.01	0.03	0.09	0.38
Passive Hndqrt to Face	−0.13	−0.03	0.12	−0.02	−0.10	0.04	0.17
Sand Throw	0.01	0.01	0.10	0.35	−0.14	−0.05	0.25
Ear Flap	−0.11	0.09	−0.13	0.37	0.01	−0.14	0.37
Temperature	0.02	0.00	0.01	0.76	−0.02	0.03	0.57
Manipulate Object	−0.06	0.04	−0.05	−0.07	0.60	−0.03	0.52
Whip Object	0.02	−0.04	0.03	0.01	0.66	−0.02	0.42
Passive Touch Head	0.03	−0.09	0.01	−0.05	−0.02	0.54	0.39
Active Touch Head	−0.04	0.01	−0.02	0.12	−0.07	0.44	0.35
Active Touch Mouth	0.07	−0.04	−0.00	−0.02	0.01	0.12	0.42
Passive Touch Mouth	0.05	−0.02	0.00	0.01	0.02	0.56	0.29

B. Interfactor correlations and reliabilities

	F1	F2	F3	F4	F5	F6
Factor 1	(0.68)					
Factor 2	−0.05	(0.63)				
Factor 3	−0.18	0.16	(0.64)			
Factor 4	0.08	−0.35	−0.11	(0.49)		
Factor 5	−0.11	−0.26	−0.16	−0.05	(0.64)	
Factor 6	−0.15	0.19	0.19	−0.11	−0.15	(0.52)

C. Summary statistics

	F1	F2	F3	F4	F5	F6
SR^2	0.07	0.05	0.05	0.05	0.05	0.04
PR^2	0.10	0.10	0.08	0.07	0.11	0.08

F, factor; h^2, community estimate; PR^2, partial r^2 statistic; SR^2, semi-partial r^2 statistic; underlined values indicate pronounced loadings (i.e., ≥ 0.3).

As can be seen in Table 6.3, Factor 1 for the adult has significant positive loadings for the variables Locomote, Active Approach, and Active Leave. Due to the configuration of the variables loading on this factor, we labelled it *Social Locomotion* as it reflects the basic social movement of elephants. As can be seen in Table 6.4, the juvenile's Factor 1 corresponds closely to that found in the adult. Factor 2 differs somewhat for the two animals. For the adult (see Table 6.3), it is marked by Active Back Into and Active Body Contact. That is to say, the adult will back into a social partner and bump her. The juvenile also backs into (Active Back Into) and body contacts (Active Body Contact) social partners (see Table 6.4, Factor 2) but she additionally orients herself such that her hindquarters are directed toward the face of her social partner (Active Hindquarters-to-Face). Factor 3 for both animals illustrates the relationships among the same variables in their passive versions. The adult (see Table 6.3) is backed into (Passive Back Into), is body contacted (Passive Body Contact), and receives the hindquarters of a social partner in a faceward orientation (Passive Hindquarters-to-Face). The juvenile, in contrast, is only backed into (Passive Back Into) and bumped (Passive Body Contact) but receives no such hindquarters orientation.

Both animals have a similar Sand factor (see Factor 4, Tables 6.3 and 6.4) in which sand is thrown repeatedly on the body (Sand throw) the ears are rhythmically flapped (Ear Flap). The loading of ambient temperature on these factors indicates the positive linear association between these behaviors and weather conditions.[2] Both animals have an Object Handling factor (see Factor 5, Tables 6.3 and 6.4) marked by manipulating an object in the trunk or mouth (Manipulate Object) and whipping the object against the body (Whip Object). Similarly, both animals have a Touch factor (see Factor 6, Tables 6.3 and 6.4) in which there is mutual touching of another animal's head and mouth (Active and Passive Touch Head, Active and Passive Touch Mouth). Active Touch Mouth, however, fails to load for the juvenile in this context.

Interfactor correlations and factor similarity

The patterns of interfactor correlations are also presented in Tables 6.3 and 6.4. Of the 30 possible correlations for each animal, only one appears to be of moderate magnitude; namely, Factors 2 and 4 for the juvenile, and this relationship is negative (see Table 6.4). Internal consistency reliabilities (Cronbach, 1951) for the primary indicators of each factor are located on the

[2] Both throwing sand on the body and ear flapping function to regulate body temperature in this species.

Table 6.5. *Coefficients of congruence among the represented factors (F)*

	Juvenile					
Adult	F1	F2	F3	F4	F5	F6
Factor 1	0.75	0.05	−0.07	−0.07	0.02	0.07
Factor 2	−0.01	0.59	0.22	−0.18	−0.03	0.01
Factor 3	−0.07	−0.06	0.69	−0.01	−0.09	0.04
Factor 4	0.04	0.04	−0.08	0.69	−0.13	−0.04
Factor 5	−0.05	−0.02	−0.07	−0.09	0.79	−0.06
Factor 6	0.01	−0.03	0.00	0.04	0.03	0.76

Underlined values are the focal diagonal elements, which, if the factors are congruent, should contain pronounced values (i.e. ≥0.6) relative to the nondiagonal elements.

diagonal of the correlation matrix and ranged from 0.49 to 0.76. Finally, coefficients of congruence are presented in Table 6.5. The congruence coefficients were of moderate magnitude, ranging from 0.59 to 0.76.

Discussion

Coherent patterns of change

Coherent and substantially overlapping change processes in each animal are immediately apparent and for the most part reflect documented behavior patterns in free-ranging elephants. Both animals had factors representing socially directed locomotion (Social Locomotion), both active and passive backing behavior (Passive Backward Approach and Active Backward Approach), temperature regulation (Sand), object manipulation (Object Handling), and social touching (Touch). There were also differences between the animals that reflect not only the nature of elephants' social organization, but also important developmental differences. At the factor level, a primary illustrative difference lies in the factors representing backward approach behavior (i.e., Passive Backward Approach and Active Backward Approach; Factors 2 and 3, Tables 6.3 and 6.4). Recall that these factors are marked differently in the two animals. In contrast to the adult, when the juvenile backed into and bumped a social partner, she did so with her hindquarters to the face of the other animal. Similarly, when the adult was backed into and bumped, she also received hindquarters in her face. The juvenile was simply backed into and bumped with no such hindquarters orientation.

The difference in these animals' hindquarter orientations was not unique to these two animals. All five subjects had similar factorial configurations. The oldest animal had active and passive backing patterns similar to those of the adult presented here (i.e., performed no hindquarters orientations, but received them). The two remaining subadults had active backing patterns similar to those of the juvenile, but also passive backing patterns similar to the adults. That is to say, they both performed hindquarters oriented backing and received it. Because the pattern of this gesture fit remarkably well to the dominance hierarchy reported in these animals (for a full report, see Hawley, 1994), a backward approach *with* a hindquarters orientation was labelled Cautious Approach[3] because animals performed it to group members of higher rank but received it from group members of lower rank. In fact, uncovering this gesture through P-technique represented an important outcome as it reflects a previously undocumented plausible marker for submissiveness in captive elephants (Hawley, 1994).

Construct validity

Also noteworthy about these two factor patterns is the lack of dual factor loadings within an animal. For example, the adult, as is shown in Table 6.3, flaps her ears (Ear Flap) in two contexts; throwing sand (Sand, Factor 4) and when manipulating objects (Object Handling, Factor 5). The general lack of dual loadings suggests that micro-level behaviors can discriminate well molar-level behavioral categories.

In addition, the micro-level variables contribute reasonably well to common factor space. On average, 41% of the variable-level variance is explained by the factors in both solutions (see the communality estimates, h^2, in Tables 6.3 and 6.4). Notable exceptions in the amount of variance in common factor space are substantively interesting. Active Hindquarters-to-Face for the adult, for example (see Table 6.3), indicates that, as the dominant animal, she did not perform this behavior in the same socially important contexts as the juvenile.

Relatedly, the molar categories showed acceptable internal consistency (i.e., alpha reliabilities; Cronbach, 1951) ranging from 0.53 to 0.76 in the adult's solution and 0.49 to 0.68 in the juvenile's solution (see Tables 6.2

[3] "Cautious" was chosen because a backwards hindquarter presentation in a nonsexual context may reflect the backing animals tendency to hide its weapons (e.g., tusks, trunk) in order to avoid negative repercussions from the dominant animal. Alternatively, a backward approach may signal intentions opposite in nature (e.g., meekness) to those of a forward advancing aggressive animal (i.e., the principle of antithesis; Darwin, 1872).

and 6.3). Measured variables are very reliably observed with greater than 0.90 observer reliability (Hawley, 1994). Considering that the observed behaviors should and do evince a sizable proportion of reliable, but unique variance, the lack of perfect communality is not attributable solely to measurement error, but, rather, uniqueness at the micro-level. That is to say, a behavior can occur outside the context of a common factor: sometimes an ear flap is just a ear flap, and reliably so. On the other hand, the degree of common variance among the variables also suggests that important molar-level constructs can be assessed and parsimoniously evaluated as reflecting important intra- and interindividual variability.

Molar level constructs provide a clearly differentiated representation of the behavioral processes of both animals. The interfactor correlations, for example, are generally very low with only one of the 30 possible interfactor correlations for both animals reaching the 0.30 level (see Tables 6.3 and 6.4), suggesting that each behavioral state, or factor, is an independent and truly separable entity. In addition, each factor explained approximately equal amounts of variance, suggesting, again, that each factor is differentially and importantly representative of the commonality among the molecular behaviors (see estimates of semi-partial and partial r^2 in Tables 6.3 and 6.4).

The degree of interindividual similarity can be seen in the coefficients of congruence between the two factor patterns (Table 6.5). Social Movement (Factor 1) is highly similar between the two animals with 0.75 congruence. Factor 2, Active Back Into for the adult and Active Cautious Approach for the juvenile, have a congruence of 0.59, perhaps dropping from the differential effect of Active Hindquarters-to-Face. Similarly, Factor 3, Passive Cautious Approach for the adult and Passive Back Into for the juvenile, showed moderate levels of similarity at 0.69, again perhaps due to the differential contribution of Passive Hindquarters-to-Face for these two animals. Sand (Factor 4) and Object Handling (Factor 5) were also similar for the two animals, with congruence of 0.69 and 0.76. respectively. Finally, Touch (Factor 6) was similar across both animals, with a 0.76 congruence coefficient.

In summary, each solution provided clear differentiable factors reflecting intraindividual variability and change and interindividual differences in their variability and change processes. Each solution was characterized by few dual factor loadings, the loadings were of moderate communality, and the resulting factors were for the most part uncorrelated. In conjunction with these aspects of validity, the multibehavioral groupings were elucidated in terms of their social context and communicative meaning. That is to say, the interpretation of each factor was suggested by the patterns of specific factor loadings and the context implied by the distinctive groupings of variables (e.g., Cautious Approach).

Conclusions

"Large-*n*" analytical techniques can be used on small subject pools to elucidate important aspects of bio-behavioral processes and development. When the occasions mode is emphasized over the subject mode of selection, P-technique factor analysis (exploratory or confirmatory) can be applied to explore the structure of behavior of an individual when certain conditions are met and particular checks are conducted. We illustrated these points on data drawn from captive elephants to highlight the potential of these techniques for broad application to observationally based studies. By doing so, complex relations among micro-level behaviors that are easily overlooked by even prolonged observation (e.g., Cautious Approach) were uncovered for each animal of the group . Only by studying individuals can the validity of such molar-categories or latent constructs such as threats, expressions of dominance, affiliation, and so forth, be explicated and compared across individuals in a first step towards sound nomothetic generalizations. In so doing, the nature of complex communication and social-interactional systems can be empirically illuminated and validated.

Acknowledgments

This research was conducted at the Max Planck Institute in conjunction with the University of California at Riverside. The authors thank J. Nesselroade, B. Schmitz, D. Todt, and K. Widaman for valuable commentary. Sincere gratitude is also extended to B. Blaszkiewitz and the staff at Tierpark, Berlin.

References

Akaike, H. (1987). Factor analysis and AIC. *Psychometrika*, **52**, 317–332.

Allport, G. W. (1937). *Personality: A Psychological Interpretation.* New York: Holt.

Anderson, T. W. (1963). The use of factor analysis in the statistical analysis of multiple time series. *Psychometrika*, **28**, 1–25.

Blurton Jones, N. (1972). *Ethological Studies of Child Behaviour.* Cambridge: Cambridge University Press.

Campbell, D. T. & Fiske, D. W. (1959). Convergent and discriminant validation by the multitrait–multimethod matrix. *Psychological Bulletin*, **56**, 81–105.

Cattell, R. B. (1952). The three basic factor-analytic research designs – their interrelations and derivatives. *Psychological Bulletin*, **49**, 499–551.

Cattell, R. B. (1963). The structuring of change by P- and incremental R-technique. In C. W. Harris (ed.), *Problems in Measuring Change*. Madison: University of Wisconsin.

Cattell, R. B., Cattell, A. K. S. & Rhymer, R. M. (1947). P-technique demonstrated in determining psycho-physical source traits in a normal individual. *Psychometrika*, **12**, 267–288.

Cronbach, L. J. (1951). Coefficient alpha and the internal structure of tests. *Psychometrika*, **16**, 297–334.

Cronbach, L. J., & Meehl, P. E. (1955). Construct validity in psychological tests. *Psychological Bulletin*, **52**, 281–302.

Daly, D. L., Bath, K. E. & Nesselroade, J. R. (1974). On the confounding of inter- and intraindividual variability in examining change patterns. *Journal of Clinical Psychology*, **30**, 33–36.

Darwin, C. (1872). *The Expression of the Emotions in Man and Animals*. Chicago: University of Chicago Press. (Reprinted 1965.)

Figueredo, A. J., Ross, D. M. & Petrinovich, L. (1992). The quantitative ethology of the zebra finch: a study in comparative psychometrics. *Multivariate Behavioral Research*, **27**, 435–458.

Golani, I. (1992). A mobility gradient in the organization of vertebrate movement: the perception of movement through symbolic language. *Behavioral and Brain Sciences*, **15**, 249–308.

Gorsuch, R. L. (1983). *Factor Analysis*. Hillsdale, NJ: Lawrence Erlbaum Associates.

Hawley, P. H. (1994). On being an elephant: A quantitative intraindividual analysis of the behavior of Elephas maximus. PhD thesis, University of California, Riverside.

Hinde, R. A. (1966). *Animal Behaviour: A Synthesis of Ethology and Comparative Psychology*. New York: McGraw Hill.

Hooker, K., Nesselroade, D. W., Nesselroade, J. R. & Lerner, R. M. (1987). The structure of intraindividual temperament in the context of mother–child dyads: P-technique factor analyses of short-term change. *Developmental Psychology*, **23**, 332–346.

Jennings, H. S. (1906). *Behavior of the Lower Organisms*. New York: Columbia University Press.

Jones, J. C., & Nesselroade, J. R. (1990). Multivariate, replicated, single-subject, repeated measures designs and P-technique factor analysis: a review of intraindividual change studies. *Experimental Aging Research*, **16**, 171–183.

Jöreskog, K. G. & Sörbom, D. (1993). *Lisrel 8 User's Guide*. Mahwah, NJ: Lawrence Erlbaum Associates.

Korth, B. (1978). A significance test for congruence coefficients for Cattell's factors matched by scanning. *Multivariate Behavioral Research*, **3**, 419–430.

Little, T. D. (1997). Mean and covariance structures (MACS) analyses of cross-cultural data: practical and theoretical issues. *Multivariate Behavioral Research*, **32**, 53–76.

Little, T. D., Lindenberger, U. & Nesselroade, J. R. (1999). On selecting indicators for multivariate measurement and modeling with latent variables: when 'good' indicators are bad and 'bad' indicators are good. *Psychological Methods*, **4,** 192–211.

Lorenz, K. (1941). Vergleichende Bewegungsstudien an Anatinen [Comparative study of movement of Anatinen]. *Journal of Ornithology*, Suppl. **89**, 194–294.

Machlis, L., Dodd, P. W. D. & Fentress, J. C. (1985). The pooling fallacy: problems arising when individuals contribute more than one observation to the data set. *Zeitschrift für Tierpsychologie*, **68**, 201–214.

Maraun, M. D. (1996). Metaphor taken as math: indeterminacy in the factor analysis model. *Multivariate Behavioral Research*, **31**, 517–538.

Molenaar, P. C. M. (1985). A dynamic factor model for the analysis of multivariate time series data. *Psychometrika*, **50**, 181–202.

Molenaar, P. C. M. (1994). Dynamic latent variable models in developmental psychology. In A. von Eye & C. C. Clogg (eds.), *Latent Variables Analysis*, pp. 155–180. Thousand Oaks, CA: Sage.

Molenaar, P. C. M., De Gooijer, J. G. & Schmitz, B. (1992). Dynamic factor analysis of nonstationary multivariate time series. *Psychometrika*, **57**, 333–349.

Morgan, C. L. (1894). Introduction to comparative psychology: New York: Schribners.

Mulaik S. (1996). Factor analysis is not just a model in pure mathematics. *Multivariate Behavioral Research*, **31**, 655–661.

Nesselroade, J. R. (1994). Exploratory factor analysis with latent variables and the study of processes of development and change. In A. von Eye & C. C. Clogg (eds.), *Latent Variables Analysis*, pp. 131–154. Thousand Oaks, CA: Sage.

Nesselroade, J. R., & Jones, C. J. (1991). Multi-modal selection effects in the study of adult development: A perspective on multivariate, replicated, single-subject, repeated measures designs. *Experimental Aging Research*, **17**, 21–27.

Nunnally, J. C. (1978). *Psychometric Theory* 2nd edition. New York: McGraw-Hill.

Parker, G. A. (1984). Evolutionarily stable strategies. In J. R. Krebs & N. B. Davies (eds.), *Behavioural Ecology: An Evolutionary Approach*, pp. 30–61. Boston: Blackwell Scientific.

Rhine, R. J., & Ender, P. B. (1983). Comparability of methods used in sampling of primate behavior. *American Journal of Primatology*, **5**, 1–15.

Roberts, M. L., & Nesselroade, J. R. (1986). Intraindividual variability in perceived locus of control in adults: P-technique factor analyses of short-term change. *Journal of Research in Personality*, **20**, 529–545.

Russell, R. L., Bryant, F. B. & Estrada, A. U. (1996). Confirmatory P-technique analyses of therapist discourse: high- versus low-quality child therapy sessions. *Journal of Consulting and Clinical Psychology*, **6**, 1366–1376.

Schmitz, B. (1987). *Zeitreihenanalyse in der Psychologie* [Time series analysis in psychology]. Weinheim: Betz Verlag.

Sebeok, T. A. (1965). Animal communication. *Science*, **147**, 1006–1014.

Senar, J. C. (1990). Agonistic communication in social species: what is communicated? *Behaviour*, **112**, 270–283.

Tucker, L. R., & Lewis, C. (1973). A reliability coefficient for maximum likelihood factor analysis. *Psychometrika*, **38,** 1–10.

Wessman, A. E. & Ricks, D. F. (1966). *Mood and Personality*. New York: Holt, Rinehart, and Winston.

Wood, P., & Brown, D. (1994). The study of intraindividual differences by means of dynamic factor models: rationale, implementation, and interpretation. *Psychological Bulletin*, **116**, 166–186.

7 Examining the relationship between environmental variables and ordination axes using latent variables and structural equation modeling

James B. Grace

Abstract

Examinations of the relationships between environmental variables and ordination results often give little consideration to the complex relationships among environmental factors. In this chapter I consider the utility of structural equation modeling (SEM) with latent variables for evaluating the relationships among environmental variables and ordination axis scores. Using an example data set, I compare the efficiency of three approaches – (1) multiple regression, (2) principle component analysis, and (3) SEM – as methods for extracting information from multivariate data. All three approaches were found to be equivalent in their ability to explain variance in response variables but differ in their ability to explain the covariation among predictor variables. In general, when sufficient theoretical knowledge exists to permit the formulation of hypotheses about the relationships among variables, structural equation modeling can provide for a more comprehensive analysis. It is suggested that the analysis of latent variables using SEM may advance our understanding of environmental effects on vegetation data in many cases.

Introduction

The search for environmental controls on communities remains a major activity of many ecologists. Considerable effort has been spent during the past few decades to develop and adapt numerical methods for analyzing the complex interrelationships among species (Goodall, 1954; Bray & Curtis, 1957; Orloci, 1967; Hill, 1973, 1974; Hill & Gauch, 1980; Gauch, 1982; ter Braak, 1986). The application of classification methods, as well as the development and adoption of ordination procedures, has led to the widespread use of these multivariate techniques by community ecologists. At the same time, considerably less attention has been paid to the investigation of interrelationships among environmental variables, in part due to limited statistical tools for analyzing such complexities. At present, our ability to comprehensively consider the relationships between environmental variables and community composition is strongly limited by a paucity of approaches for extracting insights from sets of correlated environmental variables.

A number of numerical methods for analyzing the relationships between environmental and community properties are currently available. One of the most common approaches to this problem has been through the application of regression methods, both simple regression and multiple regression. A typical use of regression is the examination of the relationships between environmental variables and ordination axes scores (so-called "indirect" analysis). Another commonly used approach is embodied in canonical correspondence analysis, which directly involves environmental data in the ordination process, resulting in an ordering of samples that is based on both species and environmental information ("direct" analysis). More advanced procedures that are increasingly used by ecologists include (1) principal components extracted from the environmental matrix, (2) canonical correlation and redundancy analysis, and (3) matrix comparison methods such as the Mantel's test, analysis of similarities, and Procrustes analysis (Legendre & Legendre, 1998).

One method for the analysis of complex multivariate relationships is structural equation modeling (SEM). While many ecologists are familiar with the special case of path analysis, the more general procedures of SEM have only recently been used in ecological studies. In this chapter I introduce the basic concepts of SEM, including the estimation of latent variables, compare SEM results with those from regression and principle component analyses, and end by considering the potential utility and limitations of this methodology.

Structural equation modeling

Historical perspective

SEM can be traced back to a method that is more familiar to most ecologists, path analysis, which was developed originally by Wright (1918) for the purpose of explicitly dealing with complex dependency relationships among variables. In the early 1970s, a generalized procedure for analyzing covariances was developed that subsumed both path analysis and factor analysis as special cases (Keesling, 1972; Jöreskog, 1973) and which came to be known as "structural equation modeling". At present, SEM is used by thousands of researchers in fields ranging from the social sciences to chemistry and biology (for a recent bibliography of applications, see Austin & Calderon, 1996). Because of the abandonment of Wrightian path analysis by most statisticians in the 1970s, the capabilities of SEM are not well represented by the discussions of path analysis sometimes found in the ecological literature. Only recently have ecologists begun to employ SEM (Johnson *et al.*, 1991; Mitchell, 1992; Pugesek & Tomer, 1995, 1996; Gough & Grace, 1999; Grace & Pugesek, 1997; Grace & Guntenspergen, 1999; Grace & Jutila, 1999; Shipley, 1999; Grace *et al.*, 2000a).

Basic terms and concepts

Basic concepts and terms relating to SEM can be found in Kline (1998), Maruyama (1998), and in other chapters in this volume. Discussions of more advanced issues can be found in Hayduk (1987, 1996), Bollen (1989), and Marcoulides & Schumacker (1996).

Briefly, SEM involves the specification of a multivariate dependence model that can be tested statistically against data. The concept of "structural" equations simply refers to the fact that the structure of dependency relationships among variables can be specified through a series of equations. The standard variable types usually employed in SEM include *indicator* variables, which are the measured variables, and *latent* variables, which are conceptual variables. A basic structural equation model includes two components: a *measurement model*, which describes how indicator variables relate to the conceptual variables; and a *structural model*, which specifies the relationships among latent variables.

The concept of latent variables is fundamental to the philosophy of SEM. It is certainly true that in many cases conceptual variables are assumed to be perfectly measured by individual indicator variables, in which case the measurement model plays a minor role in the analysis. A bivariate regression can be shown to be the simplest representation of such a model. The incorporation of latent variables in a model, however, provides the user with several advanced capabilities that may be both statistically and scientifically important. One of these capabilities is the ability to address measurement error. An often overlooked limitation of most conventional statistical procedures such as regression and analysis of variance is the assumption that independent variables are measured without error. Theoretical studies have shown that this assumption causes bias in the model parameters as well as an overestimation of error in the dependent variables (Bollen, 1989, Chapter 5). A second capability resulting from the use of latent variables is greater generality for the conclusions. The conceptual variables of most interest to scientists are often latent variables that cannot be measured perfectly by a single indicator variable. When multiple indicators are available for estimation of a latent effect, the resulting conclusions can be shown to have greater general validity (Bollen, 1989). Finally, a practical value of the use of latent variables is as an effective means of dealing with collinearity. When a set of predictor variables includes some that are highly correlated, a number of problems are created, including both variance inflation and difficulties of interpretation (Maruyama, 1998, Chapter 4). When correlated variables can be construed to be multiple measures of a more general (latent) cause, their correlations can be used to estimate the precision of that latent variable, and a structural model can be achieved that is unaffected by collinearity. Once latent variables are employed in a model, issues of both *construct validity* and *construct reliability* arise, where validity refers to the degree to which the measurement actually measures the latent variable intended by the researcher and reliability refers to the precision among indicator variables.

Another feature of SEM is that it can be used as a confirmatory procedure while applications of regression, principle components analysis, canonical correlation, Mantel's test, etc. are typically exploratory. In SEM, initial models are developed *a priori* on the basis of existing scientific knowledge and are then evaluated against the relationships in the data. *Model fit* refers to measures of how well a hypothesized model corresponds to observed covariances. In many cases, initial models are rejected in favor of revised models that are intended for further evaluation with additional data. An important principle of SEM is that model respecification should proceed conservatively and should be based on hypothesized causal relationships to avoid incorporating relationships that are due to chance features of a data set.

Methods

The data used in this paper to illustrate the statistical methods are from Grace *et al.* (2000b). They are from a study of 107 0.25 m^2 plots in a native grassland in southern Louisiana. Variables were measured at a single point in time (July 1995) and included vegetation composition, relative elevation, and several soil properties. Soil variables included pH, total carbon, total nitrogen, and extractable phosphorus, calcium, magnesium, potassium, manganese, and zinc.

Vegetation data were ordinated using nonmetric multidimensional scaling and the PC-ORD system (McCune & Mefford, 1995). Species with fewer than five occurrences were omitted before analysis. Ordination was based on Sorenson's (1948) similarity index. To simplify interpretation, the resulting ordination was rotated to maximize the relationship between axis 1 and elevation. In this chapter the relationships between environmental variables and ordination axis scores were investigated using bivariate regression, standard multiple regression, stepwise multiple regression, and principal component analysis using the SAS program (SAS Institute, 1989). Principle components analysis was performed using the standard orthogonal procedure, which is typically used in ecological applications (Legendre & Legendre, 1998). Following the extraction of principal components from the environmental matrix, components were regressed against axis scores in order to determine the components that would be retained in the final model. SEM was implemented using LISREL (Jöreskog & Sörbom, 1996). Following the formulation of an initial structural equation model, modification indices were used to guide the model refinement process until an adequate model was achieved. Further description of LISREL application procedures can be found in Pugesek & Tomer (1996) and Grace & Pugesek (1997).

Results

Ordination results, including vectors of environmental correlations, are presented in Figure 7.1. A complete and detailed description of the vegetation data are presented elsewhere (Grace *et al.*, 2000b). Rather, because the

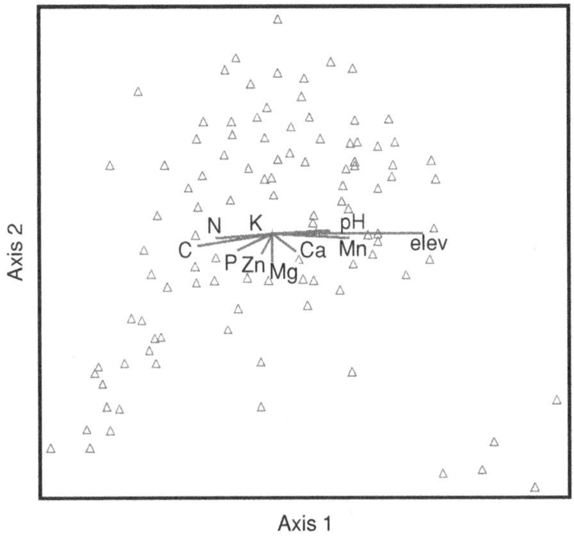

Figure 7.1. Bivariate environmental correlations with plot ordination scores (from Grace *et al.*, 2000b). The ordination of plots was obtained using nonmetric multidimensional scaling and the ordination was rotated so as to maximize the relationship between elevation and axis 1.

Table 7.1. *Correlations among environmental and axis variables (Pearson correlation coefficients). Numbers shown in bold are significant at the* $p = 0.05$ *level*

	elev	Ca	Mg	Mn	Zn	K	P	pH	C	N	Axis 1	Axis 2
elev	1.0	—	—	—	—	—	—	—	—	—	—	—
Ca	**0.42**	1.0	—	—	—	—	—	—	—	—	—	—
Mg	**0.21**	**0.79**	1.0	—	—	—	—	—	—	—	—	—
Mn	**0.71**	**0.66**	**0.42**	1.0	—	—	—	—	—	—	—	—
Zn	−0.10	**0.47**	**0.45**	**0.39**	1.0	—	—	—	—	—	—	—
K	**−0.20**	**0.40**	**0.35**	**0.25**	**0.69**	1.0	—	—	—	—	—	—
P	**−0.32**	**0.32**	**0.26**	0.08	**0.72**	**0.74**	1.0	—	—	—	—	—
pH	**0.43**	−0.02	−0.14	**0.28**	**−0.26**	**−0.46**	**−0.48**	1.0	—	—	—	—
C	**−0.44**	0.04	0.15	−0.19	**0.49**	**0.55**	**0.62**	**−0.63**	1.0	—	—	—
N	**−0.46**	−0.05	0.11	**−0.27**	**0.31**	**0.41**	**0.51**	**−0.53**	**0.81**	1.0	—	—
Axis 1	**0.77**	**0.30**	0.03	**0.55**	**−0.20**	**−0.23**	**−0.37**	**0.47**	**−0.55**	**−0.47**	1.0	—
Axis 2	0.005	**−0.26**	**−0.35**	−0.14	**−0.28**	−0.16	**−0.25**	0.11	**−0.21**	−0.13	0.17	1.0

primary purpose of this chapter is to evaluate statistical techniques, the data represented in Figure 7.1 are meant to serve as a starting point for evaluating the relationships between environmental variables and ordination results. Thus further details of the results presented in Figure 7.1 are omitted from this discussion.

Regression and principal components

Vectors showing relationships between environmental variables and ordination axes presented in Figure 7.1 represent a basic and fundamental level of evaluation of environmental effects on vegetation. In this dataset, all 10 of the environmental variables examined were found to have significant correlations with one or both axes (Table 7.1, $r \geq 0.195$). The strongest relationship found was between elevation and axis 1, while a number of other variables were also found to be related to axis 1. Five of the variables were correlated with axis 2, with magnesium having the strongest correlation. Further, 36 of the 45 possible correlations among the 10 environmental variables were found to be statistically significant (Table 7.1).

The results of standard regression are displayed diagramatically in Figure 7.2a. Five of the possible 20 partial regression coefficients were found to be statistically significant ($p < 0.05$). Sixty-five percent of the variance in axis 1 was explained by elevation, calcium (Ca), magnesium (Mg), and

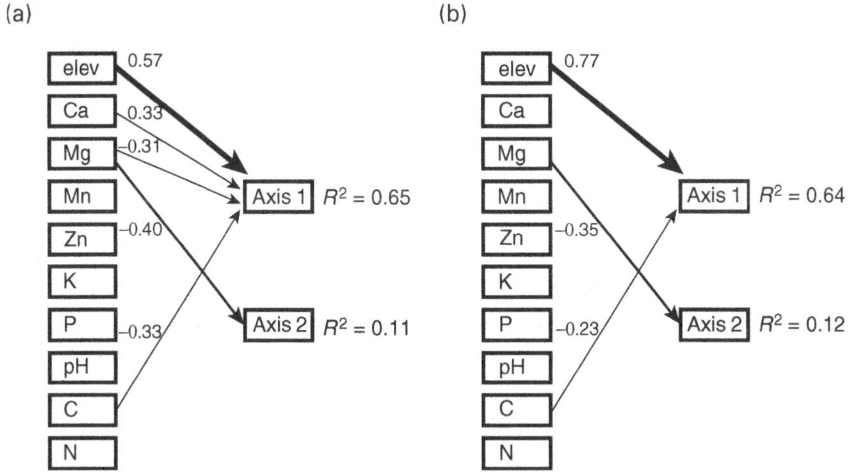

Figure 7.2. Diagramatic representation of regression results. (a) Standard regression results based on partial regression procedures. (b) Stepwise regression results based on sequential regression procedures.

Table 7.2. *Loadings resulting from principal component analysis (see Figure 7.3). Loadings greater than 0.3 are shown in bold to highlight loading patterns*

Variable	PC1	PC2	PC3	PC4	PC5
elev	−0.18	**0.44**	−0.04	**0.55**	−0.26
Ca	0.19	**0.47**	−0.29	−0.14	0.04
Mg	0.22	**0.36**	**−0.60**	−0.29	0.25
Mn	0.04	**0.52**	0.23	**0.32**	−0.11
Zn	**0.38**	0.20	**0.38**	−0.14	0.22
K	**0.41**	0.10	0.29	−0.11	−0.26
P	**0.42**	0.01	**0.32**	−0.13	−0.01
pH	**−0.33**	0.20	**0.35**	0.03	**0.77**
C	**0.40**	−0.19	−0.09	**0.41**	0.12
N	**0.35**	−0.24	−0.22	**0.52**	**0.37**

carbon (C), while 11% of the variance in axis 2 was explained by a single variable, magnesium. Stepwise regression results are presented in Figure 7.2b. For stepwise regression, the first variable loaded into the model for axis 1 was elevation, which had a partial R^2 of 0.59, while the second variable loaded into the model was carbon, which had a partial R^2 of 0.05. Additional variables did not contribute significantly to the model. Stepwise results for axis 2 included only magnesium in the model.

Principal component analysis extracted 10 components from the environmental matrix. The first five components were judged to be important on the basis of the criterion that additional components extracted less than 5% of the variance and because subsequent regression analysis showed that component 5 explained unique variance in the dependent variables. The first five components accounted for 91% of the total variance in the matrix of environmental correlations. Loadings for the first five components are given in Table 7.2. Component 1 loaded most strongly on zinc (Zn), potassium (K), pH, C, nitrogen (N), and phosphorus (P), while component 2 loaded primarily on elevation (elev), Ca, Mg, and manganese (Mn). Component 3 loaded strongly on Mg, with lesser affinities for Zn, pH, and P. Component 4 loaded strongly with elevation and N, while also loading on C and Mn. Finally, component 5 was related primarily to pH.

Regression analysis using principal components as predictors revealed that four of the first five components were significantly related to axis scores ($p < 0.05$, Figure 7.3). Component 3 was not found to contribute to

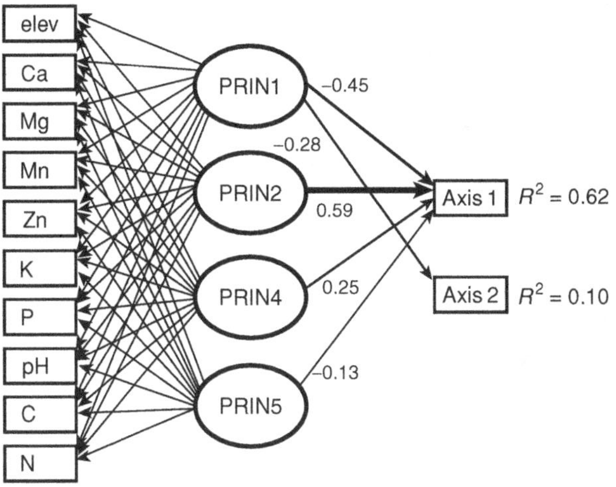

Figure 7.3. Diagramatic representation of regression using principal components. PRIN1 to PRIN5 represent principal components. Variables in ellipses represent principal components, while variables in boxes are the observed variables. For measured variables: elev refers to measured relative elevation; Ca, Mg, Mn, Zn, K, and P refer to extractable nutrient concentrations; C and N refer to total carbon and nitrogen; and pH refers to soil acidity. Components were extracted first and then separately evaluated as predictors for axes 1 and 2.

the multiple regression equation for either axis. Axis 1 was strongly related to components 1 and 2, and weakly related to components 4 and 5. Axis 2 was related only to component 1. Sixty-two percent and 10% of the variance were explained for axes 1 and 2, respectively.

Structural equation modeling

The first step in the structural equation analysis was to develop a measurement model for the environmental variables (Pugesek & Tomer, 1996). The initial measurement model (Figure 7.4) was based on an underlying hypothesis about the relationship of topography to the development of soil properties in the system being investigated. Within the study area, elevation can be seen to relate to soil properties in two important ways. First, the highest elevations at the site are associated with mima mounds, which are distinct soil mounds of disputed origin (Cain, 1974). Whether mounds are caused by differential erosion and soil particle trapping or by extrusion, the potential exists for differences in soil mineral characteristics between mounds and intermound areas. Second, areas of low elevation are more frequently flooded

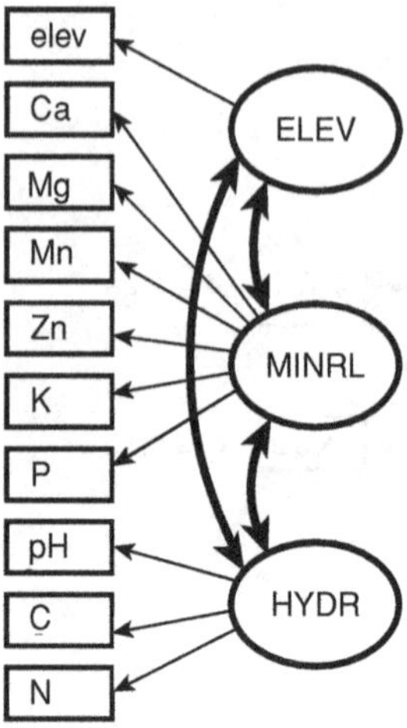

Figure 7.4. Initial measurement model. Variables shown in ellipses are latent (conceptual) variables while variables shown in boxes are indicator (observed) variables. For latent variables, ELEV refers to elevation effects, MINRL refers to effects of soil mineral composition, and HYDR refers to effects of hydric soil influences. For indicator variables: elev refers to measured relative elevation, Ca, Mg, Mn, Zn, K, and P refer to extractable nutrient concentrations; C and N refer to total carbon and nitrogen, and pH refers to soil acidity. The relationships between observed and latent variables constitute the measurement model.

and are expected to have chemically reduced soil properties and a higher accumulation of soil organic matter. For this reason, the initial model proposed an association between observed soil variables and two latent soil variables, mineral influences (MINRL) and hydric influences (HYDR). As a first approximation, it was initially hypothesized that pedogenic processes would cause covariation in the mineral components of the soil and we would expect Ca, Mg, Mn, Zn, and K to be correlated. Hydric influences were expected to affect C and N because of a greater accumulation of total carbon and nitrogen in soil organic matter under flooded conditions (Donahue *et al.*, 1977). A hydric influence on soil pH was also expected because of the development of acidic conditions in flooded soils.

Evaluation of the initial measurement model using LISREL resulted in poor model fit. A χ^2 value of 355 was obtained with 33 degrees of freedom (df) (note that degrees of freedom result from the number of unspecified relationships among variables). The observed ratio between χ^2 and degrees of freedom was 10.8, much higher than the recommended maximum of 2.0. Additional fit indices likewise showed poor model fit. The root mean square

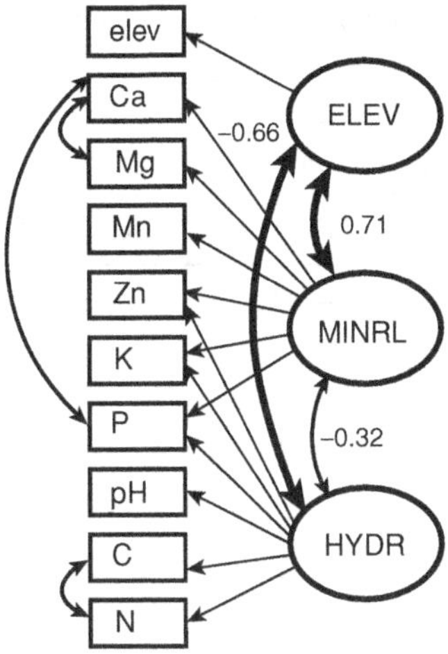

Figure 7.5. Final measurement model. Variables are as described in Figure 7.3. Double-headed arrows connecting observed variables represent correlated errors. Arrows shown represent significant paths. Numbers shown represent correlations among latent variables. Loadings between indicator and latent variables are given in Table 7.3.

error of approximation (RMSEA) value of 0.30 was substantially above the recommended maximum of 0.09. Finally, the goodness-of-fit index was 0.60, well below the recommended minimum of 0.90.

Modification indices produced by the LISREL program suggested a number of possible changes that could lead to improved model fit. Some of the changes suggested were deemed to be logically incompatible with the model. For example, it was indicated that a pathway from the latent variable ELEV to the observed variable Mn would lead to a lower χ^2 value. Some of the other modifications suggested were deemed to be biologically plausible and formed the basis for a revised measurement model (Figure 7.5).

Evaluation of the revised model led to statistical results that were deemed satisfactory. A χ^2 of 51 was obtained with 27 df (a ratio of 1.9), the RMSEA value was 0.09, and the goodness-of-fit index was 0.91. No additional modifications were suggested by the LISREL program. SEM results for the revised model are summarized in Table 7.3 and Figure 7.5. Standardized factor loadings (Table 7.3) showed that MINRL was related to Mn, Ca, Zn, K, Mg, and P while HYDR was related to P, K, Zn, C, N, and pH. Significant correlations between errors were found for Ca and Mg, Ca and P, and between C and N.

Table 7.3. *Standardized factor loadings resulting from structural equation analysis (see Figure 7.5)*

	Latent factor				
Measured variable	ELEV	MINRL	HYDR	AXIS 1	AXIS 2
elev	1.0	—	—	—	—
Ca	—	0.66	—	—	—
Mg	—	0.43	—	—	—
Mn	—	0.99	—	—	—
Zn	—	0.66	0.81	—	—
K	—	0.53	0.84	—	—
P	—	0.39	0.93	—	—
pH	—	—	−0.66	—	—
C	—	—	0.78	—	—
N	—	—	0.67	—	—
Axis 1	—	—	—	1.0	—
Axis 2	—	—	—	—	1.0

Once a satisfactory measurement model was obtained, analysis of a full model was performed. The initial structural model (proposed relationships among latent variables) was based on ordination results published elsewhere (Grace *et al.*, 2000b) and hypothesized that axis 1 of the ordination represented effects of elevation and hydric soil influences on vegetation, while axis 2 represented the effects of mineral and hydric soil factors. Results from the analysis of a full model (including both measurement and structural components) are shown in Figure 7.6. Both MINRL and HYDR were found to be strongly correlated with ELEV, while MINRL and HYDR were less strongly related to each other. Axis 1 was significantly related to ELEV and HYDR while axis 2 was related to MINRL and HYDR. Totals of 63% and 9% of the variances in axis 1 and axis 2 were explained by the model. Finally, correlations among the latent variables are presented in Table 7.4. Note that correlation between a pair of latent variables equals the sum of all direct and indirect paths between them.

Discussion

Consideration of the SEM results

Results of the SEM analyses led to the rejection of the initial measurement model and the eventual acceptance of a revised full model (Figure 7.6).

Table 7.4. *Correlations among latent variables (refer to Figure 7.6)*

	ELEV	MINRL	HYDR	AXIS 1	AXIS 2
ELEV	1.0	—	—	—	—
MINRL	0.71	1.0	—	—	—
HYDR	−0.67	−0.33	1.0	—	—
AXIS1	0.77	0.50	−0.67	1.0	—
AXIS2	0.02	−0.14	−0.21	0.07	1.0

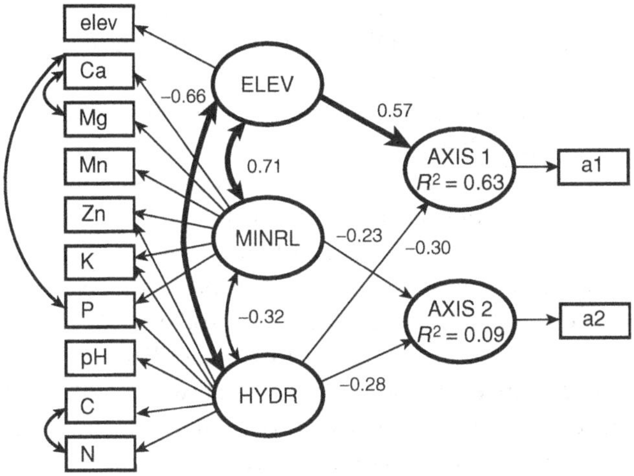

Figure 7.6. Final structural equation model. Variables are as described in Figure 7.4. Double-headed arrows connecting observed variables represent correlated errors. All arrows represent significant paths. Numbers shown represent partial regression coefficients for directional paths and correlations among latent variables for curved arrows. R^2 values represent the proportion of variance explained for the dependent latent variables. Loadings between indicator and latent variables are given in Table 7.3. Net correlations among latent variables (the sum of all direct and indirect paths) are given in Table 7.4.

First, with regard to the measurement model, three significant correlations were found among error terms. Such correlations were expected because of the nonindependence associated with testing multiple properties on a single soil sample and because of chemical associations among soil properties. Second, significant loadings of Zn, K, and P were found on HYDR, which were not proposed in the initial model. While further study of these relationships is warranted, hydric soil conditions are known to influence a large number of soil properties (Gambrell & Patrick, 1978). The availability of elements, especially Zn and P, are known to be strongly influenced by hydric conditions. On the basis of the relationships found in these data, the results suggest hydric influences on N, C, pH, P, K, and Zn. Further work is needed to determine whether these relationships are general for the study area and whether they indeed result from hydric influences on soil formation.

The results for the structural model relationships in Figure 7.6 show that 63% and 9% of the variation in axes 1 and 2 were explained by the model, virtually identical to the variance explained by the regression and PCA methods. This similarity in variance explained indicates that the latent variables MINRL and HYDR successfully represent the influences of soil properties on vegetation. For the sake of this discussion, I interpret MINRL as a representation of mineral soil influences and HYDR as a representation of the influence of hydric conditions, despite the fact that further confirmation of this interpretation is needed. Both mineral and hydric influences were strongly correlated with elevation (Figure 7.6). Higher mineral contents were generally found at higher elevations, as indicated by the positive relationship between ELEV and MINRL and the positive loadings of variables on MINRL (Table 7.3). Hydric soil influences were greatest at low elevations, as indicated by the negative correlation between ELEV and HYDR. The relatively modest correlation between hydric and mineral influences suggests that these properties relate differently to topography.

On the basis of the structural model in Figure 7.6, it appears that vegetation composition at the study site is affected by elevation, mineral soil, and hydric soil conditions. The first axis of the ordination is primarily related to unexplained elevational influences (the direct path from ELEV to AXIS1 = 0.57) and secondarily related to influences from hydric conditions (−0.30). Variation in the vegetation along axis 2 of the ordination was more or less equally related to mineral and hydric soil influences, though neither was terribly strong (−0.23 and −0.28, respectively).

The correlations among latent variables (Table 7.4) in combination with path coefficients (Figure 7.6) makes it possible to examine the complete set of relationships among latent variables. Using this approach, we can see that the total correlation between ELEV and AXIS1 (0.77) can be partitioned into influences mediated through the relationship between elevation and hydric conditions ($-0.67 \times -0.30 = +0.20$) as well as the remaining direct path (0.55), which represents elevation effects that have not been fully explained. The sum of these paths equals the net correlation

($0.20 + 0.57 = 0.77$). Further, we can see that elevation is related to axis 2 through a pair of offsetting influences. One is mediated through mineral soil influences ($0.71 \times -0.23 = -0.16$) and the other through hydric soil influences ($-0.67 \times -0.28 = +0.19$). Summing these two offsetting paths results in a nonsignificant net correlation between elevation and axis 2 that belies the underlying effects of elevation in the model.

Overall, the results of the SEM analyses are consistent with the hypothesis that the vegetation at the study site is responding primarily to hydric soil influences, mineral soil variations, and additional influences of elevation (the direct pathway). Because our initial measurement model (Figure 7.4) was rejected, we must view the measurement portion of the model in Figure 7.6 as tentative and in need of further testing. In any event, it is important to keep in mind that consistency between the statistical model and the data does not mean our scientific interpretations are correct, only that the data are not inconsistent with our interpretations.

Consideration of methodology

The results of the analyses presented in this chapter indicate that, in one sense, all the methods used can be judged to be satisfactory and equivalent. The total variance explained in axis scores was virtually identical for all methods. Thus, when the primary objective of a study is to maximize the variance explanation of the dependent variables, multiple regression is generally as satisfactory as SEM. Further, when the goal of a study is simply prediction within the parameter space included in the sample, the methods examined are equal to the task.

Regression does not attempt to explain correlations among predictors

One place where the methods examined differ is in the degree to which the correlations among environmental variables are explained in the analysis. For the simple bivariate relationships summarized in Figure 7.1, correlations among variables are ignored, except for the fact that the similarities in vector directions imply correlations among variables. For standard regression, correlations among environmental variables strongly influence the resulting partial correlations between predictors and axes. Comparing Table 7.1 with Figure 7.2, we can see that the simple correlation between elevation and axis 1 is 0.77 while the partial correlation is 0.57. A more dramatic discrepancy between simple and partial correlations can be seen for Mn, where the simple correlation is 0.55 and the partial correlation is not significant (value = 0.02). The pervasiveness of this discrepancy between simple and partial correlations is illustrated by the fact that 9 of the 10 environmental variables have significant simple correlations with axis 1 but only four have significant partial correlations when all variables are included in the model. The reliance on sequential sums of squares, such as in the stepwise regression method, results in an even greater discrepancy from simple regression results. In the sequential process, the variable with the strongest correlation is entered first and its full correlation is used to explain the dependent variable. Additional variables are judged only on their abilities to explain residual variation, leading to a very strong influence of order in the model. Thus correlations among predictors strongly influence regression results; however, explicit interpretation of correlation patterns is not included in regression analysis.

PCA provides an empirical characterization of the correlations among predictors

Both PCA and SEM attempt to explain the patterns of correlations among predictors, though they approach this in significantly different ways. PCA is based on the premise that a matrix of correlations among variables may be able to be reduced to a subset of higher-order variables or components. As reflected by the direction of arrows from component variables to observed variables in Figure 7.3, it is usually implied that the principal components can be interpreted substantiatively based on their loadings (note that these loadings are simple correlations between principal components and observed variables, and are independent of one another). In PCA the process of identifying components is directed by the data, though certainly the investigator can choose rotation options, the number of components, and variables to include in the analysis. While orthogonal rotation methods are the norm in ecological studies it can be argued that oblique methods are more reasonable and should be used more often because most ecological factors are intercorrelated to some degree in nature. In general, PCA results can be very useful in exploring relationships among correlated predictor variables and in developing hypotheses about higher-order influences. The primary limitations of PCA include (1) limited flexibility in specifying the structure of the model, (2) limits on its ability to estimate underlying latent parameters, (3) limited capacity for hypothesis testing, and (4) an inability to account for measurement error.

SEM incorporates hypotheses about underlying processes into the specification of the measurement model

The major conceptual distinction between the PCA approach and SEM is that SEM relies to a greater degree on the experience of the researcher to incorporate his or her knowledge about the underlying processes into the specification of an hypothesis about the relationships among variables. Hypothesis specification is, of course, a part of regression analysis. In the absence of an hypothesis about the direction of influence, simple correlations would be used in place of regression analysis. In the case of SEM, the incorporation of knowledge external to the data set into model specification is substantially expanded.

The development of hypotheses about the relationships among variables in the structural model is often a relatively simple transition for many ecological problems. Once researchers realize that they have the statistical freedom to specify the order of dependency among variables, they usually realize that they have sufficient information about underlying processes to develop initial hypotheses. Examples of such applications can be found in Wesser & Armbruster (1991), Mitchell (1992), Wootton (1994), and Grace & Platt (1995). In contrast, comparatively few ecological examples exist for the evaluation of hypotheses about measurement models (Pugesek & Tomer, 1996).

SEM has a number of additional strengths

The continued expansion of the use of SEM in a wide variety of fields is driven not only by the advantages of specifying and evaluating hypotheses but also by a number of statistical capabilities of the method. Unlike most other methods commonly used by biologists, including analysis of variance (ANOVA), etc., SEM provides a method for estimating and partitioning measurement error. It is not always recognized that traditional statistical methods assume that the independent variables are measured without error. When measurement error exists, as it will in many cases, the resulting parameter estimates can be shown to be biased unless measurement error is handled explicitly (Bollen, 1989, Chapter 5). In basic terms, when error in the independent variables is ignored, that error is lumped into the unexplained variance of the dependent variables, leading to both inflation of the estimated error of prediction and biased path coefficients. For this reason, efforts are being made to develop SEM approaches for use in place of ANOVA, multivariate analysis of variance (MANOVA), and multivariate analysis of covariance (MANCOVA) when measurement error is important (McArdle & Hamagami, 1996).

An additional capability of SEM is as a means of reducing the impact of collinearity among highly correlated predictor variables. When correlations among predictor variables are very high, a number of problems are created, including variance inflation and difficulties with interpretation. Within the context of regression analysis, there are limited means of dealing with such difficulties. Methods such as ridge regression can be used to reduce the impacts, but they are not highly effective (Maruyama, 1998). Two features of SEM practice can greatly reduce the potential for negative impacts of collinearity. First, highly correlated variables are often related to a common latent factor. In this case, the correlations among variables are used to estimate the latent factor, and collinearity effects do not have negative influences on either the structural or the measurement model. A second aspect of SEM practice that is very helpful in avoiding problems when interpreting the structural model is the practice of examining correlations among latent variables, indirect effects (mediated through other latent variables), and direct effects (as illustrated earlier in this chapter). By examining all these parameters, it is usually possible both to detect problems and to arrive at appropriate interpretations even when substantial collinearity exists (Maruyama, 1998).

SEM is not without limitations

At present, the family of SEM procedures and practices represents a comprehensive approach for evaluating the relationships among a set of predictor variables (Grace & Pugesek, 1998; Pugesek & Grace, 1998). This is not to say that SEM is without limitations. First and foremost, the correct analysis of complex data is inherently difficult. Vegetation scientists and ecologists in general are often faced with analyzing and interpreting large matrices of relationships. Few situations are likely to be more complicated than the case of teasing apart the effects of a set of correlated predictors on a matrix of community similarities. Because of this complexity, inadequacies of data and information can lead to erroneous conclusions even when techniques are used properly. Within the philosophy of SEM, the researcher is urged to use split data sets, overlapping data sets, and subsequent studies in order to evaluate the repeatability of results. The process of model development, evaluation, and refinement is generally seen as necessary to establish the validity of complex multivariate hypotheses.

A second limitation of multivariate analyses in general is the need for a large sample size. Small samples contain a limited amount of information and the number of parameters to be estimated is typically large. Discussion of the minimum and desired sample sizes needed for multivariate studies can be found in Hair *et al.* (1995, p. 637). As a very rough rule of thumb, a minimum of 100 samples is preferred and 50 is an absolute minimum for even the simplest of models. For very small samples, more exploratory methods such as partial least squares (Fornell & Cha, 1994) or regression and PCA are preferred.

A third limitation typically associated with most statistical studies is the assumption of linearity. Unless nonlinear (including curvilinear) relationships are explicitly considered, most methods examine only linear relationships among variables. As the complexity of the model increases, be it a structural equation model or an analysis of variance model, nonlinearities become increasingly difficult to detect and address. Approaches for dealing with this problem in SEM exist (Schumacker & Marcoulides, 1998), though they are not always simple to implement. In some cases, methods with a greater variety of ways for handling nonlinearities (Fornell & Cha, 1994) may be preferred.

The estimation of latent variables using SEM deserves further consideration by ecologists

Up to this point ecologists have had few tools for exploring the relationships among correlated predictor variables. To accomplish this using SEM will involve the use of latent variables in many cases. Many ecological concepts may be thought of as latent variables. Body size, productivity, biomass, and diversity may all best be thought of as latent variables when we are considering their relationships to other variables and to ecological processes. When we say diversity may relate to productivity we should not imagine that a single measure of productivity or diversity will precisely represent the processes of interest. In most cases, current measures of productivity act as indicators for the productivity that took place during the period of time when current patterns of diversity were being shaped. Such reasoning applies to the great majority of ecological concepts of interest. One might go so far as to say that if a variable is of general conceptual interest, it should be treated as a latent variable.

The question of how to define and measure a latent variable represents uncharted territory in ecology. There is certainly room for disagreement about what constitutes a good example of a latent variable. From a statistical standpoint, methods are available for assessing the attributes of latent variables, such as dimensionality, validity, and precision. However, most variables of general conceptual interest can be thought of as either a collection of loosely related processes or a higher level influence, depending on one's perspective and research objectives. Some may disagree with the premise that hydric influences on soil processes could constitute a single latent factor, while others may see such a claim as quite reasonable. Undoubtably, a discussion of latent variables will face the same range of viewpoints as the discussion of any general concept. Ultimately, ecologists will evaluate latent variables on the basis of their utility as a means of expressing general properties of complex data. For vegetation scientists and others trying to make sense of complex multivariate data, I suspect that the evaluation of latent variables will prove to be a useful means of extracting and interpreting the complex interrelations among variables.

Acknowledgments

This analysis was performed on data collected in collaboration with Larry Allain, US Geological Survey, and Charles Allen, University of Louisiana, Monroe. I thank Bruce Pugesek, Bill Shipley, and Bruce McCune for numerous helpful comments on the manuscript. Martha Hixon provided editorial suggestions for improving the text.

References

Austin, J. T. & Calderon, R. F. (1996). Theoretical and technical contributions to structural equation modeling: an updated annotated bibliography. *Structural Equation Modeling*, **3**, 105–175.

Bollen, K. A. (1989). *Structural Equations with Latent Variables*. New York: Wiley.

Bray, R. J. & Curtis, J. T. (1957). An ordination of the upland forest communities of southern Wisconsin. *Ecological Monographs*, **27**, 325–349.

Cain, R. H. (1974). Pimple mounds: a new viewpoint. *Ecology*, **55**, 178–182.

Donahue, R. L., Miller, R. W. & Shickluna, J. C. (1977). *An Introduction to Soils and Plant Growth*. Englewood Cliffs, NJ: Prentice-Hall, Inc.

Fornell, C. & Cha, J. (1994). Partial least squares. In R. P. Bagozzi (ed.), *Advanced Methods of Marketing Research*, pp. 52–78. Cambridge, MA: Blackwell.

Gambrell, R. P. & Patrick, W. H., Jr (1978). Chemical and microbiological properties of anaerobic soils and sediments. In D. D. Hook & R. M. M. Crawford (eds.), *Plant Life in Anaerobic Environments*, pp. 375–424. Ann Arbor, MI: Ann Arbor Science Publ.

Gauch, H. G. (1982). *Multivariate Analysis in Community Ecology*. Cambridge: Cambridge University Press.

Goodall, D. W. (1954). Objective methods for the comparison of vegetation. III. An essay in the use of factor analysis. *Australian Journal of Botany*, **1**, 39–63.

Gough, L. & Grace, J. B. (1999). Predicting effects of environmental change on plant species density: experimental evaluations in a coastal wetland. *Ecology*, **80**, 882–890.

Grace, J. B. & Guntenspergen, G. R. (1999). The effects of landscape position on plant species density: evidence of past environmental effects in a coastal wetland. *Ecoscience*, **6**, 381–391.

Grace, J. B. & Jutila, H. (1999). The relationship between species density and community biomass in grazed and ungrazed coastal meadows. *Oikos*, **85**, 398–408.

Grace, J. B. & Pugesek, B. H. (1997). A structural equation model of plant species richness and its application to a coastal wetland. *American Naturalist*, **149**, 436–460.

Grace, J. B. & Pugesek, B. H. (1998). On the use of path analysis and related procedures for the investigation of ecological problems. *American Naturalist*, **152**, 151–159.

Grace, J. B., Allain, L. & Allen, C. (2000a). Plant species richness in a coastal tallgrass prairie: the importance of environmental effects. *Journal of Vegetation Science*, **11**, 443–452.

Grace, J. B., Allain, L. & Allen, C. (2000b). Vegetation associations in a rare community type – coastal tallgrass prairie. *Plant Ecology*, **147**, 105–115.

Grace, S. L. & Platt, W. J. (1995). Neighborhood effects on juveniles in an old-growth stand of longleaf pine, *Pinus palustris*. *Oikos*, **72**, 99–105.

Hair, J. F., Jr, Anderson, R. E., Tatham, R. L. & Black, W. C. (1995). *Multivariate Data Analysis*, 4th edition. New York: Prentice-Hall.

Hayduk, L. A. (1987). *Structural Equation Modeling with LISREL: Essentials and Advances*. Baltimore, MD: Johns Hopkins University Press.

Hayduk, L. A. (1996). *LISREL: Issues, Debates, and Strategies*. Baltimore, MD: Johns Hopkins University Press.

Hill, M. O. (1973). Reciprocal averaging: an eigenvector method of ordination. *Journal of Ecology*, **61**, 237–250.

Hill, M. O. (1974). Correspondence analysis: a neglected multivariate method. *Journal of the Royal Statistical Society*, Series C, **23**, 340–354.

Hill, M. O. & Gauch, H. G. (1980). Detrended correspondence analysis, an improved ordination technique. *Vegetatio*, **42**, 47–58.

Hoyle, R. H. (ed.) (1995). *Structural Equation Modeling*. London: Sage.

Johnson, M. L, Huggins, D. G. & deNoyelles, F., Jr (1991). Ecosystem modeling with LISREL: a new approach for measuring direct and indirect effects. *Ecological Applications*, **4**, 383–398.

Jöreskog, K. G. (1973). A general method for estimating a linear structural equation system. In A. S. Goldberger & O. D. Duncan (eds.), *Structural Equation Models in the Social Sciences*, pp. 85–112. New York: Academic Press.

Jöreskog, K. G. & Sörbom, D. (1996). *LISREL 8: User's Reference Guide*. Chicago: Scientific Software International.

Keesling, J. W. (1972). Maximum likelihood approaches to causal analysis, PhD thesis. Department of Education, University of Chicago.

Kline, R. B. (1998). *Principles and Practice of Structural Equation Modeling*. New York: Guilford Press.

Legendre, P. & Legendre, L. (1998). *Numerical Ecology*. New York: Elsevier.

Marcoulides, G. A. & Schumacker, R. E. (eds.) (1996). *Advanced Structural Equation Modeling*. Mahwah, NJ: Lawrence Erlbaum Associates.

Maruyama, G. M. (1998). *Basics of Structural Equation Modeling*. Thousand Oaks, CA: Sage.

McArdle, J. J. & Hamagami, F. (1996). Multilevel models from a multiple group structural equation perspective. In G. A. Marcoulides & R. E. Schumacker (eds.), *Advanced Structural Equation Modeling*, pp. 89–124. Mahwah, NJ: Lawrence Erlbaum Associates.

McCune, B. & Mefford, M. J. (1995). PC-ORD. *Multivariate Analysis of Ecological Data*, Version 2.0. Gleneden Beach, OR: MjM Software Design.

Mitchell, R. J. (1992). Testing evolutionary and ecological hypotheses using path analysis and structural equation modeling. *Functional Ecology*, **6**, 123–129.

Orloci, L. (1967). An agglomerative method for the classification of plant communities. *Journal of Ecology*, **55**, 193–206.

Pugesek, B. H. & Grace, J. B. (1998). On the utility of path modeling for ecological and evolutionary studies. *Functional Ecology*, **12**, 853–856.

Pugesek, B. H. & Tomer, A. (1995). Determination of selection gradients using multiple regression versus structural equation models (SEM). *Biometrical Journal*, **37**, 449–462.

Pugesek, B. H. & Tomer, A. (1996). The Bumpus house sparrow data: a reanalysis using structural equation models. *Evolutionary Ecology*, **10**, 387–404.

SAS Institute (1989). *SAS/STAT User's Guide*, Version 6, 4th edition, vol. 2. Cary, NC: SAS Institute Inc.

Schumacker, R. E. & Marcoulides, G. A. (eds.) (1998). *Interaction and Nonlinear Effects in Structural Equation Modeling*. Mahwah, NJ: Lawrence Erlbaum Associates.

Shipley, B. (1999). Exploring hypothesis space: examples from organismal biology. In C. Glymour & G. Cooper (eds.), *Computation, Causation, and Search*. Cambridge, MA: AAAI/MIT Press.

Sorensen, T. A. (1948). A method of establishing groups of equal amplitude in plant sociology based on similarity of species content and its application to analyses of the vegetation on Danish commons. *Biologiske Skrifter*, **5**, 1–34.

ter Braak, C. J. F. (1986). Canonical correspondence analysis: a new eigenvector technique for multivariate direct gradient analysis. *Ecology*, **67**, 1167–1179.

Wesser, S. D. & Armbruster, W. S. (1991). Species distribution controls across a forest – steppe transition: a causal model and experimental test. *Ecological Monographs*, **61**, 323–342.

Wootton, J. T. (1994). Predicting direct and indirect effects: an integrated approach using experiments and path analysis. *Ecology*, **75**, 151–165.

Wright, S. (1918). On the nature of size factors. *Genetics*, **3**, 367–374.

8 From biological hypotheses to structural equation models: the imperfection of causal translation

Bill Shipley

Abstract

It is possible to test a multivariate biological hypothesis concerning cause–effect relationships using structural equation modeling (SEM) applied to observational data. However, to do this we must translate from the language of causality to the language of probability distributions and this process of translation is almost always imperfect. One consequence of this imperfection of causal translation is the existence of equivalent SEM models; that is, different causal models that produce exactly equivalent statistical structural equation models. In this chapter I describe how such equivalent models arise, how to find equivalent models based on path diagrams, and why their existence complicates our interpretation of standard statistical tests in SEM. I illustrate these concepts using two actual path models taken from plant ecology.

Introduction

Translation can be a tricky business. While visiting the aquatic mammals section of our local zoo with an American colleague, she asked me how to say "seal" in French. Without thinking, I told her that it was "phoque" (the "o" is pronounced "uh"). It was clear from the shocked expression on her face that she had misunderstood. The same pitfalls of translation can occur in structural equation modeling (SEM). SEM consists typically of formulating a multivariate hypothesis concerning the cause–effect relations between a set of variables and then translating this causal model into a statistical model that is capable of falsification. If one is not careful, the differences between the language of causality and the language of probability distributions can result in confusions as dramatic as my translation of "seal" and my colleague's interpretation of "phoque". These differences generate, among other things, models that make profoundly different causal predictions but exactly equivalent statistical predictions. Such sets of models are called "equivalent" models. The goal of this chapter is to describe how such statistically equivalent models arise, how to detect them, and how standard

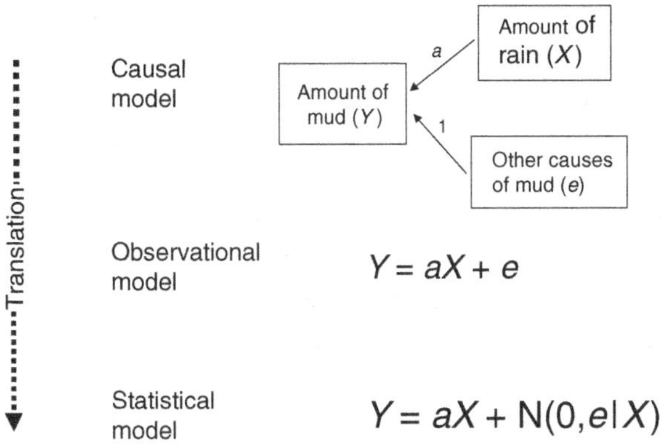

Figure 8.1. The causal model states that (1) the amount of rain causes the amount of mud, but not vice versa and (2) other things also cause mud, but not vice versa, and that there is no causal relationship between the amount of rain and the other causes of mud. The observational model states that the amount of mud observed is equivalent to the amount of rain observed times "*a*" plus the amount observed of other causes of mud. The statistical model states that the probability distribution of the amount of mud that can be observed is gaussian with an expected value of aX and a variance of $e|X$.

statistical tests in SEM should be qualified on the basis of the existence of such equivalent models. I will conclude with an analysis of two actual biological path models in which such equivalent models occur.

The first potential pitfall when translating from a causal to a statistical hypothesis is the "=" sign. Consider the following simple example (Pearl, 1998). We wish to translate the hypothesis that "rain and other sources of water cause mud" into a statistical model. Let Y represent the amount of mud that we observe at the end of the storm, let X be the amount of rain that has fallen and let "e" be all those other causes of mud (sprinklers, watering cans . . .) that are causally independent of rain. We can express this causal hypothesis either as a path diagram, an observational equation or as a statistical equation (Figure 8.1).

A very subtle, but extremely important, change has been introduced when we pass from the causal diagram to the equations. The arrow ($\rightarrow$) in the path diagram means that if we (or nature) change the amount of rain that falls to the ground, while keeping all the other causes of mud the same, then this will provoke a change in the amount of mud that we measure. It is a prediction about the effect of some physical *manipulation* on the state of some other variable in our system. Now let us pass to the equation. The "=" operator means algebraic "equality" under observational (nonmanipulative) conditions. The observational equation in Figure 8.1 means that if we observe (obtain information about) both X and e then we can infer that the value of Y is $aX + e$. Because of this algebraic equality we also know that if we observe both Y and e then we can infer that the value of X is $(Y - e)/a$. These observational inferences, along with some assumptions concerning the sampling distributions of the variables, imply a statistical relationship between the variables Y, X, and e. For example, assuming multivariate normality we can infer that the probability distribution of Y, conditional on X, is a normal distribution with a mean of $E(aX) + E(e \mid X)$ and a variance of $\mathrm{Var}(e \mid X)$. Again, the algebraic equality means that the probability distribution of X, conditional on Y, is also a normal distribution with a mean of $Y + E(e \mid Y)]/a$ and a variance of $\mathrm{Var}(e \mid Y)$. Note that, in principle, e and X need not be independent, since a structural equation is not a regression equation. It is commonly believed that the path diagram is simply a pictorial representation of the structural equations, and therefore that the arrow means the same thing as the equality sign. According to this interpretation, the path diagram is simply an imprecise mental crutch and the *real* causal model is found in the structural equations. One goal of this chapter is to convince you that the opposite is true: the real causal model is the path diagram and the structured equations are an imperfect reflection of this causal model.

Clearly, the "=" in the equation and the "→" in the path diagram cannot mean the same thing[1]. If it did then we could simply rearrange the equation and infer that the amount of rain that falls during the storm (X) is caused by the amount of mud that is measured at the end of the storm, which is absurd as a causal statement but not as an observational statement. By its very nature, a causal hypothesis implies asymmetrical relations (rain causes mud but mud does not cause rain) but we have lost this meaning when translating from the language of causes to the language of probability. The causal meaning of "→" is no more equivalent to the algebraic meaning of "=" than the French word "phoque" is equivalent to the similar sounding English word that I will decline to spell. The failure to recognize this imperfection in translation from the causal hypothesis to the statistical hypothesis can result in the scientific equivalent of a slap in the face.

[1] If we want to interpret the structural equations as a causal, rather than a statistical, model then we have to modify the meaning of "=" from an algebraic equality to a manipulative operation. I prefer to leave the meaning of "=" unchanged, which emphasizes that the structural equations are a statistical translation of the causal model but are not identical to the causal model.

The imperfection of this translation gives rise to a number of limitations in interpreting inferential tests of SEM. One limitation, which is not fully appreciated in the SEM literature, is the existence of statistically equivalent models.

How do equivalent models arise?

The traditional process of testing a SE model consists of a series of steps. First, we convert our cause–effect assumptions about a phenomenon, represented by the path diagram, into a series of linear structural equations containing a set of free parameters (variances, covariances or path coefficients). Next, we estimate the values of these free parameters in such a way as to minimize the discrepancy between the covariance structure of the data and that of the model. This estimation is done while respecting the statistical constraints imposed by the causal structure of the model; I will describe an alternative way of viewing these statistical constraints later in this chapter. The most common method is to find parameter values that minimize the maximum likelihood statistic (Bollen, 1989). Once we have these parameter estimates then we can derive the predicted covariance matrix from these fitted values and compare this predicted matrix with the observed covariance matrix obtained from our data using a fit statistic such as the maximum likelihood chi-square χ^2. Described in this way it is very difficult to see why different causal hypotheses can give exactly the same predicted covariance matrix[2]. A much easier way of understanding how a multivariate causal hypothesis is translated into a statistical hypothesis, and how this imperfect translation gives rise to observationally equivalent models, is to convert the causal hypothesis into a "directed acyclic graph"[3]. For the rest of this chapter I will simply refer to an "equivalent" model when I mean an "observationally" equivalent model.

A directed acyclic (causal) graph is simply a path diagram in which the numerical values of the (nonzero) parameters are ignored. If the

[2] One must distinguish between equivalent models, as described in this chapter, and alternative nonequivalent models that nonetheless cannot be rejected on the basis of a given significance level. This second problem is one of statistical power and can always be resolved if the sample size is large enough. Equivalent models will always give exactly the same probability level on a standard test independently of sample size.

[3] Any SEM model that does not include feedback relationships or free covariances can be expressed as an acyclic directed graph. Methods for dealing with feedback loops and free covariances exist (Andersson *et al.*, 1997b) but are more complicated and will not be discussed in this chapter.

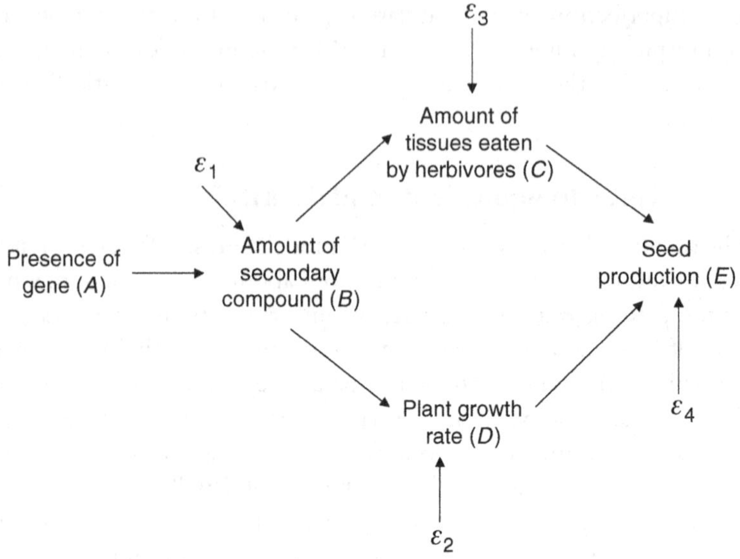

Figure 8.2. A directed acyclic graph describing the hypothesized causal relationships between five attributes of a plant. Error variables are indicated by ε.

hypothesis specifies that a variable (X) is a direct cause of another variable (Y), i.e., in which the causal effect of X on Y is not entirely mediated through another variable in the model, then draw an arrow going from X to Y. As an example, imagine that we hypothesize that the presence of a gene (A) in a plant increases the production of a toxic secondary compound (B) in its tissues. The amount of this secondary compound in the plant decreases the amount of its tissues that are eaten by a herbivore (C) but also decreases the growth rate (D) of the plant, since some of the resources of the plant are used for its production rather than for production of leaves or roots. An increase in the amount of tissues eaten by the herbivore decreases its seed production (E) and, therefore, its evolutionary fitness. An increased growth rate also increases its seed production. Figure 8.2 shows the directed graph that specifies this causal hypothesis. The independent residual or "error" variances are indicated by ε.

Notice that the description of the above biological hypothesis does not mention either "statistical associations" or "probability distributions"; rather, it refers to actual physical changes in the state of some attributes of the plant as a consequence of changing other attributes. Furthermore, these hypothesized effects are asymmetrical. For instance, according to this hypothesis, a change in the presence or absence of the gene in the plant will change the amount of the secondary compound in its tissues, but changing the amount of the secondary compound does not change the presence or absence of the gene. However, in the sort of uncontrolled observational studies for which SEM is designed, we cannot conduct controlled randomized experiments and so we have to translate, as accurately as possible, such causal assumptions into statistical consequences. This leads the thoughtful user of SEM to the following question: how accurately can a translation from a causal hypothesis to SEM be done, and what types of ambiguities remain once this most accurate translation has been obtained? If your only interest in SEM is to produce an observational (statistical) model under conditions in which no changes have been made to the system and with no connotations – explicit or implicit – of causality then you can both be thoughtful and ignore this question (Shipley & Peters, 1991). In my experience such situations are very rare in biology. Biologists are almost always ultimately concerned with causal processes and so statistical models (SE or otherwise) are viewed as tools that serve to formalize or test these causal processes.

Some theoretical work (Pearl, 1988) has shown that a graphical condition called "d-separation", when applied to such directed acyclic graphs, will generate all of the statistical information that such a statistical translation of a causal hypothesis can possess. In particular, if two variables (or sets of variables) are d-separated by fixing some set of conditioning variables to constant values in the causal system, then these two variables must also be statistically independent upon statistically conditioning on this same set of conditioning variables. These predictions of conditional independence are not restricted to any particular probability distribution or to a linear relationship between the variables[4]. This theorem is very important to biologists who wish to test their causal hypotheses using observational data. It means that, once the biologist has specified the causal hypothesis in the form of a directed acyclic graph (the path diagram), every d-separation fact that is entailed by the path diagram must be mirrored by its implied statistical independence if the causal hypothesis is correct. If any of the implied statistical independencies are not correct then the hypothesized causal hypothesis cannot be correct either. Equally important, the theorem also ensures that, if the biologist's causal hypothesis is correct, then one will never find a statistical independency in the data that is not implied by the path diagram except

[4] This is true of a causal model that does not include feedback loops. If feedback loops are present then the proof of d-separation applies only to linear models with normally distributed variables or to nonlinear models with discrete variables.

by chance[5]. In other words, d-separation gives the best (although always imperfect) statistical translation of a causal hypothesis that can be obtained without direct manipulation.

The notion of "d-separation" is very easy to grasp and, once it is understood, allows one to quickly determine whether two models are statistically equivalent. This requires a few definitions that, although unfamiliar to most readers, are not difficult to understand.

- *Directed path*: a directed path between two variables, X and Y, is the set of variables that have to be traversed in following a series of arrows (head to tail) from X to Y. For instance, there are two such directed paths from the presence of the gene (A) to the seed production (E) in Figure 8.2: $A \rightarrow B \rightarrow C \rightarrow E$ and $A \rightarrow B \rightarrow D \rightarrow E$.
- *Undirected path*: an undirected path between two variables is a set of variables that have to be traversed in going from one to the other while ignoring the direction of the arrows. For instance, there is an undirected path between C and D ($C \leftarrow B \rightarrow D$) in Figure 8.2 that is not also a directed path between C and D, since one has to violate the direction of the arrows to go from C to D.
- *Chains, forks and colliders*: a chain is a directed path between three variables with all arrows going in the same direction ($X \rightarrow Y \rightarrow Z$ or $X \leftarrow Y \leftarrow Z$). The triplet of variables A, B, and C form a chain in Figure 8.2. A fork is a pattern between three variables in which the arrows point out from the middle variable ($X \leftarrow Y \rightarrow Z$). The variables C, B, and D form a fork in Figure 8.2. A collider is a pattern between three variables in which the arrows point into the middle variable ($X \rightarrow Y \leftarrow Z$). The variables C, E, and D form a collider on E in Figure 8.2.

An undirected path p between two variables X and Z is d-separated (or "blocked") by a set of conditioning variables Q if and only if:

1. p contains a chain ($X \rightarrow Y \rightarrow Z$) or a fork ($X \leftarrow Y \rightarrow Z$) such that the variable Y is a member of the conditioning set Q, or
2. p contains at least one collider ($X \rightarrow Y \leftarrow Z$) such that Y is not a member of the conditioning set Q and no descendant of Y (i.e., a

[5] By "chance" I mean either the usual random sampling fluctuations that exist in any statistical test, or due to the fact that alternative paths between two variables exactly cancel out.

Table 8.1. *The d-separation relationships and resulting probabilistic independencies that are implied by the path model in Figure 8.2*

Independent d-separation fact	Pearson (partial) correlation
A d-separated from C given B	$\rho(A, C/B) = 0$
A d-separated from D given B	$\rho(A, D/B) = 0$
C d-separated from D given B	$\rho(C, D/B) = 0$
B d-separated from E given both C and D	$\rho(B, E/\{C, D\}) = 0$
A d-separated from E given both C and D	$\rho(A, E/\{C, D\}) = 0$

variable that is, itself, caused by Y) is a member of the conditioning set Q.

Finally, two variables are d-separated (or blocked) given a conditioning set of variables Q in a directed graph if and only if every undirected path between the two variables is d-separated given Q.

These definitions are best understood with an example, but after a few minutes of practice even novices can quickly determine whether two variables are blocked. Given the causal graph in Figure 8.2, there are only five d-separation facts that are required to completely specify the probability distribution that results from this causal hypothesis. This number is equal to the number of pairs of variables in the directed graph that do not have an arrow between them. If we assume that the variables in Figure 8.2 are normally distributed and that the functional relations between them are linear, then these five d-separation facts can be equated with Pearson partial correlations, as shown in Table 8.1. These independent d-separation facts can be used to test structural equations models not involving latent variables without resorting to maximum likelihood methods and can even be used to perform nonparametric tests of such models (Shipley, 2000a).

Given the set of d-separation facts that are generated by a directed acyclic graph, it can be proven (Verma & Pearl, 1990) that two different causal models are observationally equivalent if they possess the same set of d-separation facts. By "equivalent" I mean both that they predict exactly the same independence facts ("independence equivalent") and that they predict exactly the same covariance structure ("covariance equivalent"). A structural equation model that includes free covariances (curved double-headed arrows) is not a directed acyclic graph. In such cases, d-separation still predicts independence equivalence but not covariance equivalence. In such cases, models that are "independence equivalent" might still imply different covariance matrices and so give different probability estimates using the standard maximum likelihood χ^2 statistic. Even simpler, any two causal models represented by directed acyclic graphs are observationally equivalent if they have the same set of variables, the same set of edges (i.e., arrows in which their directions are ignored), and the same set of "unshielded" colliders. An unshielded collider is a set of three variables that form a collider and in which the two variables at the tails of the arrows do not also have an arrow between them[6]. Arrows coming from the mutually independent error variables are not included in the definition of such unshielded colliders. For instance, the variables *C*, *E* and *D* in Figure 8.2 form an unshielded collider because they form a collider ($C \rightarrow E \leftarrow D$) and *C* and *D* do not have an arrow from one to the other. In practice this means that if one can change the direction of any arrows in the causal model without removing or adding any unshielded colliders, the two models are observationally equivalent. If you test them using standard SEM techniques you will find that they give the same values of their test statistic and the same probability levels, and this will be true independently of the sample size.

If we take Figure 8.2 as an example, then which alternative models will, or will not, be observationally equivalent[7]? First, note that the set of variables $\{C \rightarrow E \leftarrow D\}$ form an unshielded collider. Therefore no model that changes the direction of either of these two arrows will be equivalent to the model in Figure 8.2. If we change only the direction of the arrow between *B* and *C* (Figure 8.3a) then we will have introduced a new unshielded collider in the triplet of variables $\{A \rightarrow B \leftarrow C\}$. This new model will not be observationally equivalent to Figure 8.2; the same applies if we change only the direction of the arrow between *B* and *D*. On the other hand, if we change only the direction of the arrow between *A* and *B*, then no new unshielded colliders are introduced (Figure 8.3b) and so this model is equivalent to Figure 8.2. Similarly, if we change the directions of the arrows between *A* and *B* and also between *B* and *C*, then no new unshielded colliders are introduced (Figure 8.3d). This model is therefore also observationally equivalent to Figure 8.2; the same applies if we change the direction of both the arrow between *A* and *B* and that between *B* and *D* (Figure 8.3c). A number of algorithms have been devised that will list all statistical models that are observationally equivalent to a given causal

[6] An unshielded noncollider is a set of three variables that form a chain or a fork and in which the two variables at the ends of the triplet do not have an arrow between them.

[7] The arrows associated with the error terms (ε) have no effect on the equivalence of models. They will often be ignored in this chapter, where their presence is obvious.

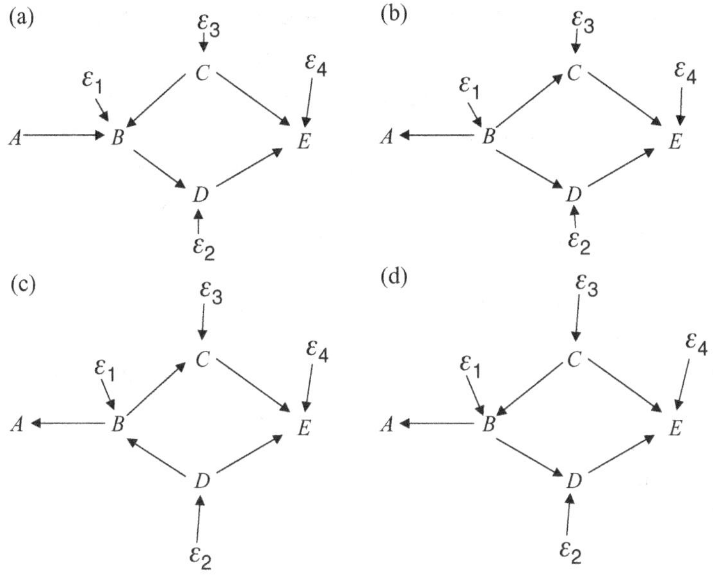

Figure 8.3. The first model (a) is not observationally equivalent to the model in Figure 8.2. The next three models (b, c and d) all make different causal predictions but exactly equivalent observational predictions. These three models are observationally equivalent to the model in Figure 8.2 and would give exactly the same probability levels if tested with SEM. Error variables are indicated by ε.

model (Chickering, 1995; Meek, 1995; Andersson *et al.*, 1997a), although none of these has yet been incorporated into publicly available computer programs.

Using equivalent models in the design stage

Since one common use of SEM is to test an *a priori* causal hypothesis, it is important to consider equivalent models before collecting data, since it will be impossible to statistically differentiate between one's preferred causal explanation and those different causal explanations that are observationally equivalent. Perhaps some equivalent models can be excluded by other (nonstatistical) means? The temporal order with which the data are collected can often exclude some equivalent models, since causal effects cannot travel backwards in time. Since the presence of unshielded colliders is critical in determining observational equivalence, one might be able to include some new variable into the model that produces such unshielded collisions.

Equivalent models and hypothesis testing in SEM

Other chapters have described the logic of hypothesis testing in SEM and the conditions needed to draw statistical inferences. Imagine that we have already specified our causal hypothesis (Figure 8.2), obtained the maximum likelihood parameter estimates of our free parameters, verified that the residuals are multivariate normal and the relationships linear, and calculated the resulting test statistic and its probability level. How does the presence of equivalent models complicate our biological conclusions?

If the probability level is below our chosen significance level, then we are justified in rejecting our causal hypothesis. Yet the existence of equivalent models means that we are also justified in rejecting all of these equivalent models as well. This is a bonus that is not often recognized. On the other hand, if the probability level is above our chosen significance level, then we are in a state of ignorance. We do not have good evidence to reject our causal hypothesis, but we do not have any good (statistical) reason to accept it either. First, we might not have sufficient statistical power to detect some small but real difference between our observed and predicted covariance matrices. This is a serious problem but can always be solved in principle by increasing our sample size. Second, the presence of observationally equivalent models makes our nonsignificant result even more difficult to interpret since, even with an infinite sample size, we would not be able to distinguish between them. In other words, we are really testing the entire set of equivalent models against all other (nonequivalent) models. We thought that we were testing a single causal hypothesis (Figure 8.2) but, because of the imperfect translation of causal statements into probability distributions, we now realize that we are actually testing only those causal assumptions, common to all of the equivalent models, that can be converted from statements of causality to statements of conditional probability distributions (Figure 8.3). This is true even though the different equivalent models can make very different causal statements. In order to differentiate between these equivalent models we must either introduce nonstatistical information based on our biological knowledge, introduce new variables into the model that will create other non-shielded colliders or perturb our system in some way and retest with data from this perturbed system.

An alternative notation for equivalent models

Other chapters have already introduced the notion of parameter underidentification in SEM. A parameter in a set of structural equations is "underidentified" if there is more than one value of the parameter for which the likelihood is maximized. In such cases the data do not yield enough information to allow the fitting procedure to choose one value as preferable to another. When this happens, most SEM programs will warn the user not to trust the resulting parameter estimates or the test statistics. However, SEM also suffers from another form of under-identification that is both much more common and much less acknowledged; Scheines *et al.* (1998) called this "causal under-identification". If a structural equation model possesses equivalent models (and most do), then the data do not yield enough information to allow us to choose one equivalent model as preferable to another as a causal description. Remember that a structural equation model is an imperfect statistical translation of a specific causal hypothesis. This means that your structural equation model may well be over-identified in its parameters, allowing an inferential test to be performed, but still be under-identified in its causal assumptions. Since the main purpose of SEM is to allow falsifiable tests of causal hypotheses using observational data, this is not a minor limitation! In this section I wish to introduce a way of drawing path diagrams, adapted from Spirtes *et al.* (1993), that incorporates information concerning equivalent models and that therefore make such causal under-identification more explicit. I will use the causal model shown in Figure 8.2 as an example.

Steps:

1. Remove all directions from your path diagram; i.e., remove the arrowheads from your arrows (Figure 8.4a).
2. Draw an open circle at each end of each line connecting two variables (Figure 8.4b). The circle means that we do not know whether or not there should be an arrowhead at that position.
3. For each set of unshielded colliders in the original path diagram, replace the circles on each side of the collider variable with arrowheads pointing into it (Figure 8.4c).
4. For each set of unshielded noncolliders in the original path diagram, keep the circles on each side of the noncolliding middle variable, but also draw a line under the middle variable indicating the triplet in question. The underline means that there cannot be arrowheads pointing into this variable from the other two variables in the triplet (Figure 8.4d).

The resulting "path diagram"[8] summarizes what causal knowledge can, and cannot, be tested using SEM. If there is an arrowhead pointing into a variable, then this means that such a directed path exists in every

[8] Technically, this is called a partially directed graph.

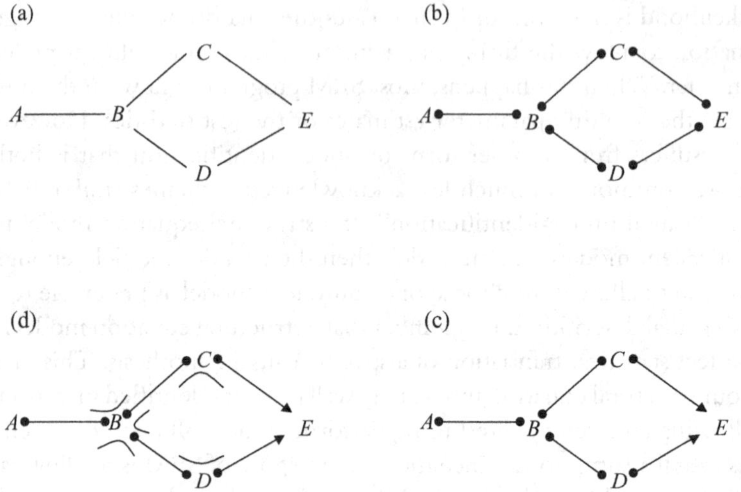

Figure 8.4. Steps involved in constructing a partially directed acyclic graph that shows all the different causal models that are observationally equivalent to the model in Figure 8.2. Error variables are not shown in this figure to improve clarity.

model equivalent to your original causal model. If there is an underlined variable, then this means that no model equivalent to your original causal model can have arrowheads pointing into this variable from both directions. It also means that the data do not allow us to distinguish between cases in which no arrowheads, or only one arrowhead, point into it[9]. In fact, the notation can be simplified even further by omitting the underlining and simply including the rule that any variable that is not an unshielded collider is understood to be a noncollider.

In the next section I will illustrate these ideas using two biological path diagrams taken from evolutionary plant biology.

Seed dispersal in St Lucie's Cherry

Jordano (1995) published a study in which various attributes of individual trees of *Prunus mahaleb* were related to the number of seeds per tree that

[9] This notation permits double-headed arrows between two variables. Such double-headed arrows do not represent feedback loops, but rather the presence of a latent variable not included in the original path diagram. When such double-headed arrows are permitted, all equivalent models are independence equivalent (they make the same d-separation predictions), but are not necessarily covariance equivalent and could therefore produce different maximum likelihood χ^2 statistics.

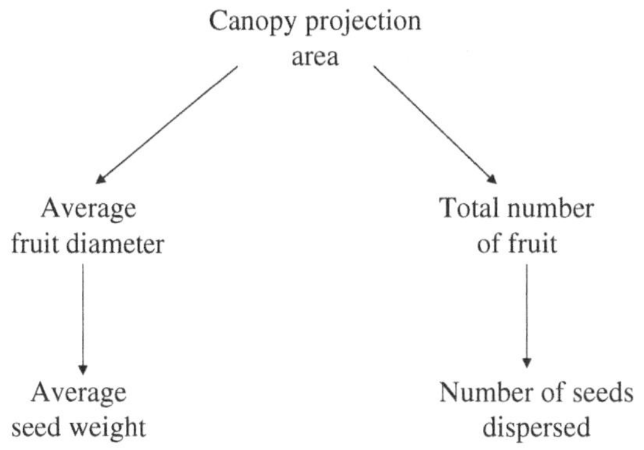

Figure 8.5. A path model proposed in Shipley (1997) relating five attributes of St Lucie's Cherry.

were dispersed by birds. Here, I will concentrate on five attributes: canopy projection area, the total number of ripe fruit produced, the average fruit diameter, the average seed weight, and the total number of seeds dispersed at the end of the growing season. I (Shipley, 1997) gave one model, based on these data, which was obtained by exploratory path analysis (Figure 8.5).

There are no sets of variables in this model that form unshielded colliders. The partially directed path diagram corresponding to Figure 8.5 is shown in Figure 8.6, along with the five path models that are observationally equivalent[10]. On the other hand, logic dictates that the number of seeds that are dispersed at the end of the growing season cannot cause the number of fruit that were produced during the growing season, and so we can reject the model in Figure 8.6 having this orientation. This is an example of using nonstatistical knowledge to exclude some equivalent models. Unless we have sufficiently good biological knowledge to exclude other causal orderings, the remaining four equivalent models must be considered to be equally well supported by the available data. In other words, it is not possible, even with an infinite sample size, to extract more causal information from such data than is represented by the partially directed path diagram.

An interspecific model of gas exchange in leaves

Evolutionary biologists often invoke causal explanations for patterns of covariation between the attributes of organisms. Such causal explanations often involve trade-offs between traits due to conflicting selection pressures. One

[10] There are other, nonequivalent, models that would pass the usual maximum likelihood χ^2 test because of the rather small sample size of this data set.

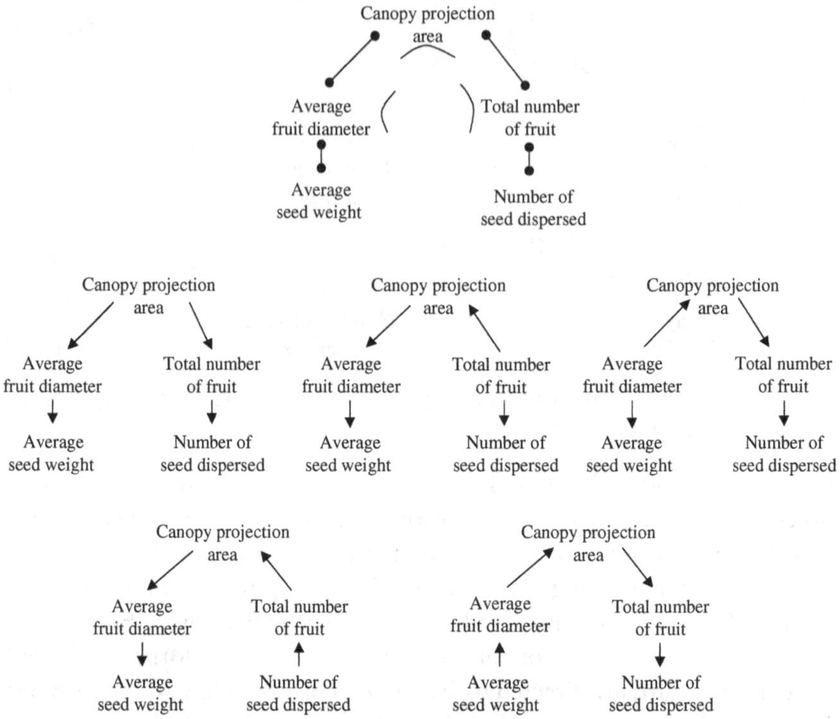

Figure 8.6. The partially directed acyclic graph of Figure 8.5 (top) and the five observationally equivalent models of Figure 8.5. Error variables are not shown in this figure to improve clarity.

example is the model of stomatal regulation and gas exchange in leaves due to conflicting requirements for maximizing net photosynthetic rate while minimizing water loss (Cowan & Farqahar, 1977). One prediction of this model is that the observed rate of net photosynthesis is limited below its physiological maximum by the behavior of the stomates in order to maximize carbon gain while minimizing water loss. This, and other biological considerations, led Martin Lechowicz and I (Shipley & Lechowicz, 2000) to propose an interspecific path model relating specific leaf mass (leaf dry weight divided by leaf area), leaf nitrogen concentration, stomatal conductance to water, net photosynthetic rate, and internal leaf CO_2 concentration (Figure 8.7a). Note that, although in this path diagram there are two arrows that collide at the internal leaf CO_2 concentration, the triplet of variables {Stomatal conductance → Leaf CO_2 concentration ← Net photosynthesis} does not form an *unshielded* collider, since there is also an arrow from Stomatal Conductance to Net Photosynthesis. Therefore, the partially directed

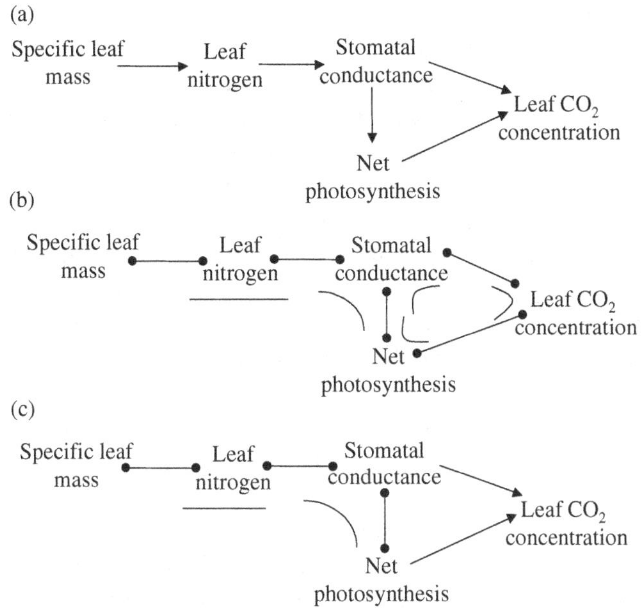

Figure 8.7. (a) A proposed path model relating specific leaf area, leaf nitrogen concentration and leaf gas exchange. (b) The partially directed acyclic graph showing all equivalent models. (c) A reduced partially directed acyclic graph after removing impossible causal relationships, based on the physics of gaseous diffusion. Error variables are not shown in this figure to improve clarity.

path diagram for this causal model cannot place any restrictions at all on the directions of the arrows between these three variables (Figure 8.7b). On the other hand, nonstatistical knowledge from basic physics can resolve this ambiguity. The concentration of CO_2 within the leaf is determined by the rate at which it is entering the leaf, determined by the degree of stomatal opening as measured by stomatal conductance, and by the rate at which it is being removed from the internal air by photosynthesis. This gives the second partially directed path diagram (Figure 8.7c).

Conclusions

SEM does not test causal models; rather, it tests statistical consequences of the causal model. These statistical consequences are obtained after translating the causal hypothesis into a statistical model. This process of translation is imperfect. The imperfection arises because the inherently asymmetrical relationships of the causal process are replaced by inherently symmetrical relationships involving probability distributions. One consequence of this imperfect translation is that different causal models can have exactly the same statistical consequences, giving rise to observationally equivalent models. This is called "causal under-identification". Because of such causal under-identification, one can rarely test one causal model against all others. Instead, one tests the set of equivalent models against the set of all nonequivalent models. The lack of recognition of this process of imperfect translation, when going from a causal model to a SE model, has resulted in a great deal of confusion amongst users of SEM in other disciplines. This confusion has often resulted in protracted battles concerning the validity of SEM as a way of testing causal hypotheses with observational data[11]. Since biologists are just beginning to use SEM as a methodological tool, an appreciation of the relationship between causal models and SEM should allow us both to avoid such arguments and increase the acceptance of SEM as a valid addition to our statistical toolbox.

The notion of d-separation and equivalent models leads to many other useful results for biologists, including exploratory methods of path analysis. These topics, and much else, are dealt with by Shipley (2000b).

Acknowledgements

Suggestions by Judea Pearl and Adrian Tomer improved this chapter. Financial support was provided by the Natural Sciences and Engineering Research Council of Canada.

References

Andersson, S. A., Madigan, D. & Perlman, M. D. (1997a). On the Markov equivalence of chain graphs, undirected graphs and acyclic digraphs. *Scandinavian Journal of Statistics*, **24**, 81–102.

Andersson, S. A., Madigan, D., Perlman, M. D. & Richardson, T. S. (1997b). Graphical Markov models in multivariate analysis. Technical Report, Indiana University and University of Washington.

Bollen, K. A. (1989). *Structural Equations with Latent Variables*. New York: Wiley.

Chambers, W. V. (1991). Inferring formal causation from corresponding regressions. *Journal of Mind and Behavior*, **12**, 49–70.

[11] Closely related to the confusion generated by unrecognized equivalent models is the long-standing debate, familiar to biologists, of whether researchers should concentrate on causal models or phenomenological (predictive) models (Shipley & Peters, 1991). The distinction between causal, observational, and statistical models should clear up this confusion.

Chickering, D. M. (1995). A transformational characterization of Bayesian network structures. In P. Besnard & S. Hanks (eds.), *Uncertainty in Artificial Intelligence* 11, pp. 87–98. San Francisco: Morgan Kaufmann.

Cowan, I. R. & Farquhar, G. D. (1977). Stomatal function in relation to leaf metabolism environment. In D. H. Jennings (ed.) *Integration of Activity in the Higher Plant*, pp. 471–505. Cambridge: Cambridge University Press.

Jordano, P. (1995). Frugivore-mediated selection on fruit and seed size: birds and St. Lucie's cherry, *Prunus mahaleb. Ecology*, **76**, 2627–2639.

Meek, C. (1995). Causal inference and causal explanation with background knowledge. In P. Besnard & S. Hanks (eds.), *Uncertainty in Artificial Intelligence 11*, pp. 403–410. San Francisco: Morgan Kaufmann.

Pearl, J. (1988). *Probabilistic Reasoning in Intelligent Systems. Networks of Plausible Inference*. San Mateo, CA: Morgan Kaufmann.

Pearl, J. (1998). Graphs, causality, and structural equation models. *Sociological Methods and Research*, **27**, 226–284.

Scheines, R., Spirtes, P., Glymour, C., Meek, C. & Richardson, T. (1998). The TETRAD project: Constraint based aids to causal model specification. *Multivariate Behavioral Research*, **33**, 65–117.

Shipley, B. (1997). Exploratory path analysis with applications in ecology and evolution. *American Naturalist*, **149**, 1113–1138.

Shipley, B. (2000a). A new inferential test for path models based on directed acyclic graphs. *Structural Equation Modeling*, **7**, 206–218.

Shipley, B. (2000b). *Cause and Correlation in Biology. A User's Guide to Path Analysis, Structural Equations, and Causal Inference.* Cambridge: Cambridge University Press.

Shipley, B. & Lechowicz, M. (2000). The functional co-ordination of leaf morphology, nitrogen concentration, and gas exchange in 40 wetland species. *Functional Ecology*, **7**, 183–194.

Shipley, B. & Peters, R. H. (1991). The seduction by mechanism: a reply to Tilman. *American Naturalist*, **138**, 1276–1282.

Sprites, P., Glymour, C. & Scheines, R. (1993). *Causation, Prediction, and Search. Springer-Verlag Lecture Notes in Statistics*, **81**. New York: Springer-Verlag.

Verma, T. & Pearl, J. (1990). Equivalence and synthesis of causal models. In P. Bonissone, M. Henrion, L. N. Kanal & J. F. Lemmer (eds.), *Uncertainty in Artificial Intelligence* **6**, pp. 255–268. Cambridge, MA: Elsevier Science.

9 Analyzing dynamic systems: a comparison of structural equation modeling and system dynamics modeling

Peter S. Hovmand

Abstract

Structural equation modeling (SEM) is a method of testing systems of equations against observed covariance matrices while system dynamic modeling (SDM) simulates systems of differential equations representing nonlinear feedback loops. SEM has typically been constrained to linear systems of equations, but more recent developments have allowed SEM to represent nonlinear as well as nonrecursive (feedback) relationships. In this sense, SEM path diagrams with nonrecursive relationships and SDM causal loop diagrams look similar. This chapter analyzes the relationship between system dynamics modeling and structural equation modeling. A formal notion of equivalence between system dynamic models and structural equation models is developed. This is then analyzed through an illustration comparing the implied covariance matrices from one system dynamics model against the implied covariance matrices of two "equivalent" LISREL models. The first LISREL model tests whether an "equivalent" model can explain the entire family of covariance matrices generated by a system dynamics model, while the second LISREL model is developed to fit the data. The results suggest that structural equation models are able to explain the covariance matrices of a system dynamics model with changes in feedback loop dominance, but only by sacrificing a substantial portion of the latent structure representing feedback loops.

Introduction

Changing over time and involving a variety of feedback mechanisms, many biological systems are inherently dynamic. This chapter examines the relationship between two methods that are often used to model dynamic biological systems: structural equation modeling (SEM) and system dynamic modeling (SDM). Structural equation models generally rely on systems of linear equations to describe the relationships between all variables (Bollen, 1989, p. 3), while formal system dynamics models use systems of coupled differential equations to model the feedback mechanisms (Richardson, 1996, p. 48). To the extent that these two approaches are both trying to explain similar dynamic phenomena, it is important to understand the relationship between them. Specifically, it is critical to understand the extent to which these two methods yield equivalent models, i.e., the notion that two different models can both explain the same covariance structure. Model equivalence has become an area of increasing concern in structural equation modeling because equivalent models may "entail different and potentially incompatible, controversial or even opposite explanations of the studies phenomenon" (Raykov & Penev, 1999, p. 200). Failure to consider model equivalence can lead to serious misinterpretations of models. This chapter develops and demonstrates a notion of equivalence between system dynamics models and structural equation models.

One of the main challenges with comparing SEM and SDM is that, while the former is amenable to analytical solutions, the latter is not. Determining whether a system of linear equations is solvable and then finding the solution can generally be done analytically. In SEM, this allows one to calculate the implied covariance matrix and establish analytically the conditions under which two different models are equivalent. The problem with applying a similar approach to examining the equivalence between models in SDM and SEM is that systems of differential equations are, in general, nonlinear and therefore not amenable to the same type of analytical solution. In fact, most efforts at analyzing systems of differential equations involving nonlinear relationships require numerical methods to approximate the solution. The real problem is then to find a way of relating models in SDM into SEM, and to do so in such a way that one is able to investigate claims about particular models and sets of models. The general strategy in this chapter will be first to extend notions of equivalence and implied covariance from SEM into SDM and then to demonstrate their application by using computer simulations of a dynamic system.

System dynamics

The idea of modeling biological systems with systems of differential equations is not new, nor restricted to system dynamics. Von Bertalanffy (1968), for example, clearly saw that the underlying structure of living systems could be represented by systems of differential equations in his development of a general system theory. The Lotka–Volterra equations are frequently cited as examples of a simple system of nonlinear equations for representing predator–prey populations (Epstein, 1997). The main barrier to using systems of differential equations has been the difficulty of finding solutions to the system.

Working independently and from a paradigm of control systems theory in electrical engineering (Richardson, 1991), Jay W. Forrester pioneered a set of heuristics for identifying and simulating dynamic behavior with systems of differential equations. The key concept in system dynamics was, according to Forrester, the existence of feedback loops between the variables. Summarizing his method formally in *Principles of Systems*, Forrester (1968) emphasized an endogenous perspective, meaning that one should seek to explain problems by looking at feedback loops (versus exogenous influences).

System dynamics modeling is notably different from many quantitative methods in that there is a lack of emphasis on such activities as parameter estimation and statistical measures of model fit. Instead, SDM involves using a set of modeling heuristics (e.g., Levine *et al.*, 1992, p. 217), which if followed lead one to conclude that a given model is a valid representation of a specific problem. One of the major reasons for this is that most preliminary sensitivity analyses of dynamic systems demonstrate that most parameters are insensitive to even major variations (Levine & Lodwick, 1992, p. 121). Included as a step in developing a system dynamics model, a sensitivity analysis essentially involves varying each parameter (e.g., through Monte Carlo techniques) and assessing whether the qualitative nature of the time-series curves have changed. Parameters that alter the qualitative behavior of the system are frequently referred to as the leverage points of the system, and in some cases efforts are made to carefully estimate their values.

Some work has been done to examine the extent to which structural equation models can be converted into dynamic models. Levine (1992), for example, applied Roberts' pulse process technique to path diagrams. One limitation is that the pulse process is explicitly discrete. Another is that while the method does allow one to observe behavior over time, it does not include lags or delays as one might expect in a dynamic model. In order for these effects to be incorporated, the model must explicitly include a lagged variable.

Definitions of equivalence

Hayduk (1996) defines two notions of equivalence for SEM: (1) broad equivalence where two models produce the same set of covariance matrices when the coefficients of the parameters are allowed to vary; and (2) narrow equivalence where two models explain equally well the same sample covariance matrix. These two notions can be extended to include system dynamic models, as they too are essentially systems of equations involving a number of parameters.

To abstract Hayduk's notions of equivalence, let us begin by letting θ represent the ordered set of parameters that describes a specific system of equations or model. Then two systems represented by θ_1 and θ_2 are said to be identical if θ_1 and θ_2 are identical. Two systems represented by θ_1 and θ_2 are said to have an identical pattern of parameters if θ_1 and θ_2 have the same pattern of fixed and free parameters.

Now let $\aleph(\theta)$ be the set of all θ_α values such that θ_α is pattern identical with θ. Then two models represented by θ_1 and θ_2 are defined as broadly equivalent if, for each θ_α element of $\aleph(\theta_1)$, there exits a θ_β such that it is an element of $\aleph(\theta_2)$ and $\mathbf{\Sigma}(\theta_\alpha) = \mathbf{\Sigma}(\theta_\beta)$, where $\mathbf{\Sigma}(\theta)$ is the covariance matrix implied by the set of parameters θ. Two models represented by θ_1 and θ_2 are defined as narrowly equivalent if $\mathbf{\Sigma}(\theta_1) = \mathbf{\Sigma}(\theta_2) = \mathbf{S}$, where $\mathbf{S}$ is the sample covariance matrix.

From this, the equivalence between SDM and SEM can be defined in more precise terms. Let Ω be a subset of $\aleph(\theta_1)$ where θ_1 represents a system dynamics model. Then SDM is said to be broadly equivalent to SEM within Ω if for all θ_α values within Ω there exists a structural equation model represented by θ_2 such that θ_α is broadly equivalent to θ_2. And SDM is said to be narrowly equivalent to SEM within Ω if for all θ_α values within Ω there exists a structural equation model represented by θ_2 such that θ_α is narrowly equivalent to θ_2. Determining the type of equivalence between SDM and SEM (if any) within Ω can thereby be restated as a problem of calculating and comparing the implied covariance matrices from SDM and SEM.

Implied covariance

SEM is organized around being able to calculate the implied covariance matrix for a specific set of parameters. SDM is not. In general, system dynamic models are too complicated to solve analytically as they can easily involve nonlinear systems of ordinary and partial differential equations. This means that, in general, it is not possible to specify some function g such that $\theta_2 = g(\theta_1)$, where θ_1 represents the parameters to the system dynamics model while θ_2 represents the parameters to some corresponding structural equation model. And it is therefore not possible to generally determine $\mathbf{\Sigma}(g(\theta_1))$ analytically.

Instead of trying to identify a function such as g or solving the system of differential equations analytically, one might employ the use of numerical methods on computers to approximate the solution to the system of differential equations. Then by systematically varying selected parameters from a given probability distribution across a number of simulations, one can generate a distribution of observed data and thereby approximate the implied covariance matrix. One thereby knows *a priori* from the simulation that for any given system dynamics model θ_1, $\mathbf{\Sigma}(\theta_1) \approx \mathbf{S}$.

Thus, for a given structural equation model θ_2, one can calculate the implied covariance matrix $\mathbf{\Sigma}(\theta_2)$ and compare that with the sample covariance matrix $\mathbf{S}$. Since $\mathbf{S}$ is known from the simulation to be approximately equal to $\mathbf{\Sigma}(\theta_1)$, this means that a test of $\mathbf{\Sigma}(\theta_2)$ against $\mathbf{S}$ is essentially a test against $\mathbf{\Sigma}(\theta_1)$. Thus one can test whether the two approaches yield the same covariance matrix for a given θ_1 and θ_2.

Model selection

There is an indefinite number of subsets Ω that one might want to consider with respect to the equivalence between SDM and SEM. Of particular interest to understanding the dynamics of biological systems are models that have several feedback mechanisms resulting in complex dynamic behavior. One of the simplest examples for such a model is a system archetype called "fixes that fail" (Senge, 1990).

In "fixes that fail" there is an initial problem to which a fix is applied, but the fix has unintended consequences that have a delayed effect on the original problem. Consequently, there is a short-term benefit in applying the fix that decreases the perceived problem. Unfortunately later, the unintended consequences catch up and start increasing the problem. With the fix having been previously successful, more of the fix is applied in an attempt to respond to the increase in the original problem. But at this point, the unintended effects are increasing as rapidly as the fix can be applied, and the problem begins to grow continuously (see Figure 9.1).

"Fixes that fail" can be applied to a variety of biological phenomena. For example, it is commonly known that there is an increasing number of pests that threaten crop yields that developed a resistance to conventional insecticides. In "fixes that fail", initial insecticide applications (fix) have great short-term benefits as pests populations (problem) are dramatically reduced, with immediate increases in crop yields. But pests develop a resistance over time to the chemicals (unintended consequence). As this happens, farmers begin using increasing amounts of insecticides with decreasing benefits.

There are two basic feedback loops in "fixes that fail". In feedback loop L1, the problem increases the fix, which has the effect of decreasing

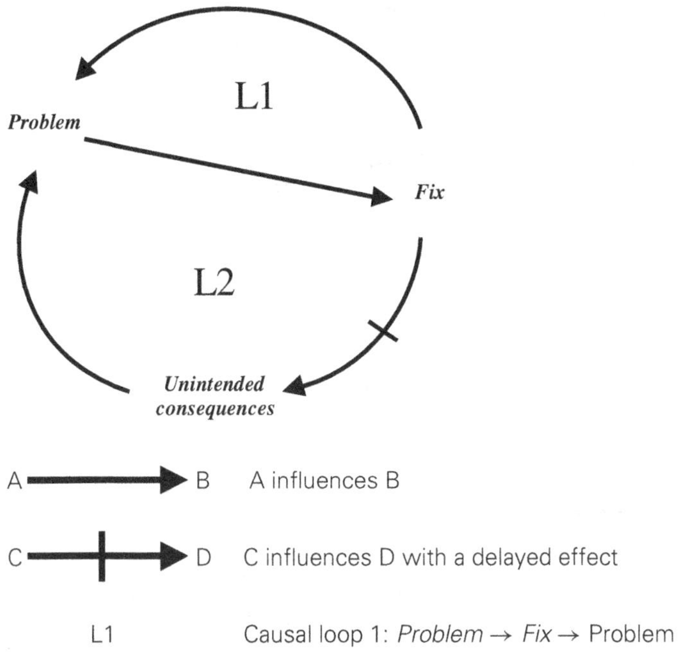

Figure 9.1. "Fixes that fail" archetype.

the problem. In feedback loop L2, the problem increases the fix, which then has a delayed effect of the unintended consequences that eventually increase the problem.

A central feature in system dynamics is the analysis of feedback loop dominance. A feedback loop L dominates the behavior of a given variable x within a time interval T when the behavior of x can be largely explained in terms of the behavior of L. This means that the behaviors of other (nondominant) feedback loops do not affect the behavior of x within the time interval T. Hence, removing or deactivating the nondominant feedback loops will not affect the behavior of x within T.

In "fixes that fail", feedback loop dominance oscillates between L1 and L2. Initially, L1 and L2 are competing, but L1 soon takes over until the unintended effects catch up in their impact on the original problem. At that point, L2 takes over, which spurs an increase in problems and pushes the dominance back to L1. But now, as the unintended consequences are consistently increasing the problems, L2 catches up with L1 and they return to a "stable" dynamic where the growth of the original problem is kept linear.

The specific system dynamics model for "fixes that fail" to be analyzed is illustrated in Figure 9.2. Since SDM generally involves simulations

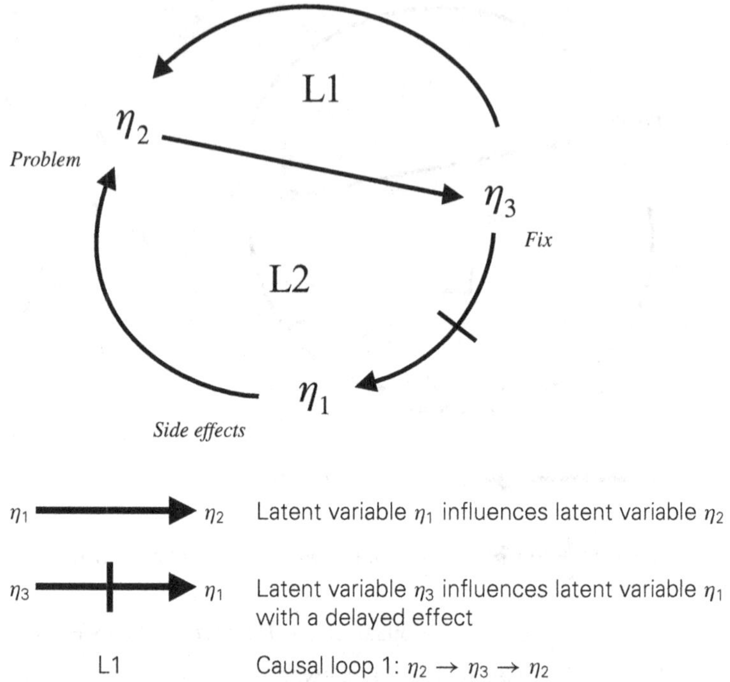
L1
η_2
Problem
η_3
Fix
L2
η_1
Side effects
η_1 η_2 Latent variable η_1 influences latent variable η_2
η_3 η_1 Latent variable η_3 influences latent variable η_1 with a delayed effect
L1 Causal loop 1: $\eta_2 \rightarrow \eta_3 \rightarrow \eta_2$

Figure 9.2. SDM causal loop diagram.

as one of the main analytical tools, the values of all latent variables are known and indicator variables are usually excluded from causal loop diagrams.

There are a number of ways that one might infer a structural equation model from a system dynamics model. It is arguably the similarity between the pictorial representations of SDM causal loop diagrams and SEM path diagrams that has led to instances where SDM is seen as a method equivalent to SEM. So one of the more obvious approaches is to preserve the latent structure while appending a measurement model, as has been done in Figure 9.3. If one considers the structural equation model as pictured in Figure 9.3 as representing an underlying causal structure between the latent variables, then this mapping from SDM to SEM arguably preserves the semantics of the system dynamics model.

Developing the corresponding equations is not as easy, as there is a trade-off between preserving the semantics of the latent structure and preserving the mathematical relationships between the latent variables. For example, SDM has no problem including nonlinear interactions, while SEM is generally restricted to linear equations. Although developing equations that represent the nonlinear interactions in SEM is possible, the resulting

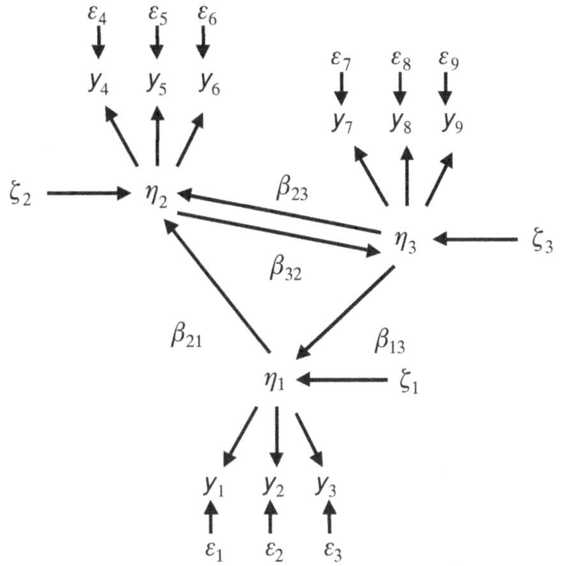

Figure 9.3. SEM path diagram. η, latent variable; ζ, latent variable error term; y, observed variable; ε, observed variable error term; β_{23}, coefficient, influence of η_3 on η_2.

path diagram would no longer correspond as nicely to the system dynamics model. One would have improved the mathematical correspondence between the models, but compromised the semantic correspondence. To keep matters simple in this chapter, the SEM equations are taken from the implied path diagram. A comparison of the basic equations is listed in Table 9.1. Let equations describing the system dynamics model in Table 9.1 be represented by θ_1 and the implied structural equation model by θ_2.

Now there are several points worth noting. First, system dynamics models are time variant and generate time-series data, so one can usually only talk about the state of the system at a given time. In this system dynamics model, the state of the system is specific to the elapsed time of the simulation, represented by the variable t_s where t_s being equal to zero refers to the initial starting time of the simulation. In this model, the units of time are arbitrary, but they are determined ultimately by the time horizon of the phenomenon under study. Second, as with most simulations, the sources of variation need to be explicitly included. For analysis of "fixes that fail", two sources have been specified: normally distributed measurement error ε_i for each observed variable y_i and normally distributed sampling times t_s around a mean sampling time $\bar{t}_s$, meaning that there is then one implied covariance matrix for each $\bar{t}_s$.

Table 9.1. *Model equations*

	θ_1 System dynamics model	θ_2 Implied structural equation model
Latent model	$\frac{\partial \eta_1}{\partial t} = \xi_2 \eta_3 + \eta_{1,t=0}$	$\eta_1 = \beta_{13}\eta_3 + \zeta_1$
	$\frac{\partial \eta_2}{\partial t} = -\eta_2 \eta_3 + \eta_1 + \xi_1$	$\eta_2 = \beta_{21}\eta_1 + \beta_{23}\eta_3 + \zeta_2$
	$\frac{\partial \eta_3}{\partial t} = -1.1 \times \eta_3 + \eta_2 + \eta_{3,t=0}$	$\eta_3 = \beta_{32}\eta_2 + \zeta_3$
Measurement model	$y_1(t_s) = \eta_1(t_s) + \varepsilon_1$	$y_1 = \eta_1 + \varepsilon_1$
	$y_2(t_s) = \eta_1(t_s) + \varepsilon_2$	$y_2 = \eta_1 + \varepsilon_2$
	$y_3(t_s) = \eta_1(t_s) + \varepsilon_3$	$y_3 = \eta_1 + \varepsilon_3$
	$y_4(t_s) = \eta_2(t_s) + \varepsilon_4$	$y_4 = \eta_2 + \varepsilon_4$
	$y_5(t_s) = \eta_2(t_s) + \varepsilon_5$	$y_5 = \eta_2 + \varepsilon_5$
	$y_6(t_s) = \eta_2(t_s) + \varepsilon_6$	$y_6 = \eta_2 + \varepsilon_6$
	$y_7(t_s) = \eta_3(t_s) + \varepsilon_7$	$y_7 = \eta_3 + \varepsilon_7$
	$y_8(t_s) = \eta_3(t_s) + \varepsilon_8$	$y_8 = \eta_3 + \varepsilon_8$
	$y_9(t_s) = \eta_3(t_s) + \varepsilon_9$	$y_9 = \eta_3 + \varepsilon_9$
Fixed parameters in the latent model	$E(\xi_1) = 0$	
	$Var(\xi_1) = 0$	
	$E(\xi_2) = 0.25$	
	$Var(\xi_2) = 0$	
Parameters that are fixed for each simulated study	$\forall i, 1 \le i \le p,\ E(\varepsilon_i) = 0$	
	$\forall i, 1 \le i \le p,\ Var(\varepsilon_i) = 1.0$	
	$E(t_s) = k\frac{t_{stop} - t_{start}}{N_{studies}}$	
	$Var(t_s) = 0.01$	

One necessary but not sufficient condition of SDM and SEM being broadly equivalent within Ω is that the family of implied covariance matrices from a single system dynamic model θ_1 considered over a time interval can be modeled by the family of covariance matrices implied by $\aleph(\theta_2)$ for some structural equation model represented by θ_2.

This criterion would be supported in the "fixes that fail" example if a series of covariance matrices were estimated over a given time interval and, for each implied covariance matrix, there existed a structural equation model in $\aleph(\theta_2)$ with the same implied covariance matrix. Failure to find support for this criterion in one example would provide support for rejecting the claim that SDM and SEM are broadly equivalent within Ω.

Testing this criterion involves generating this series of implied covariance matrices from the system dynamics model θ_1 via a series of simulations and using them for the sample covariance matrix for a given structural equation model θ_2. If one can estimate the free parameters of θ_2 and satisfy criteria for model fit, then one could arguably claim that both θ_1 and θ_2 yielded the same implied covariance matrices.

One sufficient condition for SDM and SEM being narrowly equivalent within Ω is that, for a given sample covariance matrix, there exists a system dynamics model θ_1 and a structural equation model θ_2 such that they both explain the same given sample covariance matrix. Testing this condition would involve finding a structural equation model θ_2 that was both identified and satisfied goodness-of-fit criteria over a given time interval.

Method

There are a variety of sophisticated modeling packages for both system dynamics and structural equation modeling. For this demonstration, a software package called Powersim (1997) was used to simulate the "fixes that fail" model. The model was developed and tested using conventional SDM methods.

Systems dynamic models are "in flux" when feedback loop dominance with respect to a given variable changes, which corresponds to changes in the qualitative time-series behavior of a specific variable. Consequently it is important to determine the intervals where feedback loop dominance is stable and points where the dominance of feedback loops change. To do this, feedback loop dominance was analyzed with respect to the variable problem (see Figure 9.1, also represented as η_2 in Figure 9.2). This is a reasonable representation of the overall system's behavior, since the variable η_2 is contained in both feedback loops, L1 and L2.

In SDM, the analysis of feedback loops is complicated by the fact that a variable may be affected by a number of loops simultaneously and it is generally impossible to understand and analyze these effects without the use of simulations. Ford (1999) provided a formal method for identifying feedback loop dominance with respect to a specific variable x within a given time interval T by analyzing changes in the expression $\partial(|\partial x/\partial t|)/\partial t$ as loops are systematically turned off in the model.

The next step was to estimate the implied covariance matrices across time. This was accomplished by estimating the covariance matrix at a particular time $\bar{t}_s$ between 0.0 and 10.0. There was one simulated study for each covariance matrix at $\bar{t}_s$, with each study involving one run for each case. For each case, a time t was randomly selected from a normal distribution around $\bar{t}_s$. Powersim was then used to simulate that case from 0.0 to 10.0, with the data from that case being recorded at time t. When all the cases for a study had been simulated and data recorded, the observed covariance matrix was calculated and used as the estimate for the covariance matrix at $\bar{t}_s$.

A critical issue became selecting the appropriate minimum number of studies within a time interval of feedback loop dominance. It turned out to be too difficult to analyze the changes in covariance structure across time when there were too few studies for a given time interval of feedback loop dominance. So a minimum number of studies per time interval was set to five, meaning that the total number of studies depended on the smallest time interval for a given phase of feedback loop dominance. For the "fixes that fail" model, this turned out to require about 100 studies in order to simulate the time period from 0.0 to 10.

The number of cases per study was determined somewhat arbitrarily and based on balancing computing time against minimizing the variability in the covariance matrices due to sampling error. The results are based on 1000 cases per study.

Initial simulations used the existing features of Powersim to save data to a file for each study, which was then merged using a macro in SPSS. With 100 000 runs of the Powersim model, this approach became impractical. So a Visual Basic program was developed to manage the Powersim runs, collect the data for each study, calculate the covariance matrix, and save the data and covariance matrix in a format compatible with LISREL.

LISREL was then used to specify the structural equation model and estimate the parameters against the series of covariance matrices by setting the number of repetitions equal to the number of simulated studies (i.e., RP = 100). The goodness-of-fit statistics from each repetition were then saved in an output file and analyzed using Excel.

Results

The simulation time ran from 0.0 to 10.0. Each case had the same latent variable time series η_1, η_2, and η_3 (plotted in Figure 9.4). The analysis of the loop dominance was with respect to η_2, where $\partial(|\partial\eta_2/\partial t|)/\partial t$ was plotted against time (see Figure 9.5). Loop dominance changes with respect to η_2

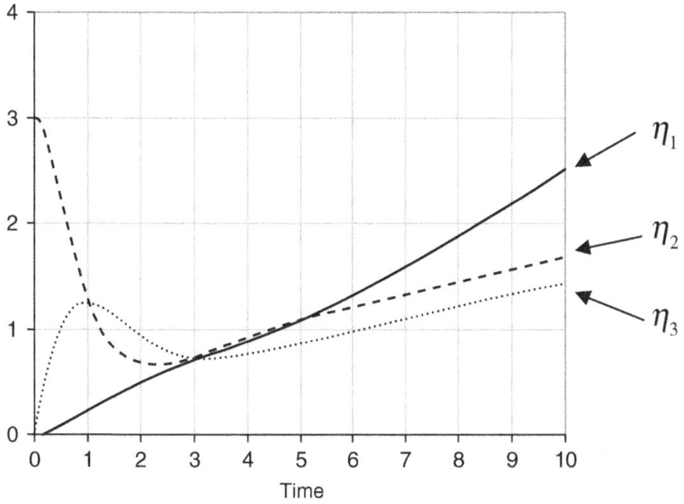

Figure 9.4. Latent variables η_1, η_2, and η_3 in system dynamics model.

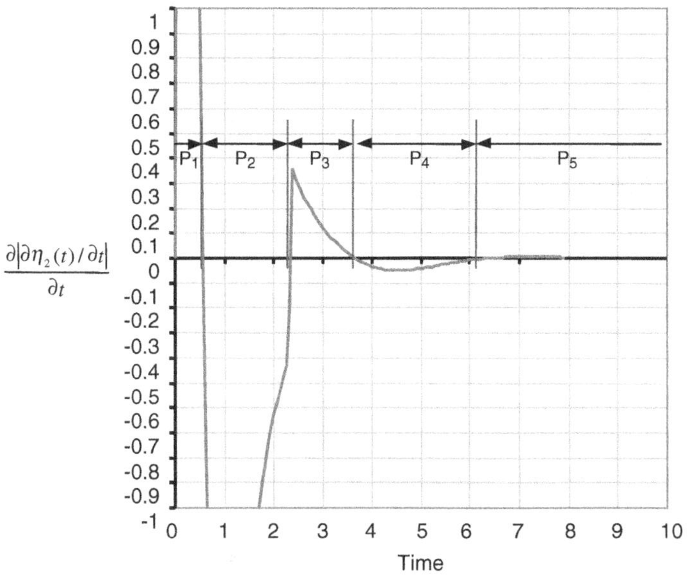

Figure 9.5. Analysis of loop dominance. P1, phase 1 of loop dominance.

Table 9.2. *Phases of loop dominance*

Phase of loop dominance	Time interval	Dominant loop(s)
P1	0.0 to 0.6	L1 and L2
P2	0.6 to 2.4	L1
P3	2.4 to 3.7	L2
P4	3.7 to 6.3	L1
P5	6.3 to 10.0	L1 and L2

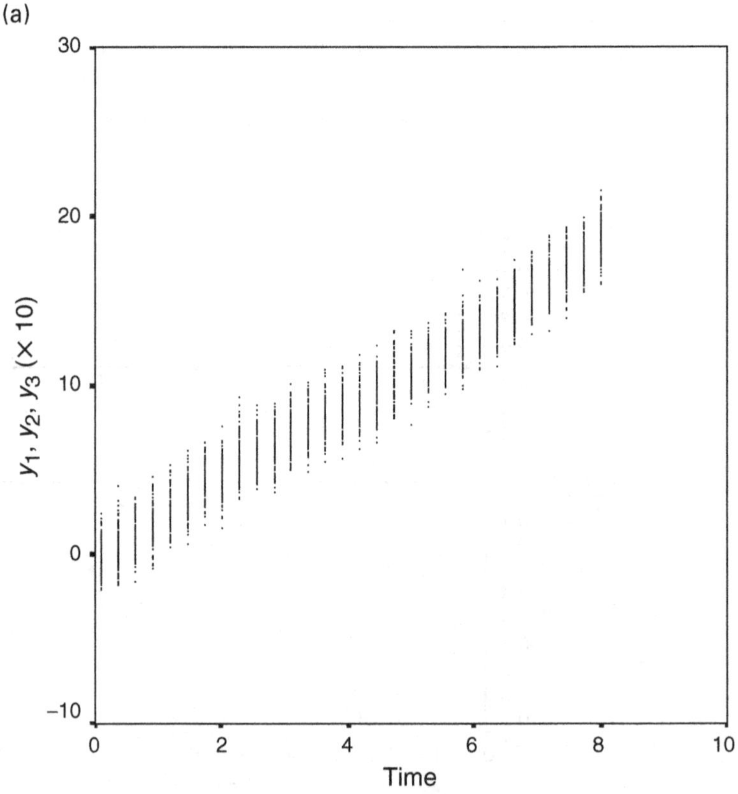

Figure 9.6. (a) Estimates y_1, y_2, and y_3 of η_1 from system dynamics model (1% of cases represented). (b) Estimates y_4, y_5, and y_6 of η_2 from system dynamics model (1% of cases represented). (c) Estimates y_7, y_8, and y_9 of η_3 from system dynamics model (1% of cases represented).

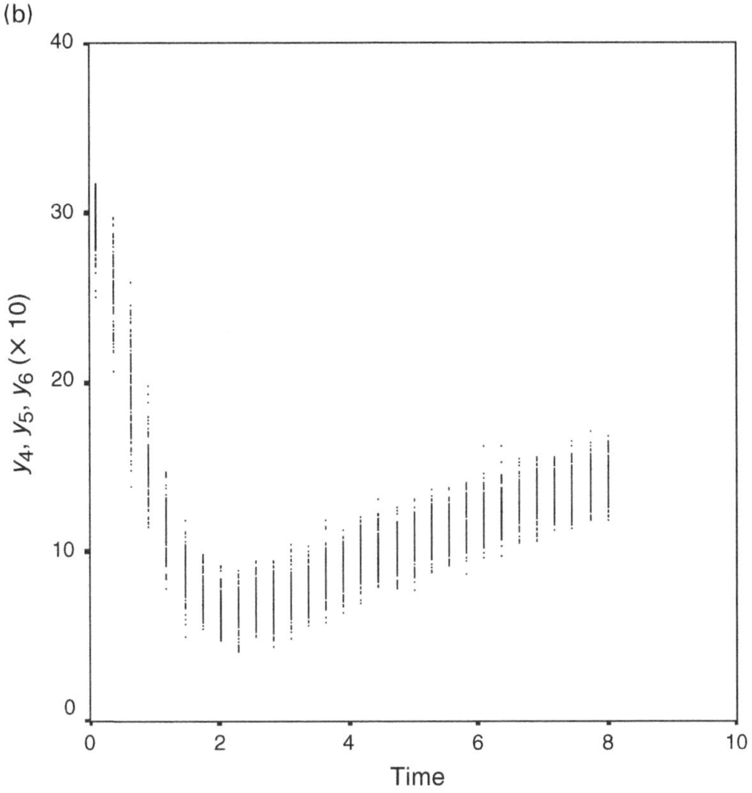

Figure 9.6. *(cont.)*

whenever $\partial(|\partial\eta_2/\partial t|)/\partial t$ equals zero. The time intervals for each phase as well as the loops that are dominating the behavior of η_2 are listed in Table 9.2.

There were three indicator variables for each latent variable. Each case generated a different set of values for the observed variables $y_1, \ldots, y_9$ (see Figure 9.6a–c).

Two LISREL models were developed. Model 1 was the general structural equation model depicted in Figure 9.3, and was used to analyze whether a single structural equation model could explain all the covariance matrices of a dynamic system model over time (see Figure 9.7). Chi-square (χ^2) and root mean square error of approximation (RMSEA) statistics for model 1 indicated that it did not fit at all during the first two phases of

(c)

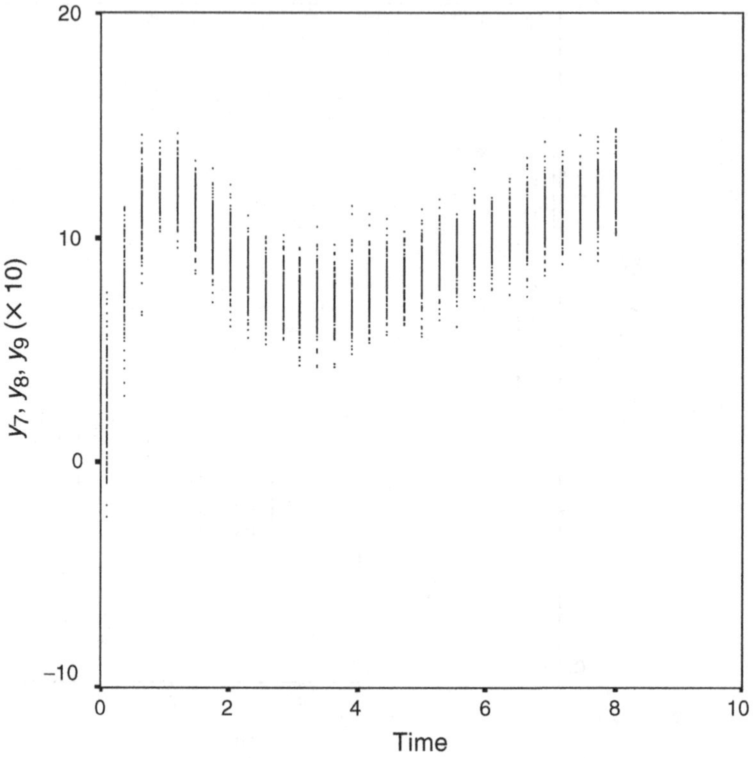

Figure 9.6. *(cont.)*

feedback loop dominance (see Figures 9.8 and 9.9). Near the transition into phase P3, the fit statistics improved substantially and remained fairly good for the remaining phases and transitions.

Model 2 was developed as a submodel of Model 1 with the goal of maximizing fit across the entire simulation time domain (see Figure 9.10). Attempts at including a feedback relationship resulted in a high proportion of models being rejected on the grounds that they did not converge or had inadmissible solutions, in which case χ^2 and RMSEA are reported as negative in Figures 9.11 and 9.12. Model 2 was rejected in the transitions into and out of phase P3 because the solutions did not converge and were not admissible. The results are summarized in Table 9.3.

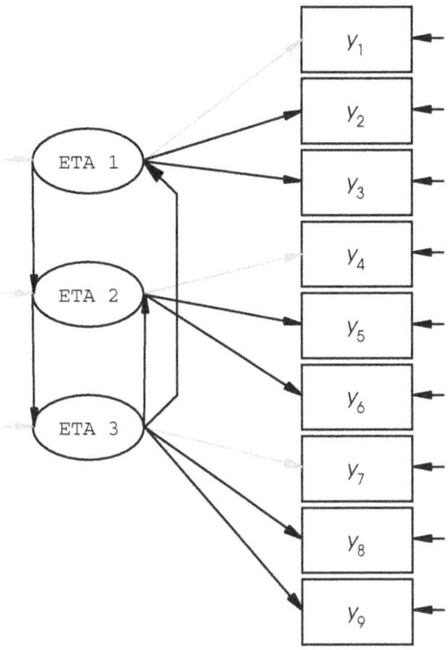

Latent variable η_1

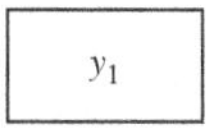

Observed variable y_1

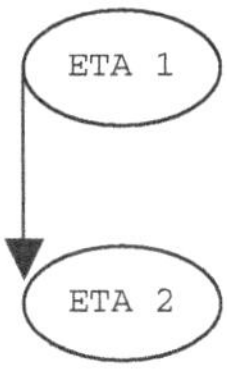

Latent variable η_1 influences latent variable η_2

LISREL syntax for model 1

```
1129c LOOP 1+2
! Developed from t=5 covariance matrix
DA NI=13 NO=1000 MA=CM RP=100
LA
s1 s2 s3 y1 y2 y3 y4 y5 y6 y7 y8 y9 s4
CM fi=cov1129c.dat
SE
4 5 6 7 8 9 10 11 12 /
```

Figure 9.7. Conceptual diagram for model 1.

```
MO NY=9 NE=3 LY=FU,FI BE=FU,FI PS=sy,fi TE=sy,fi
LE
'ETA 1' 'ETA 2' 'ETA 3'
FR TE(1,1) TE(2,2) TE(3,3) TE(4,4) TE(5,5) TE(6,6) TE(7,7)
  TE(8,8) TE(9,9)
FR LY(1,1) LY(2,1) LY(3,1) LY(4,2) LY(5,2) LY(6,2) LY(7,3)
  LY(8,3) LY(9,3)
FR BE(2,1) BE(3,2) BE(1,3)
FR BE(2,3)
PD
OU ME=ul RS be=s1129LI2.bes gf=s1129L12.gfs
```

Figure 9.7. *(cont.)*

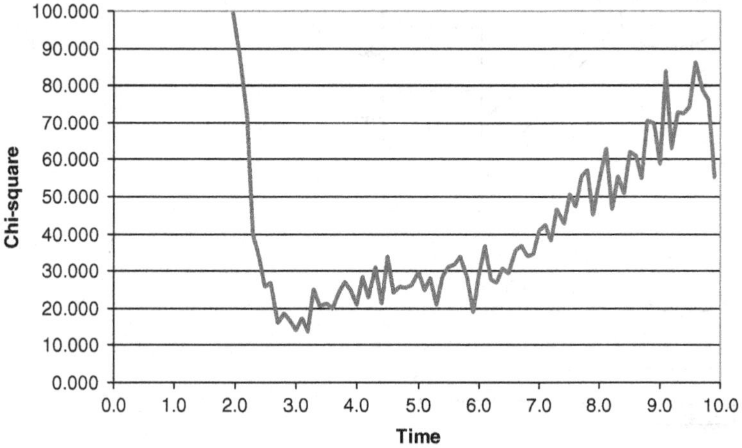

Figure 9.8. Chi-square statistic for model 1.

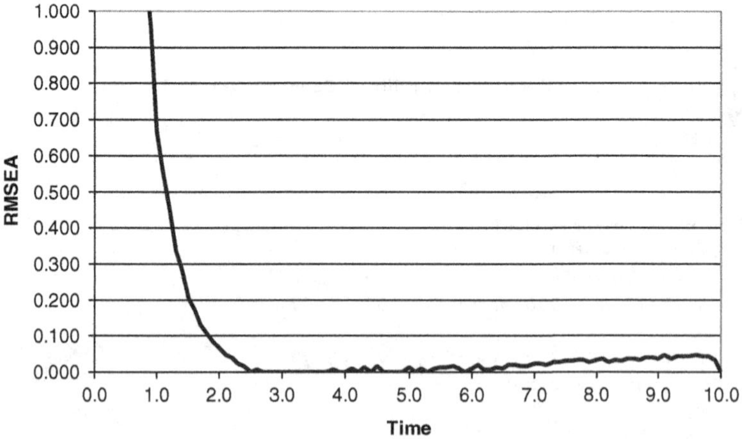

Figure 9.9. RMSEA for model 1.

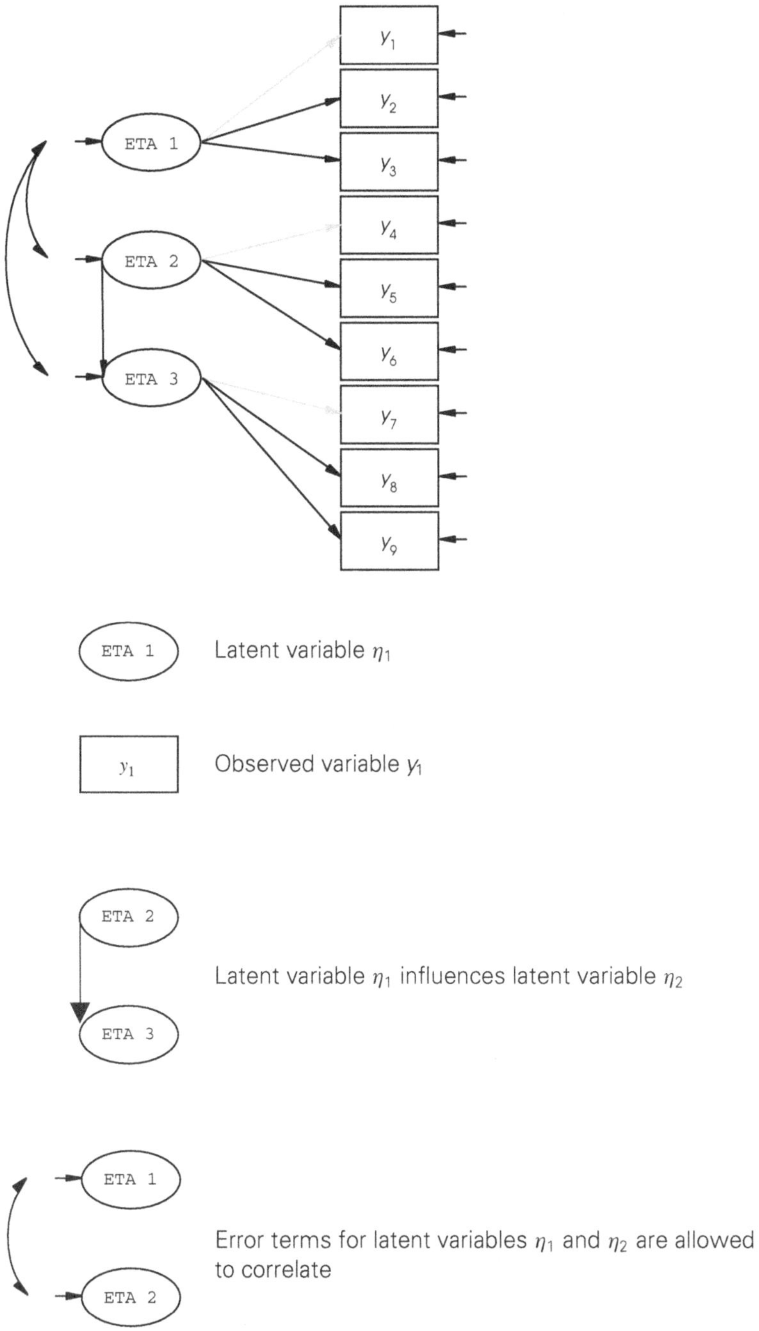

Figure 9.10. Conceptual diagram for model 2.

LISREL syntax for analyzing model 2

```
1129c LOOP 2
! Developed from t=5 covariance matrix
DA NI=13 NO=1000 MA=CM RP=100
LA
s1 s2 s3 y1 y2 y3 y4 y5 y6 y7 y8 y9 s4
CM fi=cov1129c.dat
SE
4 5 6 7 8 9 10 11 12 /
MO NY=9 NE=3 LY=FU,FI BE=FU,fi PS=sy,fi TE=sy,fi
LE
'ETA 1' 'ETA 2' 'ETA 3'
FR TE(1,1) TE(2,2) TE(3,3) TE(4,4) TE(5,5) TE(6,6) TE(7,7)
  TE(8,8) TE(9,9)
FR LY(1,1) LY(2,1) LY(3,1) LY(4,2) LY(5,2) LY(6,2) LY(7,3)
  LY(8,3) LY(9,3)
FR PS(1,3)
FR PS(1,2)
FR PS(1,1) PS(2,2) PS(3,3)
FR PS(3,2)
PD
OU ME=ul RS be=s1129LI2.bes gf=s1129L12.gfs
```

Figure 9.10. (*cont.*)

Table 9.3. *Summary of results*

Phase of loop dominance	Time interval	Dominant loop(s)	Model 1 fit	Model 2 fit
P1	0.0 to 0.6	L1 and L2	No	Yes
Transition	0.6	–	No	Yes
P2	0.6 to 2.4	L1	No	Yes
Transition	2.4	–	Yes	No[a,b]
P3	2.4 to 3.7	L2	Yes	Yes
Transition	3.7	–	Yes	No[a,b]
P4	3.7 to 6.3	L1	Yes	Yes
Transition	6.3	–	Yes	Yes
P5	6.3 to 10.0	L1 and L2	Yes[c]	Yes

[a] Fails to converge.
[b] Not admissible after 50 iterations.
[c] RMSEA increases up to about 0.045 and then decreases again.

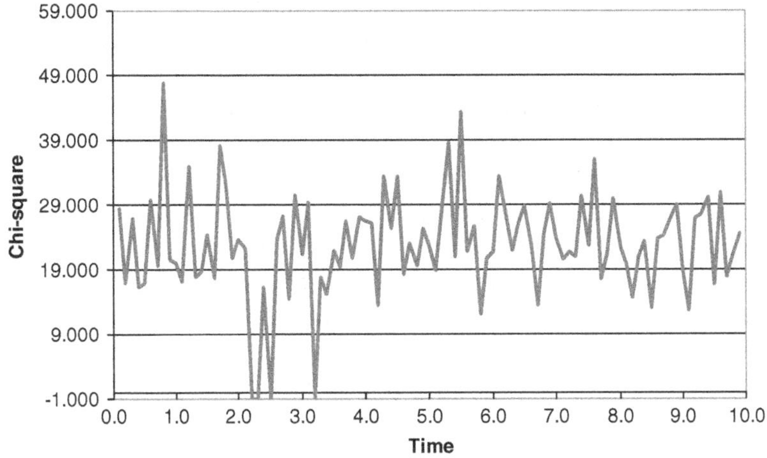

Figure 9.11. Chi-square statistic for model 2.

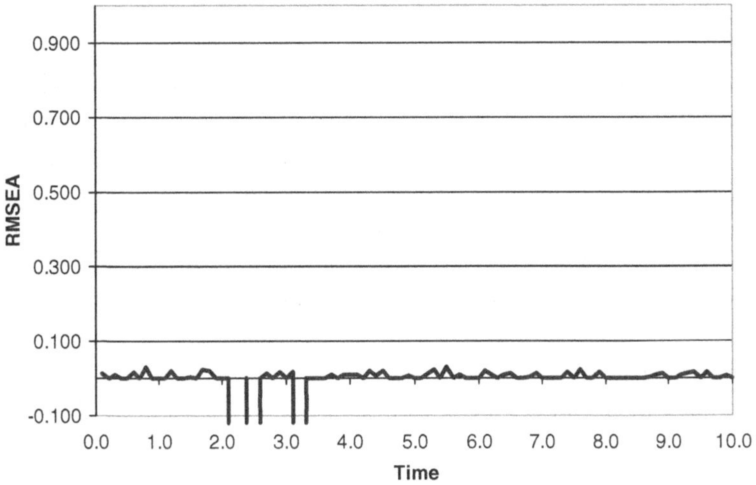

Figure 9.12. RMSEA for model 2.

Discussion

Model 1 is arguably the structural equation model that is most similar to the "fixes that fail" system dynamics model; yet it rather dramatically failed to explain the implied covariance structure during the first two phases P1 and P2 of the system dynamics model. This lends support to rejecting the claim that SDM and SEM are in general broadly equivalent. Model 1 was able to explain the implied covariance matrices during the subsequent phases, but these phases could arguably be considered quasi-static. Hence SDM and SEM do not appear to be broadly equivalent with respect to Ω when Ω includes periods of dynamic behavior.

Model 2 did considerably better at fitting the overall family of implied covariance matrices, with the exception of the transitions into and out of phase P3. If one looks at Figure 9.4, one can note that not only are the slopes relatively flat, but, at these points, the slopes are changing signs. In essence, measurement error begins to dominate the covariance matrices during these transitions and essentially washes out the effects. Excluding the transitions around P3, model 2 would support the claim that SDM and SEM can be narrowly equivalent.

There are several additional points worth making. In order for model 2 to achieve the fit across the entire time domain, a substantial amount of structural relationships between the latent variables had to be removed. Model 2 provides little insight into the structure of the "fixes that fail" model.

One way of thinking of model 2 is to consider it as a model of the transient response and not the underlying structure that has that particular transient response as a solution. Another way to think about this is with the example of growth curves. One way to model a growth curve is to write growth as a function of time, e.g., $g(t) = e^t$. This would be the curve that we observe across time. But this expression is really the model for the *solution* to the underlying relationship $g'(t) = g(t)$. The underlying relationship is what causes the observed behavior. Modeling the solution to a system of equations is not the same thing as modeling the underlying system of equations.

This suggests that SEM is best for (1) representing the solution to a set of equations describing dynamic behavior or (2) representing the underlying structure when the relationships are relatively stable (time invariant). In contrast, SDM is best suited to modeling the underlying structure of dynamic behavior.

There are also situations when the two approaches can be used in complementary fashion. Formal system dynamic models involve simulating the time-series solution to systems of coupled differential equations by successively estimating the behavior in small intervals of time. But this makes system dynamics models sensitive to the size of the time intervals used in the simulation. Large intervals increase the error accumulated during the simulation, while small intervals increase the number of steps to be calculated and overall computation time. So there is an optimum time interval for each model. System dynamic models that need to represent behavior involving different time scales (e.g., minutes and years) run into problems because the best time interval for one part of the model will create major problems for another part. For example, system dynamics models would have a very difficult time representing population growth (measured on the scale of months or years) by modeling the dynamics of metabolism (measure on the scale of seconds or minutes).

The solution is often to consider one part of the model static with respect to the rest of the model. In the population growth example, one might consider the dynamics of metabolism to be relatively time invariant with respect to population size. This does not mean that metabolism does not change, but instead that metabolism should be considered a function of other variables (excluding time). SEM is an excellent method for estimating these time-invariant relationships, and system dynamics models implicitly specify such relationships through table functions.

Conclusion

In this chapter I have suggested some ways that SDM and SEM might be related by expanding the concepts of broad and narrow equivalence to include SDM. While one example is not enough to make general claims about the relationship between SDM and SEM, it does offer a demonstration of how one might go about developing a better sense of which models do and do not work and under what conditions. What has not been addressed is how structural equation models correspond to system dynamic models. This direction, however, is arguably a bit easier as models in SEM are amenable to analytical solutions, which implies that one can probably find simple expressions mapping models from SEM into SDM. Also not addressed are methods of estimating model parameters in SDM by optimizing a fit function. While the problems associated with solving nonlinear systems of differential equations might have made this task unfeasible, this is likely to improve dramatically with continued advances in computing power and interconnectivity of software packages. This suggests that, in time, the overlap between SDM and SEM will grow substantially.

References

Bollen, K. A. (1989). *Structural Equations with Latent Variables*. New York: Wiley.

Epstein, J. M. (1997). *Nonlinear Dynamics, Mathematical Biology, and Social Science*. Reading, MA: Addison-Wesley Publishing.

Ford, D. N. (1999). A behavioral approach to feedback loop dominance analysis. *System Dynamics Review*, **1**(15), 3–36.

Forrester, J. W. (1968). *Principles of Systems*, 2nd preliminary edition. Cambridge, MA: Wright-Allen Press.

Hayduk, L. (1996). Equivalence, loops, and longitudinal models. In *LISREL Issues, Debates and Strategies*, pp. 79–120. Baltimore, MD: Johns Hopkins University Press.

Levine, L. R. (1992). An introduction to qualitative dynamics. In R. L. Levine & H. E. Fitzgerald (eds.), *Analysis of Dynamic Psychological Systems*, vol. 1 *Basic Approaches to General Systems, Dynamic Systems, and Cybernetics*, pp. 267–327. New York: Plenum Press.

Levine, L. R. & Lodwick, W. (1992). Parameter estimation and assessing the fit of dynamic models. In R. L. Levine & H. E. Fitzgerald (eds.), *Analysis of Dynamic Psychological Systems*, vol. 2 *Methods and Applications*, pp. 119–150. New York: Plenum Press.

Levine, L. R., Sell, M. V. & Rubin, B. (1992). System dynamics and the analysis of feedback processes in social and behavioral systems. In R. L. Levine & H. E. Fitzgerald (eds.), *Analysis of Dynamic Psychological Systems*, vol. 1 *Basic Approaches to General Systems, Dynamic Systems, and Cybernetics*, pp. 145–266. New York: Plenum Press.

Powersim [Computer program]. (1997). Isdalstø: Powersim.

Raykov, T. & Penev, S. (1999). On structural equation model equivalence. *Multivariate Behavioral Research*, **34**, 199–244.

Richardson, G. P. (1991). *Feedback Thought in the Social Science and Systems Theory*. Philadelphia: University of Pennsylvania Press.

Richardson, G. P. (1996). System dynamics: simulation for policy analysis from a feedback perspective. In G. P. Richardson (ed.), *Modeling for Management. I: Simulation in Support of Systems Thinking*, pp. 47–72. Brookfield: Dartmouth Publishing.

Senge, P. M. (1990). *The Fifth Discipline: The Art and Practice of the Learning Organization*. New York: Doubleday.

Von Bertalanffy, L. (1968). *General System Theory: Foundations, Development, Applications*, revised edition. New York: George Braziller.

10 Estimating analysis of variance models as structural equation models

Michael J. Rovine and Peter C. M. Molenaar

Abstract

In this chapter we describe how to estimate a number of different analysis of variance (ANOVA) models through the use of structural equations modeling (SEM). The three kinds of model we concentrate on are the one-way ANOVA model, the repeated measures ANOVA model, and the random coefficients model.

Introduction

Due to its flexibility, structural equation modeling (SEM) has the capability to estimate each of the analysis of variance (ANOVA) models in a number of different ways. To be practical, we will describe what we feel is the most straightforward estimation method for each of these models. For the ordinary ANOVA model, we will exemplify the method based on dummy-coding described by Huitema (1980) and Jöreskog & Sörbom (1993). We will describe a method for estimating any repeated ANOVA by expressing the model as a regression in which the repeated measures are transformed into a set of trend variables (Hertzog & Rovine, 1985). We will present a method for estimating random coefficients models based on the general linear mixed model (GLMM; Laird & Ware, 1982). Rovine & Molenaar (1998) have described SEM strategies for estimating GLMM models with patterned covariance matrices and for models in which the parameters representing individual curves are to be estimated (Rovine & Molenaar, 2000).

We will use the LISREL program to estimate these models and will refer to the LISREL parameterization of SEM throughout the chapter. While we could have used any SEM program to estimate these models, we believe that there is some intuitive appeal of the matrix language of LISREL and similar programs.

The LISREL model

The very general LISREL model (Jöreskog & Sörbom, 1993) allows one to model covariance structures implied by models that include both manifest (observed) and latent (unobserved) variables. The model can be thought of as a combination of confirmatory factor analysis (Long, 1983) and regression (Pedhazur, 1997). The general model is parameterized so that certain of the LISREL matrices (the measurement models) hold the parameters related to the confirmatory factor analysis while other matrices (the structural regression models) consist of the parameters for the regressions to be estimated. By selecting certain of the LISREL matrices one can, for example, estimate an ordinary regression model. By selecting a different matrix, one can estimate a simple confirmatory factor model. By using the measurement model to define latent variables, one can estimate regression relationships among those latent variables. In addition to modeling covariance structures, LISREL provides the capacity for modeling mean structures. This capacity will, of course, be important for estimating ANOVA models.

The LISREL model is divided into an x-side and a y-side. On the x-side of the LISREL model,

$$\boldsymbol{x} = \boldsymbol{\tau}_x + \boldsymbol{\Lambda}_x \boldsymbol{\xi} + \boldsymbol{\delta}, \tag{10.1}$$

where $\boldsymbol{\Lambda}_x$ is a matrix of factor loadings (regression weights) relating $\boldsymbol{\xi}$, a vector of factor scores, to the vector of observed variables, $\boldsymbol{x}$; $\boldsymbol{\delta}$ is a vector of residuals representing the portion of each observed variable not determined by the latent variables. $\boldsymbol{\tau}_x$ is a vector of constant intercept terms. On the x-side, the expectation for the covariance matrix of the observed variables takes on the form

$$\mathrm{E}\left[(\boldsymbol{x} - \boldsymbol{\mu}_x)(\boldsymbol{x} - \boldsymbol{\mu}_x)^{\mathrm{T}}\right] = \boldsymbol{\Lambda}_x \boldsymbol{\Phi} \boldsymbol{\Lambda}_x^{\mathrm{T}} + \boldsymbol{\Theta}_\delta, \tag{10.2}$$

where $\boldsymbol{\Phi}$ is the covariance matrix of the latent variables, $\boldsymbol{\Theta}_\delta$ is the covariance matrix of the residuals ($\boldsymbol{\delta}$), and $\boldsymbol{\mu}_x$ is the vector of means of the observed variables. The superscript T indicates a transposed matrix.

On the y-side of the model,

$$\boldsymbol{y} = \boldsymbol{\tau}_y + \boldsymbol{\Lambda}_y \boldsymbol{\eta} + \boldsymbol{\epsilon}, \tag{10.3}$$

where $\boldsymbol{\Lambda}_y$ is a matrix of factor loadings (regression weights) that relate a vector of factor scores, $\boldsymbol{\eta}$, to a vector of observed scores, $\boldsymbol{y}$. $\boldsymbol{\epsilon}$ is a vector of residuals, and $\boldsymbol{\tau}_y$ is a vector of intercept terms.

The y-side differs from the x-side in that regression relationships can exist among the latent variables. In addition, if the x-side latent variables are considered exogenous and the y-side latent variables are considered endogenous, the regressions of y-side variables onto x-side variables can be estimated in a $\mathbf{\Gamma}$ matrix. These regression relationships among the latent variables can be expressed as

$$\eta = \alpha + \mathbf{B}\eta + \mathbf{\Gamma}\xi + \zeta, \tag{10.4}$$

where the regression weights in $\mathbf{\Gamma}$ allow exogenous x-side latent variables, ξ, to predict the η. Regression relationships among the η appear in the $\mathbf{B}$ matrix. ζ are the residuals for these structural regression equations, and α is a vector of constant intercept terms for the latent variables.

In the case in which only the y-side of the LISREL model is used,

$$\eta = \alpha + \mathbf{B}\eta + \zeta. \tag{10.5}$$

When only the y-side of the LISREL model is used the expectation for the covariance matrix of the observed variables is

$$\mathbf{E}\left[(y - \mu_y)(y - \mu_y)^{\mathrm{T}}\right] = \mathbf{\Lambda}_y(\mathbf{I} - \mathbf{B})^{-1}\mathbf{\Psi}(\mathbf{I} - \mathbf{B})^{-1^{\mathrm{T}}}\mathbf{\Lambda}_y^{\mathrm{T}} + \mathbf{\Theta}_\epsilon, \tag{10.6}$$

where $\mathbf{\Theta}_\epsilon$ is a covariance matrix of residuals (ϵ); $\mathbf{B}$ is matrix of regression weights among the latent variables; and $\mathbf{\Psi}$ is a covariance matrix of the regression equation residuals (ζ).

As a simple case let us begin with a set of regression equations among manifest variables. Consider the following set of equations in which regression coefficients are designated as β_i and intercept terms as α_k:

$$\begin{aligned}
y_1 &= \alpha_1 + \beta_1 x_1 + \beta_2 x_2 + \epsilon_1,\\
y_2 &= \alpha_2 + \beta_3 x_1 + \beta_4 x_2 + \epsilon_2,\\
y_3 &= \alpha_3 + \beta_5 y_1 + \beta_6 y_2 + \epsilon_3,\\
y_4 &= \alpha_4 + \beta_7 y_3 + \epsilon_4.
\end{aligned}$$

To express this set of equations as a LISREL model, we first make a distinction between the x manifest variables and the y manifest variables. Since the $\mathbf{\Gamma}$ and $\mathbf{B}$ express regression relationships among latent variables, we use the respective measurement models to set each manifest variable equivalent to a latent variable. We accomplish this by setting $\mathbf{\Lambda}_x = 0$ and $\mathbf{\Theta}_\delta$ (the covariance matrix of the x-side measurement errors) $= 0$ on the x-side, and by setting $\mathbf{\Lambda}_y = 0$ and $\mathbf{\Theta}_\epsilon = 0$ on the y-side. We can allow means other than 0 for our manifest variables by estimating parameters in $\boldsymbol{\tau}_x$ and $\boldsymbol{\tau}_y$.

We will consider this option for particular models. If $\boldsymbol{\tau}_x = 0$ and $\boldsymbol{\tau}_y = 0$ then equations 10.1 and 10.3 become

$$\boldsymbol{x} = \boldsymbol{\xi}$$

and

$$\boldsymbol{y} = \boldsymbol{\eta}.$$

We can then rewrite the regression equations as

$$\eta_1 = \alpha_1 + \gamma_{11}\xi_1 + \gamma_{12}\xi_2 + \zeta_1,$$
$$\eta_2 = \alpha_2 + \gamma_{21}\xi_1 + \gamma_{22}\xi_2 + \zeta_2,$$
$$\eta_3 = \alpha_3 + \beta_{31}\eta_1 + \beta_{32}\eta_2 + \zeta_3,$$
$$\eta_4 = \alpha_4 + \beta_{43}\eta_3 + \zeta_4.$$

In terms of equation 10.4 this becomes

$$\begin{bmatrix}\eta_1\\ \eta_2\\ \eta_3\\ \eta_4\end{bmatrix} = \begin{bmatrix}\alpha_1\\ \alpha_2\\ \alpha_3\\ \alpha_4\end{bmatrix} + \begin{bmatrix}\gamma_{11} & \gamma_{12}\\ \gamma_{21} & \gamma_{22}\\ 0 & 0\\ 0 & 0\end{bmatrix}\begin{bmatrix}\xi_1\\ \xi_2\end{bmatrix} + \begin{bmatrix}0 & 0 & 0 & 0\\ 0 & 0 & 0 & 0\\ \beta_{31} & \beta_{32} & 0 & 0\\ 0 & 0 & \beta_{43} & 0\end{bmatrix}\begin{bmatrix}\eta_1\\ \eta_2\\ \eta_3\\ \eta_4\end{bmatrix} + \begin{bmatrix}\zeta_1\\ \zeta_2\\ \zeta_3\\ \zeta_4\end{bmatrix}.$$

Estimating an ANOVA model as SEM through the use of dummy-coding

Consider the data appearing in Table 10.1 (p. 241). Subjects have been randomly assigned to one of three groups, and we are interested in determining whether the groups differ on the dependent variable, biology score. We test that by determining whether mean differences exist among the three groups. To express that model as a regression, we first dummy code the groups as follows:

Group	v_1	v_2
1	1	0
2	0	1
3	0	0

The following regression equation

$$\hat{y} = \beta_0 + \beta_1 v_1 + \beta_2 v_2 \tag{10.7}$$

is equivalent to a one-way ANOVA. For each group the predicted value of y is the mean of that group. For the three groups, equation 10.7 becomes

$$\bar{y}_{\text{Group1}} = \beta_0 + \beta_1,$$
$$\bar{y}_{\text{Group2}} = \beta_0 + \beta_2,$$
$$\bar{y}_{\text{Group3}} = \beta_0.$$

These three equations can be solved to yield:

$$\beta_0 = \bar{y}_{\text{Group3}},$$
$$\beta_1 = \bar{y}_{\text{Group1}} - \bar{y}_{\text{Group3}},$$
$$\beta_2 = \bar{y}_{\text{Group2}} - \bar{y}_{\text{Group3}}.$$

The multiple correlation coefficient squared (R^2) from this equation can be used to calculate the F-statistic for the one-way ANOVA. The F-statistic can be computed as:

$$F = \frac{R^2/(G-1)}{(1-R^2)/(N-G)}, \tag{10.8}$$

with df $= (G-1), (N-G)$, where G is the number of groups in the analysis and N is the sample size. The tests of the regression coefficients represent either planned or follow-up comparisons on the means.

ANOVA as SEM

We will first estimate the ANOVA as suggested by Huitema (1980) and Jöreskog & Sörbom (1993). The regression weights will be estimated in the $\mathbf{\Gamma}$ matrix. The $\mathbf{\Gamma}$ matrix is used to regress *latent* endogenous y-side variables onto *latent* exogenous x-side variables. Since the variables in the ANOVA are manifest, we can use the x- and y-side measurement models (equations 10.1 and 10.3) to set each manifest variable equivalent to a corresponding latent variable.

Consider equation 10.1. If we set $\boldsymbol{\tau}_x = 0$, $\mathbf{\Theta}_\epsilon = 0$, and $\mathbf{\Lambda}_x = \mathbf{I}$, where $\mathbf{I}$ is the identity matrix, then equation 10.1 becomes

$$\boldsymbol{x} = \boldsymbol{\xi}.$$

Similarly, in equation 10.3, we set $\boldsymbol{\tau}_y = 0$, $\boldsymbol{\Theta}_\epsilon = 0$, and $\boldsymbol{\Lambda}_y = \mathbf{I}$, where $\mathbf{I}$ is the identity matrix, then equation 10.3 becomes

$$\boldsymbol{y} = \boldsymbol{\eta}.$$

The regression relationships are modeled in equation 10.4. Since we are regressing y-side variables onto x-side variables, we set $\mathbf{B} = 0$. In their example, Jöreskog & Sörbom were estimating the regression weights (i.e., β_1 and β_2) as a means of determining the squared multiple correlation coefficient (R^2), which was then used to calculate the omnibus F-test. They did not estimate the intercept (β_0).

To model the regression intercept, we have a number of choices. If we input the means vector along with the covariance matrix of the manifest variables, we could use the $\boldsymbol{\alpha}$ vector in equation 10.4. Equation 10.4 becomes

$$\boldsymbol{\eta} = \boldsymbol{\alpha} + \boldsymbol{\Gamma}\boldsymbol{\xi} + \boldsymbol{\zeta}. \tag{10.9}$$

In terms of equation 10.7 we have three variables, y, v_1, and v_2. Having used the measurement model to set each manifest variable equivalent to a latent variable with no error we have

$$y_1 = \eta_1 = y,$$
$$x_1 = \xi_1 = v_1,$$
$$x_2 = \xi_2 = v_2,$$

and equation 10.9 becomes

$$[y_1] = [\alpha_1] + [\gamma_1 \ \gamma_2]\begin{bmatrix}\xi_1 \\ \xi_2\end{bmatrix} + [\zeta_1],$$

which is equivalent to equation 10.7. However, one problem remains. At this point we have modeled only the mean of the dependent variable. That we have not explicitly included parameters for the means of the independent variables will result in a lack of fit if those means are indeed different from zero. In addition, although the model is saturated in the means, we would see degrees of freedom other than zero. We can model the means of the independent variables by including the $\boldsymbol{\tau}_x$ vector in the x-side measurement model. The measurement model would then be

$$\boldsymbol{x} = \boldsymbol{\tau}_x + \boldsymbol{\xi}.$$

The model would now have the correct degrees of freedom (i.e., zero), and all parameters would be correctly estimated.

Table 10.1. *Simulated data for the one-way ANOVA*

bio	*int*	d_1	d_2	d_3
15	1	1	0	0
19	1	1	0	0
21	1	1	0	0
27	1	1	0	0
35	1	1	0	0
39	1	1	0	0
23	1	1	0	0
38	1	1	0	0
33	1	1	0	0
50	1	1	0	0
20	1	0	1	0
34	1	0	1	0
28	1	0	1	0
35	1	0	1	0
42	1	0	1	0
44	1	0	1	0
46	1	0	1	0
47	1	0	1	0
40	1	0	1	0
54	1	0	1	0
14	1	0	0	1
20	1	0	0	1
30	1	0	0	1
32	1	0	0	1
34	1	0	0	1
42	1	0	0	1
40	1	0	0	1
38	1	0	0	1
54	1	0	0	1
56	1	0	0	1

An example

In Table 10.1 we include an imaginary data set modeled after Huitema (1980). Subjects are in one of three groups (d_1–d_3). With this data set we are interested in testing whether the groups differ on the dependent variable, *bio*. Jöreskog & Sörbom (1993) estimated the R^2 in a similar data set and

Table 10.2. *SAS PROC REG results for one-way ANOVA*

Dependent variable: *bio*

		Analysis of variance			
Source	df	Sum of squares	Mean square	F-value	Prob > F
Model	2	420.00000	210.00000	1.604	0.2198
Error	27	3536.00000	130.96296		
C total	29	3956.00000			

Root MSE	11.44391	R^2	0.1062
Dep. mean	35.00000	Adj. R^2	0.0400
CV	32.69687		

		Parameter estimates			
Variable	df	Parameter estimate	Standard error	T for H_0: Parameter = 0	Prob > $\|T\|$
Int	1	36.000000	3.61888053	9.948	0.0001
d_1	1	−6.000000	5.11786993	−1.172	0.2513
d_2	1	3.000000	5.11786993	0.586	0.5626

df, degrees of freedom; MSE, mean square error; C, centered; CV, coefficient of variation; T, T-value; H_0, null hypothesis.

used that to construct the F-test for the ANOVA. Here we will, in addition, estimate the intercept (*int*) of the model.

As a comparison with the SEM results, we first present the results of the ANOVA computed using SAS PROC REG (SAS, 1989). These results appear in Table 10.2.

We will present three different but equivalent parameterizations of this model.

Model 1

In the first parameterization, we define *bio* as a y-side variable and the dummy-code variables v_1 and v_2, along with the intercept, *int*, as the x-side variables. The input matrix will be the sums of squares and cross-products (SSCP) matrix. By using this input matrix we can treat the intercept as if it were an additional regression predictor. The SSCP coefficients related to the intercept (the last row of the SSCP matrix) give N times the mean of *bio*, v_1, and v_2 along with the *N* of the sample. The LISREL input file that we used to estimate this model is:

```
One-way ANOVA - Model 1
DA NI=4 NO=30 MA=CM
CM FU
*
41132.00000   300.00000   400.00000   1060.00000
  300.00000    10.00000     0.00000     10.00000
  400.00000     0.00000    10.00000     10.00000
 1060.00000    10.00000    10.00000     30.00000
LA
Y D1 D2 INT
MO NY=1 NX=3
OU ND=3
```

Notice that even though we include the SSCP matrix as input, we tell the program that the input matrix is a covariance matrix (MA=CM). This "tricks" the program into allowing us to use the intercept as if it were an ordinary variable. Part of the output from this program appears in Table 10.3.

The regression coefficients appear in the $\mathbf{\Gamma}$ matrix. From these coefficients we conclude that the mean differences between groups 1 and 3 (−6.0) and groups 2 and 3 (4.0) are not significantly different from 0. The $\mathbf{\Phi}$ matrix is an SSCP matrix of the predictor variables and the constant related to the intercept. The sum of the squared equation residuals appears in the $\mathbf{\Psi}$ matrix. The parameter estimates are identical to the SAS PROC GLM results. The LISREL standard errors are a tiny bit larger than those of SAS (this result tends to hold across a wide range of structural models). We cannot use the equation R^2. Since we are using an SSCP matrix masquerading as a covariance matrix, the sums of squares used to calculate the R^2 are uncorrected (i.e., not centered).

Model 2

In the second model, we will use only the *y*-side of the LISREL model, and compute the regression coefficients in the $\mathbf{B}$ matrix. Note that in this model the intercept does not appear explicitly in the input matrix as in model 1. In this model we estimate the intercept in $\boldsymbol{\alpha}$, the vector of latent variable means.

Table 10.3. *LISREL results for the one-way ANOVA (model 1)*

LISREL estimates (maximum likelihood) given as:

Estimate
(standard error)
t-value

	d_1	d_2	*Int*
Γ			
γ	−6.000	4.000	36.000
	(4.940)	(4.940)	(3.493)
	−1.215	0.810	10.307
Φ			
d_1	10.000		
	(2.774)		
	3.606		
d_2	—	10.000	
	(1.961)	(2.774)	
	0.000	3.606	
Int	10.000	10.000	30.000
	(3.922)	(3.922)	(8.321)
	2.550	2.550	3.606

Ψ

γ
3172.000
(879.755)
3.606

Squared multiple correlations for structural equations

γ
0.923

Goodness-of-fit statistics
χ^2 with 0 df = 0.00 (p = 1.000)

If $\eta_1 = bio$, $\eta_2 = v_1$, and $\eta_3 = v_2$, then in terms of equation 10.5, the model is

$$\begin{bmatrix} \eta_1 \\ \eta_2 \\ \eta_3 \end{bmatrix} = \begin{bmatrix} \alpha_1 \\ \alpha_2 \\ \alpha_3 \end{bmatrix} + \begin{bmatrix} 0 & \beta_{12} & \beta_{13} \\ 0 & 0 & 0 \\ 0 & 0 & 0 \end{bmatrix} \begin{bmatrix} \eta_1 \\ \eta_2 \\ \eta_3 \end{bmatrix} + \begin{bmatrix} \zeta_1 \\ 0 \\ 0 \end{bmatrix},$$

or

$$\eta_1 = \alpha_1 + \beta_{12}\eta_2 + \beta_{13}\eta_3 + \zeta_1$$
$$\eta_2 = \alpha_2$$
$$\eta_3 = \alpha_3.$$

The input file for this model is

```
One-way ANOVA - Model 2
DA NI=3 NO=30 MA=CM
CM FU
*
126.8505747  -1.8390805   1.6091954
 -1.8390805   0.2298851  -0.1149425
  1.6091954  -0.1149425   0.2298851
ME FU
*
35.3333333 0.3333333 0.3333333
MO NY=3 NE=3 LY=ID TE=ZE AL=FR BE=FU,FI PS=SY,FI

PA BE
*
0 1 1
0 0 0
0 0 0
PA PS
*
1
0 1
0 1 1
PA AL
1 1 1
MA BE
*
0 .1 .1
0 0 0
```

```
0 0 0
MA PS
*
136
0 .10
0 0 .10
MA AL
*
36 .33 .33
OU ND=3
```

For this model the input is a covariance matrix along with a vector of means. Part of the output from this program appears in Table 10.4. The regression estimates appear in the **B** matrix, the intercept along with the means of the dummy code variables appear in the $\boldsymbol{\alpha}$ vector. $\boldsymbol{\Psi}$ now contains a covariance matrix of the predictors along with the variance of the regression residuals (ψ_{11}). The estimates resulting from model 2 are identical to those of model 1, except in this case each equation R^2 is correct.

A different parameterization for the ANOVA model

We could also estimate the intercept by adding the constant 1 to each case in the raw data file (a fourth "variable") and using a SSCP matrix as input. Using this strategy we would use LISREL's y-side model. Consider equation 10.5. This model again expresses regression relationships among latent variables. As before, we turn the equation into a regression among manifest variables by using the y-side measurement model, equation 10.3, to set each manifest variable equivalent to a latent variable.

We set $\boldsymbol{\alpha} = 0$ and equation 10.5 becomes

$$\eta = \mathbf{B}\eta + \zeta, \tag{10.10}$$

which will be used to estimate the regression coefficients.

In terms of equation 10.7 we have three variables, y, v_1, and v_2. We add a fourth "variable", the constant 1, that will be used to model the intercept resulting in

$$\begin{aligned} y_1 &= y \\ y_2 &= 1 \\ y_3 &= v_1 \\ y_4 &= v_2. \end{aligned} \tag{10.11}$$

Table 10.4. *LISREL results for the one-way ANOVA (Model 2)*

LISREL estimates (maximum likelihood) given as:

Estimate
(Standard error)
t-value

	Var 1	Var 2	Var 3
β			
Var 1	—	−6.000	4.000
		(4.677)	(4.677)
		−1.283	0.855
Var 2	—	—	—
Var 3	—	—	—
Covariance matrix of η			
Var 1	126.851		
Var 2	−1.839	0.230	
Var 3	1.609	−0.115	0.230
Ψ			
Var 1	109.379		
	(28.724)		
	3.808		
Var 2	—	0.230	
		(0.060)	
		3.808	
Var 3	—	−0.115	0.230
		(0.048)	(0.060)
		−2.408	3.808
Squared multiple correlations for structural equations			
	0.138	—	—
α			
	36.000	0.333	0.333
	(3.326)	(0.089)	(0.089)
	0.823	3.744	3.744
Goodness-of-fit statistics			
χ^2 with 0 df = 0.00 (p = 1.000)			

Remembering that $[\eta_1, \eta_2, \eta_3, \eta_4] = [y_1, y_2, y_3, y_4]$, the matrices represented by equation 10.9 become

$$\begin{bmatrix} \eta_1 \\ \eta_2 \\ \eta_3 \\ \eta_4 \end{bmatrix} = \begin{bmatrix} 0 & \beta_{12} & \beta_{13} & \beta_{14} \\ 0 & 0 & 0 & 0 \\ 0 & 0 & 0 & 0 \\ 0 & 0 & 0 & 0 \end{bmatrix} \begin{bmatrix} \eta_1 \\ \eta_2 \\ \eta_3 \\ \eta_4 \end{bmatrix} + \begin{bmatrix} \zeta_1 \\ \zeta_2 \\ \zeta_3 \\ \zeta_4 \end{bmatrix},$$

which is equivalent to the set of equations

$$\begin{aligned} \eta_1 &= \beta_{12}\eta_2 + \beta_{13}\eta_3 + \beta_{14}\eta_4 + \zeta_1 \\ \eta_2 &= \zeta_2 \\ \eta_3 &= \zeta_3 \\ \eta_4 &= \zeta_4. \end{aligned}$$

From equation 10.11 this becomes

$$\begin{aligned} y_1 &= \beta_0 y_2 + \beta_1 y_3 + \beta_2 y_4 + \zeta_1 = \beta_0 + \beta_1 v_1 + \beta_2 v_2 + \zeta_1 \\ y_2 &= 1 = \zeta_2 \\ y_3 &= v_1 = \zeta_3 \\ y_4 &= v_2 = \zeta_4. \end{aligned}$$

Although the input matrix is an SSCP matrix, we will tell LISREL that the input matrix is a covariance matrix; otherwise, LISREL would convert the matrix to a covariance matrix and the inclusion of a constant would cause the resulting covariance matrix to be nonpositive definite. The only drawback to this "trick" is that, as with model 1, the R^2 will be incorrectly calculated. To correctly calculate R^2, we must remove the constant and use the covariance matrix (not the SSCP matrix) as input. So, completing the analysis would require two computer runs.

The LISREL input for this parameterization is

```
One-way ANOVA - Model 3
DA NI=3 NO=30 MA=CM
CM FU
*
126.8505747   -1.8390805    1.6091954
 -1.8390805    0.2298851   -0.1149425
  1.6091954   -0.1149425    0.2298851
ME FU
*
35.3333333 0.3333333 0.3333333
```

```
MO NY=1 NX=2 NE=1 NK=2 LY=ID LX=ID TE=ZE TD=ZE AL=FR  C
       GA=FU,FI PS=DI PH=SY,FI TX=FR
PA GA
1 1
PA PS
1
PA AL
1
PA PH
1
1 1
PA TX
1 1
MA PS
120
MA PH
.2
-.01 .2
OU ND=3
```

The parameter estimates resulting from this file appear in Table 10.5. Once again, those estimates match the results of the previous two models.

Repeated measures ANOVA as SEM

In this section we describe how to model repeated measures ANOVA within the SEM framework. A simple and relatively straightforward way to place this repeated measures ANOVA within the SEM framework involves creating trend variables to model the within factors, and treating those trend variables as dependent variables in an ANOVA in which the between factors appear as the grouping factors. We model the between factors (e.g. by dummy coding) as described above.

A one-way repeated measures ANOVA

We first consider a one-way repeated measures ANOVA in which individuals have been measured repeatedly on three different occasions, yielding three levels of the repeated measures factor. With this analysis we wish to determine whether differences in the occasion means of the dependent variable exist. We transform the dependent variable into trend variables. For this transformation we are using orthogonal polynomial coefficients (Kirk,

Table 10.5. *LISREL results for the one-way ANOVA (model 3)*

LISREL estimates (maximum likelihood) given as:
Estimate
(Standard error)
t-value

(i) $\boldsymbol{\Gamma}$

	Var 2	Var 3
Var 1	−6.000	4.000
	(4.677)	(4.677)
	−1.283	0.855

(ii) Covariance matrix of $\boldsymbol{\eta}$ and $\boldsymbol{\xi}$

	Var 1	Var 2	Var 3
Var 1	126.851		
Var 2	−1.839	0.230	
Var 3	1.609	−0.115	0.230

(iii) $\boldsymbol{\Phi}$

	Var 2	Var 3
Var 2	0.230	
	(0.060)	
	3.808	
Var 3	−0.115	0.230
	(0.048)	(0.060)
	−2.408	3.808

(iv) $\boldsymbol{\Psi}$

Var 1
109.379
(28.724)
3.808

(v) Squared multiple correlations for structural equations

Var 1
0.138

Table 10.5. (*cont.*)

(vi) τ

	Var 2	Var 3
	0.333	0.333
	(0.089)	(0.089)
	3.744	3.744

(vii) α

	Var 1
	35.333
	(2.091)
	16.894

(viii) Goodness-of-fit statistics
χ^2 with 0 df = 0.0 ($p = 1.000$)

1982); though this is one of a number of options. We use the following set of coefficients:

Trend	$(c_1$	c_2	$c_3)$	/	$\sqrt{\sum c_j^2}$
Linear	(−1	0	1)	/	$\sqrt{2}$
Quadratic	(1	−2	1)	/	$\sqrt{6}$
Sum	(1	1	1)	/	$\sqrt{3}$,

to create a set of trend scores for each case as

$$y_{\text{trend}} = c_1 \times y_{\text{occ1}} + c_2 \times y_{\text{occ2}} + c_3 \times y_{\text{occ3}},$$

which would create the trend scores

$$\begin{aligned} y_{i_{\text{linear}}} &= -1 \times y_{i_{\text{occ1}}} + 0 \times y_{i_{\text{occ2}}} + 1 \times y_{i_{\text{occ3}}} \\ y_{i_{\text{quadratic}}} &= 1 \times y_{i_{\text{occ1}}} + -2 \times y_{i_{\text{occ2}}} + 1 \times y_{i_{\text{occ3}}} \\ y_{i_{\text{sum}}} &= 1 \times y_{i_{\text{occ1}}} + 1 \times y_{i_{\text{occ2}}} + 1 \times y_{i_{\text{occ3}}}. \end{aligned}$$

We can now consider regressing each of these variables onto a constant (i.e., 1). Take the linear trend score. By regressing the linear trend score onto a constant we have the regression equation

$$y_{i_{\text{linear}}} = \beta_0(1) + \epsilon_i. \tag{10.12}$$

We now invoke the general solution to the general linear model. For any regression equation given by

$$y_i = \beta_0(1) + \beta_1 x_{1i} + \beta_2 x_{2i} + \cdots + \beta_p x_{pi} + \epsilon_i. \tag{10.13}$$

If $\mathbf{X}$ represents the design matrix (the matrix of values of the independent variables including a column of 1s if the intercept (β_0) is to be included), and $\boldsymbol{Y}$ is a vector of the values of the dependent variable, the vector of regression weights $\boldsymbol{\beta}$ can be calculated as

$$\boldsymbol{\beta} = (\mathbf{X}^{\mathrm{T}}\mathbf{X})^{-1}\,(\mathbf{X}^{\mathrm{T}}\boldsymbol{Y}). \tag{10.14}$$

For equation 10.12, $\mathbf{X}$ is a column of 1s and $\boldsymbol{Y}$ is the vector of linear quadratic trend scores and equation 10.14 becomes

$$\boldsymbol{\beta}_0 = \left(\sum_{i=1}^{N} 1\right)^{-1} \sum_{i=1}^{N} y_{\text{linear}} = \frac{\sum_{i=1}^{N} y_{\text{linear}}}{N},$$

or the mean of the linear trend variable. We can see how to interpret the mean of the linear trend variable (along with other variables) if we take the mean of each of these trend variables. We see that:

$$\begin{aligned} \bar{y}_{\text{linear}} &= -1 \times \bar{y}_{\text{occ1}} + 0 \times \bar{y}_{\text{occ2}} + 1 \times \bar{y}_{\text{occ3}} \\ \bar{y}_{\text{quadratic}} &= 1 \times \bar{y}_{\text{occ1}} + -2 \times \bar{y}_{\text{occ2}} + 1 \times \bar{y}_{\text{occ3}} \\ \bar{y}_{\text{sum}} &= 1 \times \bar{y}_{\text{occ1}} + 1 \times \bar{y}_{\text{occ2}} + 1 \times \bar{y}_{\text{occ3}}. \end{aligned}$$

The first two equations represent a set of contrasts on the occasion means. The third equation represents (three times) the grand mean across all three occasions. We are now in a position to estimate the one-way repeated measures ANOVA model. In this model we regress each of the trend variables (including the sum) onto the constant. This results in three regression equations, which test respectively whether the average linear trend, the average quadratic trend, and the grand mean (i.e., the sum) are equal to zero. By using orthonormalized coefficients (dividing each coefficient, c_j, in a particular trend by $(\sum c_j^2)^{1/2}$), we could calculate the sums of squares for the respective linear and quadratic trend scores and then pool these sums of squares. We would then have the sums of squares for the omnibus test of whether any differences exist among the three means, and we could use these pooled sums of squares to construct the F-test for the overall test of mean differences. By creating the trend variables, we also have a set of contrasts that, depending on our hypothesis, can be used either as planned or follow-up contrasts.

To estimate this model in LISREL we use the same strategy as in our first example. This model differs from the previous model in two ways: (1) we now have a set of regression equations rather than a single regression equation and (2) our only "independent" variable is the constant, 1 (i.e., we are only estimating the intercept of the model). Using the second parameterization described above, these equations can be expressed in the LISREL model as

$$y = \eta$$

and

$$\eta = \mathbf{B}\eta + \zeta$$

in which the structural regression equation becomes

$$\begin{bmatrix} \eta_1 \\ \eta_2 \\ \eta_3 \\ \eta_4 \end{bmatrix} = \begin{bmatrix} 0 & 0 & 0 & \beta_{14} \\ 0 & 0 & 0 & \beta_{24} \\ 0 & 0 & 0 & \beta_{34} \\ 0 & 0 & 0 & 0 \end{bmatrix} \begin{bmatrix} \eta_1 \\ \eta_2 \\ \eta_3 \\ \eta_4 \end{bmatrix} + \begin{bmatrix} \zeta_1 \\ \zeta_2 \\ \zeta_3 \\ \zeta_4 \end{bmatrix},$$

where $\eta_1 = y_{\text{linear}}$, $\eta_2 = y_{\text{quadratic}}$, $\eta_3 = y_{\text{sum}}$, and η_4 is the constant, 1. The measurement model is again used to define each latent variable, η, as equivalent either to a manifest variable or to the constant. The regression weights β_{14}, β_{24}, and β_{34} estimate, respectively, the means of the linear and quadratic trend scores and (three times) the grand mean.

A two-way repeated measures ANOVA

Now consider a more complex design in which individuals who have been randomly assigned to one of three groups are measured repeatedly on four occasions yielding a 4 Time × 3 Group ANOVA in which we call Time a "within" factor and Group a "between" factor. We are interested in whether differences exist among the Time × Group interactions means and among the main effect means for Time and Group, respectively.

We again use orthogonal polynomial coefficients (Kirk, 1982) to create trend scores. For four occasions of measurement, this would entail creating linear, quadratic, and cubic trend scores along with a mean (or sum) score using the following set of coefficients:

Trend	c_1	c_2	c_3	c_4	$\sqrt{\sum c_j^2}$
Linear	−3	−1	1	3	$\sqrt{20}$
Quadratic	1	−1	−1	1	2
Cubic	−1	3	−3	1	$\sqrt{20}$
Sum	1	1	1	1	2

to create the new variables:

$$y_{\text{new}} = c_1 \times y_{\text{occ1}} + c_2 \times y_{\text{occ2}} + c_3 \times y_{\text{occ3}} + c_4 \times y_{\text{occ4}}.$$

The within effects are again contained in the new dependent variables. The between effect, Group, is dummy-coded as in the example above. The resultant model is equivalent to four regression equations in which each of the respective trend variables are regressed onto the vectors representing the Group factor. We also estimate the intercept in this model. The regression equations represented by this model are

$$\begin{aligned} y_{\text{linear}} &= \beta_0 + \beta_1 v_1 + \beta_2 v_2 + \zeta_1 \\ y_{\text{quadratic}} &= \beta_0 + \beta_1 v_1 + \beta_2 v_2 + \zeta_2 \\ y_{\text{cubic}} &= \beta_0 + \beta_1 v_1 + \beta_2 v_2 + \zeta_3 \\ y_{\text{sum}} &= \beta_0 + \beta_1 v_1 + \beta_2 v_2 + \zeta_4. \end{aligned} \tag{10.15}$$

In this set of equations the between effects, which in this case include the test of group differences pooled across time, are estimated in the equation for y_{sum}. In the equation for each of the trend variables, the intercept tests the main effect for that trend while the regression weights related to the dummy-coded variables test the Trend × Group interaction.

As described for the one-way repeated measures ANOVA, the sums of squares represented in the regression equations can be pooled to yield the omnibus sums of squares necessary to construct the standard F-tests.

In terms of the LISREL model we will again be using the measurement model to set each latent variable equivalent to a corresponding manifest variable. In terms of equation 10.10, $\eta_1 = y_{\text{linear}}$, $\eta_2 = y_{\text{quadratic}}$, $\eta_3 = y_{\text{cubic}}$, $\eta_4 = y_{\text{sum}}$, $\eta_5 =$ constant, $\eta_6 = v_1$, and $\eta_7 = v_2$. The structural equation becomes

$$\begin{bmatrix}\eta_1\\\eta_2\\\eta_3\\\eta_4\\\eta_5\\\eta_6\\\eta_7\end{bmatrix}=\begin{bmatrix}0&0&0&0&\beta_{15}&\beta_{16}&\beta_{17}\\0&0&0&0&\beta_{25}&\beta_{26}&\beta_{27}\\0&0&0&0&\beta_{35}&\beta_{36}&\beta_{37}\\0&0&0&0&\beta_{45}&\beta_{46}&\beta_{47}\\0&0&0&0&0&0&0\\0&0&0&0&0&0&0\\0&0&0&0&0&0&0\end{bmatrix}\begin{bmatrix}\eta_1\\\eta_2\\\eta_3\\\eta_4\\\eta_5\\\eta_6\\\eta_7\end{bmatrix}+\begin{bmatrix}\zeta_1\\\zeta_2\\\zeta_3\\\zeta_4\\\zeta_5\\\zeta_6\\\zeta_7\end{bmatrix},$$

where each row of regression weights (β_{ij}) represents the set of weights corresponding to one of the trend scores, with β_{i5} corresponding to the intercept for that equation, and k_{i6} and β_{i7} corresponding to the dummy-coded variables.

An example

As an example of this model, consider the following data set: 120 race horses have been measured repeatedly on four occasions on a variable called Adjustment (HMADJ1, HMADJ2, etc.). This variable represents the adjustment behavior of race horses after they have been sold to a new owner. These horses have been divided into three groups (40 each) on the basis of their overall level of performance before the sale (Group) resulting in a 4 Time × 3 Group repeated measures ANOVA design. The data are available from the authors on request. To provide a reference for the LISREL analysis, we first analyzed these data using SAS PROC GLM. A partial listing of the results appears in Table 10.6.

We analyzed these data first using the *repeated* option of PROC GLM. The omnibus test for differences in the between effect Group means appears first. Next come the test of differences among the within effects means for the effects, Time and Time × Group. Only the main effect for Time is statistically significant.

Next come the tests of the polynomial trend scores. TIME.1, .2, and, .3 are the respective linear, quadratic, and cubic trend scores. The test represented by "MEAN" is a test of whether the mean of that particular trend score is statistically different than 0. The test represented by "GROUP" is a test of whether there are group differences in that trend (i.e., the Group × Trend interaction). Only the test of the quadratic trend score shows a result significantly different from zero.

Table 10.6.

A. Analysis using the PROC GLM repeated *option*

Source	df	Type III SS	Mean square	F-value	Pr>F
Repeated measures analysis of variance tests of hypotheses for between-subjects effects					
Group	2	8.267	4.133	0.04	0.9645
Error	117	13373.900	114.307		
Repeated measures analysis of variance univariate tests of hypotheses for within-subject effects					
Time	3	1347.2000000	449.0666667	27.39	0.0001
Time × Group	6	86.5000000	14.4166667	0.88	0.5102
Error (Time)	351	5755.3000000	16.3968661		
Repeated measures analysis of variance analysis of variance of contrast variables					
Contrast variable: TIME.1					
Mean	1	42.6666667	42.6666667	1.76	0.1872
Group	2	15.3033333	7.6516667	0.32	0.7299
Error	117	2835.630000	24.2361538		
Contrast variable: TIME.2					
Mean	1	1280.53333	1280.53333	98.25	0.0001
Group	2	66.51667	33.25833	2.55	0.0823
Error	117	1524.95000	13.03376		
Contrast variable: TIME.3					
Mean	1	24.0000000	24.0000000	2.01	0.1586
Group	2	4.6800000	2.3400000	0.20	0.8220
Error	117	1394.720000	11.9206838		

SS, Sum of squares; TIME.N represents the n-th degree polynomial contrast for TIME.

B. Analysis using trend scores and effect coding

(i) Dependent variable: HMADJLIN

	Analysis of variance				
Source	df	Sum of squares	Mean square	F-value	Prob > F
Model	2	15.30333	7.65167	0.316	0.7299
Error	117	2835.63000	24.23615		
C total	119	2850.93333			
Root MSE	4.92302	R^2	0.0054		
Dep mean	−0.59628	Adj. R^2	−0.0116		
CV	−825.61603				

Table 10.6. (*cont.*)

		Parameter estimates			
Variable	df	Parameter estimate	Standard error	T for H_0: Parameter = 0	Prob > |T|
Int	1	−0.596285	0.44940844	−1.327	0.1872
W_1	1	0.4370	0.495662	0.63555952	0.780
W_2	1	−0.331683	0.63555952	−0.522	0.6027

(ii) Dependent variable: HMADJQUA

		Analysis of variance			
Source	df	Sum of squares	Mean square	F-value	Prob > F
Model	2	66.51667	33.25833	2.552	0.0823
Error	117	1524.95000	13.03376		
C total	119	1591.46667			

Root MSE	3.61023	R^2	=	0.0418
Dep. mean	3.26667	Adj. R^2	=	0.0254
CV	110.51725			

		Parameter estimates			
Variable	df	Parameter estimate	Standard error	T for H_0: Parameter = 0	Prob > |T|
Int	1	3.266667	0.32956740	9.912	0.0001
W_1	1	−0.691667	0.46607869	−1.484	0.1405
W_2	1	−0.341667	0.46607869	−0.733	0.4650

(iii) Dependent variable: HMADJCUB

		Analysis of variance			
Source	df	Sum of squares	Mean square	F-value	Prob > F
Model	2	4.68000	2.34000	0.196	0.8220
Error	117	1394.72000	11.92068		
C total	119	1399.40000			

Root MSE	3.45263	R^2	=	0.0033
Dep. mean	−0.44721	Adj. R^2	=	−0.0137
CV	−772.03250			

Table 10.6. (*cont.*)

		Parameter estimates			
Variable	df	Parameter estimate	Standard error	*T* for H_0: Parameter = 0	Prob > |*T*|
Int	1	−0.447214	0.31518095	−1.419	0.1586
W_1	1	−0.201246	0.44573317	−0.451	0.6525
W_2	1	0.268328	0.44573317	0.602	0.5483

(iv) Dependent variable: HMADJSUM

	Analysis of variance				
Source	df	Sum of squares	Mean square	*F*-value	Prob > *F*
Model	2	8.26667	4.13333	0.036	0.9645
Error	117	13373.90000	114.30684		
C Total	119	13382.16667			

Root MSE	10.69144	R^2	0.0006
Dep. mean	35.58333	Adj. R^2	−0.0165
CV	30.04619		

		Parameter estimates			
Variable	df	Parameter estimate	Standard error	*T* for H_0: Parameter = 0	Prob > |*T*|
Int.	1	35.583333	0.97599026	36.459	0.0001
W_1	1	0.366667	1.38025866	0.266	0.7910
W_2	1	−0.233333	1.38025866	−0.169	0.8660

C, centered; df, degrees of freedom; MSE, mean standard error; CV, coefficient of variation.

Modifying the strategy described above somewhat, we next effects coded the factor Group, as follows:

Group	w_1	w_2
1	1	0
2	0	1
3	−1	−1

and included the results regressing each of the trend variables and the sum variable onto the effects coded variables, w_1 and w_2. The results also appear in Table 10.6. By comparing the result with the *repeated* options test of trends, we can see that the analyses are the same. For example, consider the trend variable HMADJLIN. This variable is equivalent to TIME.1. Notice that the test for the R^2 when regressing HMADJLIN onto w_1 and w_2 is the same as the test of the Group effect for TIME.1. We can compare the test of regressing the HMADJSUM onto the w_1 and w_2 and see that test is equivalent to the between Group effect. Since we have effects coded, each intercept tests whether the mean of the trend variable is different from 0. We can compare this with the test of the "MEAN" of the corresponding TIME. variable and see that they are indeed the same.

In Table 10.7 we present the LISREL results for the same analysis. We used the following LISREL input file the linear, quadratic, and cubic trend variables:

```
Repeated measures Anova (within effects)
DA NI=6 NO=120 MA=CM
CM FU
*
  23.9574230   3.5388751  -1.6075630   -0.8198916   0.2217277
-0.0563715
   3.5388751  13.3736695   0.8737577    1.8893557  -0.5798319
-0.4621849
  -1.6075630   0.8737577  11.7596639   -0.8756367  -0.0450972
0.1127429
  -0.8198916   1.8893557  -0.8756367  112.4551821   0.1680672
-0.0336134
   0.2217277  -0.5798319  -0.0450972    0.1680672   0.6722689
0.3361345
  -0.0563715  -0.4621849   0.1127429   -0.0336134   0.3361345
0.6722689
ME FU
*
-0.5963 3.2667 -0.4472 35.5833 0 0
LA
HMADJLIN HMADJQUA HMADJCUB HMADJSUM W1 W2
SE
1 2 3 5 6 /
MO NY=5 NE=5 LY=ID TE=ZE AL=FR BE=FU,FI PS=SY,FI
```

Table 10.7. *LISREL results for the repeated measure ANOVA*

LISREL estimates (maximum likelihood) given as:

Estimate
(Standard error)
t-value

(i) Within effects

	HMADJLIN	HMADJQUA	HMADJCUB	W_1	W_2
$\boldsymbol{\beta}$					
HMADJLIN	—	—	—	0.496 (0.630) 0.787	−0.332 (0.630) −0.526
HMADJQUA	—	—	—	−0.692 (0.462) −1.497	−0.342 (0.462) −0.739
HMADJCUB	—	—	—	−0.201 (0.442) −0.455	0.268 (0.442) 0.607
W_1	—	—	—	—	—
W_2	—	—	—	—	—
$\boldsymbol{\Psi}$					
HMADJLIN	23.829 (3.089) 7.714				
HMADJQUA	— — —	12.815 (1.661) 7.714			
HMADJCUB	—	—	11.720 (1.519) 7.714		
W_1	—	—	—	0.672 (0.087) 7.714	
W_2	—	—	—	0.336 (0.069) 4.879	0.672 0.672 7.714
$\boldsymbol{\alpha}$					
	−0.596 (0.447) −1.333	3.267 (0.328) 9.955	−0.447 (0.314) −1.425	0.000 (0.075) 0.000	0.000 (0.075) 0.000

Goodness-of-fit statistics
χ^2 with 3 df = 7.455 (p = 0.0587)

Table 10.7. (*cont.*)

(ii) Between effects

	HMADJSUM	ν_1	ν_2
$\boldsymbol{\beta}$			
HMADJSUM	—	0.367	−0.233
		(1.369)	(1.369)
		0.268	−0.170
ν_1	—	—	—
ν_2	—	—	—
$\boldsymbol{\Psi}$			
HMADJSUM	112.386		
	(14.570)		
	7.714		
ν_1	—	0.672	
		(0.087)	
		7.714	
ν_2	—	0.336	0.672
		(0.069)	(0.087)
		4.879	7.714
$\boldsymbol{\alpha}$			
	35.583	0.000	0.000
	(0.972)	(0.075)	(0.075)
	36.615	0.000	0.000

Goodness-of-fit statistics
χ^2 with 0 df = 0.00 (p = 1.000)

```
PA BE
*
0 0 0 1 1
0 0 0 1 1
0 0 0 1 1
0 0 0 0 0
0 0 0 0 0
PA PS
*
1
0 1
0 0 1
0 0 0 1
0 0 0 1 1
```

```
PA AL
1 1 1 1 1
MA BE
*
0 0 0 .1 .1
0 0 0 .1 .1
0 0 0 .1 .1
0 0 0 0 0
0 0 0 0 0
MA PS
*
20
0 10
0 0 5
0 0 0 .67
0 0 0 .33 .67
MA AL
*
0 0 0 0 0
OU ND=3
```

Looking at Table 10.7, we see that the regression parameter estimates corresponding to w_1 and w_2 appear in the **B** matrix, while the intercept appears in the $\boldsymbol{\alpha}$ vector.

To test the Group effect, we used the dependent variable HMADJSUM. We used the following LISREL input file:

```
Repeated measures Anova (between effects)
DA NI=6 NO=120 MA=CM
CM FU
*
   23.9574230   3.5388751  -1.6075630   -0.8198916   0.2217277
-0.0563715
    3.5388751  13.3736695   0.8737577    1.8893557  -0.5798319
-0.4621849
   -1.6075630   0.8737577  11.7596639   -0.8756367  -0.0450972
0.1127429
   -0.8198916   1.8893557  -0.8756367  112.4551821   0.1680672
```

```
-0.0336134
    0.2217277 -0.5798319 -0.0450972  0.1680672 0.6722689
0.3361345
   -0.0563715 -0.4621849  0.1127429 -0.0336134 0.3361345
0.6722689
ME FU
*
-0.5963 3.2667 -0.4472 35.5833 0 0
LA
HMADJLIN HMADJQUA HMADJCUB HMADJSUM W1 W2
SE
4 5 6 /
MO NY=3 NE=3 LY=ID TE=ZE AL=FR BE=FU,FI PS=SY,FI
PA BE
*
0 1 1
0 0 0
0 0 0
PA PS
*
1
0 1
0 1 1
PA AL
1 1 1
MA BE
*
0 .1 .1
0 0 0
0 0 0
MA PS
*
50
 0 .67
 0 .33 .67
MA AL
*
0 0 0 0 0
OU ND=3
```

The results are included in Table 10.7. We see again that the results agree with the SAS PROC GLM results. The regression effects again appear in the **B** matrix. The α vector now contains the test of the grand mean.

Estimating a random coefficients model as SEM

The final model we will describe is the random coefficients model based on the GLMM as described by Laird & Ware (1982). As an example of this type of model, we will estimate the linear growth curve model.

Consider again animals who are measured on some variable of interest repeatedly on four occasions. In the previous analyses, we were primarily interested in determining whether differences existed in the occasion means of that variable. Although as part of the analysis we implicitly calculated animal trend scores, we were not interested in those trend scores per se. Had we been interested, we could have considered those trends scores as indicative of animal differences. Now suppose that we believe each animal's scores follow a straight line, and, furthermore, that we are interested in predicting the line for each animal. We could then imagine that each horse could be described by some best-fitting straight line through their data points. The observed data points would differ from these predicted values. We could imagine an animal's scatterplot in which the "errors" represent the distances between that predicted straight line and the observed values as if we had performed a regression on each animal. Each animal would then have a separate regression line (represented by its slope and intercept) and a separate set of "errors" around that line. Once we have a complete set of these slopes and intercepts, we could then imagine calculating the average slope and the average intercept for the group. The test of whether slope $= 0$ would represent a specific hypothesis on the group means.

In terms of the GLMM, we could model an animal's observed data as

$$y_i = \mathbf{X}_i\boldsymbol{\beta} + \mathbf{Z}_i\boldsymbol{\gamma}_i + \epsilon_i \tag{10.16}$$

where for the i-th animal, $\mathbf{X}_i$ is a design matrix for the fixed effects, $\boldsymbol{\beta}$ is a vector of the fixed effects, $\mathbf{Z}_i$ is an $n_i \times g$ design matrix for these random effects, and $\boldsymbol{\gamma}_i$ is a $g \times 1$ vector of random effects. For a model that assumes animals follow a straight line, one can think of this equation as breaking down an animal's observed values into (1) the group line (indicated by $\mathbf{X}_i\boldsymbol{\beta}$), (2) the degree that the animal's line differs from the group line (indicated by $\mathbf{Z}_i\boldsymbol{\gamma}_i$), and (3) an animal's deviations around their own line (indicated by ϵ_i). The γ_i are the deviations from the subgroup curve and are assumed to be independently distributed across animals with a distribution $\gamma_i \sim N(0, \sigma^2\mathbf{D})$ where $\mathbf{D}$ is an arbitrary "between" covariance matrix, σ is the standard deviation, and the ϵ_i are the errors representing the degree to which the animal deviates at each occasion from their own predicted line. The errors are assumed to have the distribution $\epsilon_i \sim N(0, \sigma^2\mathbf{W}_i)$ where $\mathbf{W}_i$ is an arbitrary "within" covariance matrix. For the models to be considered, each animal will be measured on the same occasions and the random effects design matrix will be the same for all animals so we drop the index for $\mathbf{Z}$ and n.

Equation 10.16 is the general linear mixed model proposed by Laird & Ware (1982) and based on the work of Harville (1974, 1976, 1977). The linear growth curve presented above represents one submodel of this very general useful model.

Consider again the data used in the 4 Time × 3 Group repeated measures ANOVA example. Each animal has been repeatedly measured on four occasions and is a member of one of three groups. For group 1, the design matrix and parameters related to the fixed effects can be written as

$$\mathbf{X}_i\beta = \begin{bmatrix} 1 & 1 & 0 & 0 & 0 & 0 \\ 1 & 2 & 0 & 0 & 0 & 0 \\ 1 & 3 & 0 & 0 & 0 & 0 \\ 1 & 4 & 0 & 0 & 0 & 0 \end{bmatrix} \begin{bmatrix} \beta_1 \\ \beta_2 \\ \beta_3 \\ \beta_4 \\ \beta_5 \\ \beta_6 \end{bmatrix} \tag{10.17}$$

and the parameters, β_1 and β_2 represent the respective slopes and intercepts for group 1. By multiplying out the above matrix we see that the predicted values for the group line are $\beta_1 + \beta_2$, $\beta_1 + 2\beta_2$, $\beta_1 + 3\beta_2$, and $\beta_1 + 4\beta_2$, yielding the group 1 straight line.

The design matrices for groups 2 and 3 are

$$\begin{bmatrix} 1 & 1 & 1 & 1 & 0 & 0 \\ 1 & 2 & 1 & 2 & 0 & 0 \\ 1 & 3 & 1 & 3 & 0 & 0 \\ 1 & 4 & 1 & 4 & 0 & 0 \end{bmatrix} \text{ and } \begin{bmatrix} 1 & 1 & 0 & 0 & 1 & 1 \\ 1 & 2 & 0 & 0 & 1 & 2 \\ 1 & 3 & 0 & 0 & 1 & 3 \\ 1 & 4 & 0 & 0 & 1 & 4 \end{bmatrix},$$

respectively. As a result, β_3 and β_4 are the respective differences in intercept and slope between groups 1 and 2 and β_5 and β_6 are the respective differences in intercept and slope between groups 1 and 3.

Each animal's data will be modeled as differing from the group line through the random effects, which can be written as

$$\mathbf{Z}\gamma_i + \epsilon_i = \begin{bmatrix} 1 & 1 \\ 1 & 2 \\ 1 & 3 \\ 1 & 4 \end{bmatrix} \begin{bmatrix} \gamma_{1i} \\ \gamma_{2i} \end{bmatrix} + \begin{bmatrix} \epsilon_{1i} \\ \epsilon_{2i} \\ \epsilon_{3i} \\ \epsilon_{4i} \end{bmatrix}, \tag{10.18}$$

where γ_{1i} and γ_{2i} represent the respective i-th animal's deviations from their group's intercept and slope, and the ϵ_i are the within-animal errors (the residuals around each animal's own regression line). The total covariance matrix for the i-th animal is

$$\sigma^2 \mathbf{V}_i = \sigma^2 (\mathbf{Z}\ \mathbf{D}\ \mathbf{Z}^{\mathrm{T}} + \mathbf{I}) \tag{10.19}$$

This is the covariance structure for the random coefficients model (Jennrich & Schlucter, 1986). For the linear growth curve model of interest we must estimate $\boldsymbol{\beta}$, σ^2, and $\mathbf{D}$. Once these values are known, we can estimate the values of the random effects.

Using SEM to estimate the linear curve model

Rovine and Molenaar (1998) have shown how to use the following method to estimate a number of different ANOVA-type models including the random coefficients model (Rovine & Molenaar, 2000) described here.

The SEM strategy to be used consists of placing the design matrices corresponding to both the fixed and random effects into $\mathbf{\Lambda}_y$, differentiating between the fixed and random effects in the $\mathbf{\Psi}$ matrix in which we estimate the covariance matrix of the random effects, and estimating variance of the "within" errors (the deviations around the individual lines) in the $\mathbf{\Theta}_\epsilon$ matrix.

We start with the y-side LISREL model in which the measurement model (equation 10.3) is

$$y = \boldsymbol{\tau}_y + \mathbf{\Lambda}_y \boldsymbol{\eta} + \boldsymbol{\epsilon}$$

and the structural regression equation (equation 10.5) is

$$\boldsymbol{\eta} = \boldsymbol{\alpha} + \mathbf{B}\boldsymbol{\eta} + \boldsymbol{\zeta}.$$

Most typically, the measurement model is used to perform confirmatory factor analysis or to indicate latent variables to be used in structural regressions. On the y-side the $\mathbf{\Lambda}_y$ matrix functions as the matrix of factor loadings and the $\mathbf{\Psi}$ matrix as the covariance matrix of the regression errors. In this application, however, we will use $\mathbf{\Lambda}$ and Ψ to define the fixed and random effects.

Considering the three group model, for each group we concatenate the fixed and random effects design matrices into a single matrix. In terms of the LISREL model, this is akin to treating each effect as a "factor". The "factor scores" related to the fixed effects are constant, so the variance of each fixed effect "factor" is zero. On the other hand, we estimate the covariance matrix of the "factors" associated with the random effect. Since $\mathbf{\Psi}$ is the covariance matrix of the factors, we partition $\mathbf{\Psi}$ into $\mathbf{\Psi}_{\mathrm{f}} = 0$, the covariance matrix of the fixed effects, $\mathbf{\Psi}_{\mathrm{r}}$, the covariance matrix of the random effects, and $\mathbf{\Psi}_{\mathrm{fr}} = 0$, the covariances between the fixed effect and random effect "factors".

In terms of the LISREL matrices we set $\mathbf{B} = 0$ and $\boldsymbol{\tau}_y = 0$ in equations 10.3 and 10.5. $\mathbf{\Psi}$ thus becomes the covariance matrix of $\boldsymbol{\zeta}$, which is now equivalent to the covariance matrix of $\boldsymbol{\eta}$. By partitioning $\mathbf{\Psi}$ as

$$\begin{bmatrix} \mathbf{\Psi}_{\mathrm{f}} & \mathbf{\Psi}_{\mathrm{fr}} \\ \mathbf{\Psi}_{\mathrm{fr}} & \mathbf{\Psi}_{\mathrm{r}} \end{bmatrix} = \begin{bmatrix} 0 & 0 \\ 0 & \mathbf{\Psi}_{\mathrm{r}} \end{bmatrix},$$

we define $\boldsymbol{\eta}$ as a constant vector (no variance) for the fixed effects. Equation 10.3 becomes

$$y = \mathbf{\Lambda}_y \boldsymbol{\eta} + \epsilon, \tag{10.20}$$

where $\mathbf{\Lambda}_y$ is a partitioned matrix $(\mathbf{\Lambda}_{\mathrm{f}} | \mathbf{\Lambda}_{\mathrm{r}})$ of the design matrices corresponding to the fixed and random effects. Since $\boldsymbol{y}$ is a vector of the observed values for the i-th subject on the repeatedly measured variables, $\mathbf{\Theta}_\epsilon$ is the covariance matrix of the errors for i-th subject.

The expectation of equation 10.5 becomes

$$\mathrm{E}(\boldsymbol{\eta}) = \boldsymbol{\alpha} \tag{10.21}$$

and the $\boldsymbol{\alpha}$ will contain the parameter estimates of the regression model.

In the example described above, we have three different groups each of which will have different values for the coding vectors. We will use the multiple group option in LISREL to apply the coding vectors for each group. Within each of the groups, the values of the coding vectors in the design matrix will be constant.

For group 1, the matrix specification corresponding to equation 10.11 is

$$\mathbf{y}_i = \mathbf{X}\beta + \mathbf{Z}\gamma_i + \epsilon_i$$

$$= \begin{bmatrix} 1 & 1 & 0 & 0 & 0 & 0 & 1 & 1 \\ 1 & 2 & 0 & 0 & 0 & 0 & 1 & 2 \\ 1 & 3 & 0 & 0 & 0 & 0 & 1 & 3 \\ 1 & 4 & 0 & 0 & 0 & 0 & 1 & 4 \end{bmatrix} \begin{bmatrix} \beta_1 \\ \beta_2 \\ \beta_3 \\ \beta_4 \\ \beta_5 \\ \beta_6 \\ \gamma_{1i} \\ \gamma_{2i} \end{bmatrix} + \begin{bmatrix} \epsilon_{1i} \\ \epsilon_{2i} \\ \epsilon_{3i} \\ \epsilon_{4i} \end{bmatrix}, \qquad (10.22)$$

in which the first six columns of the design matrix correspond to the fixed effects and the last two correspond to the random effects. Groups 2 and 3 replace $\mathbf{X}|\mathbf{Z}$ with the respective design matrices,

$$\begin{bmatrix} 1 & 1 & 1 & 1 & 0 & 0 & 1 & 1 \\ 1 & 2 & 1 & 2 & 0 & 0 & 1 & 2 \\ 1 & 3 & 1 & 3 & 0 & 0 & 1 & 3 \\ 1 & 4 & 1 & 4 & 0 & 0 & 1 & 4 \end{bmatrix} \text{ and } \begin{bmatrix} 1 & 1 & 0 & 0 & 1 & 1 & 1 & 1 \\ 1 & 2 & 0 & 0 & 1 & 2 & 1 & 2 \\ 1 & 3 & 0 & 0 & 1 & 3 & 1 & 3 \\ 1 & 4 & 0 & 0 & 1 & 4 & 1 & 4 \end{bmatrix}.$$

By adding the random effects to the design matrix, $\mathbf{\Lambda}_y$, we now distinguish between the fixed and random effects in the $\mathbf{\Psi}$ matrix (the covariance matrix of the η values) by partitioning the $\mathbf{\Psi}$ matrix according to the corresponding columns in $\mathbf{\Lambda}_y$ that represent the fixed and random effects. $\mathbf{\Psi}$ has the form

$$\begin{bmatrix} 0 & 0 & 0 & 0 & 0 & 0 & 0 & 0 \\ 0 & 0 & 0 & 0 & 0 & 0 & 0 & 0 \\ 0 & 0 & 0 & 0 & 0 & 0 & 0 & 0 \\ 0 & 0 & 0 & 0 & 0 & 0 & 0 & 0 \\ 0 & 0 & 0 & 0 & 0 & 0 & 0 & 0 \\ 0 & 0 & 0 & 0 & 0 & 0 & 0 & 0 \\ 0 & 0 & 0 & 0 & 0 & 0 & \psi_{55} & \psi_{56} \\ 0 & 0 & 0 & 0 & 0 & 0 & \psi_{65} & \psi_{66} \end{bmatrix} \qquad (10.23)$$

$$\mathbf{\Psi} = \begin{bmatrix} \mathbf{\Psi}_{\mathrm{f}} & \mathbf{\Psi}_{\mathrm{fr}} \\ \mathbf{\Psi}_{\mathrm{fr}} & \mathbf{\Psi}_{\mathrm{r}} \end{bmatrix} = \begin{bmatrix} \mathbf{0} & \mathbf{0} \\ \mathbf{0} & \mathbf{\Psi}_r \end{bmatrix}$$

In this matrix the first six rows and columns correspond to the fixed effects and the last two the random effects that have the covariance matrix.

To provide the correct error structure for the data, we assume: (1) that the covariance matrix of the γ_i, $\sigma^2\mathbf{D}$, is equal across all subjects; (2) that the within-subject errors are uncorrelated ($\mathbf{W}_i = \mathbf{I}$); and (3) that the within-subject error variances are all equal. These assumptions yield (in terms of the LISREL matrices) the total covariance matrix for the i-th subject:

$$\sigma^2\mathbf{V}_i = \mathbf{Z}_i\mathbf{\Psi}\mathbf{Z}_i^{\mathrm{T}} + \mathbf{\Theta}_\epsilon, \tag{10.24}$$

where $\mathbf{\Psi} = \sigma^2\mathbf{D}$ and $\mathbf{\Theta}_\epsilon = \sigma^2\mathbf{I}$.

Once we have estimated Ψ and $\mathbf{\Theta}_\epsilon$, we can determine $\mathbf{D}$ and σ^2. We are then able to estimate the individual random effects (Jones, 1993). The best linear unbiased prediction (BLUP) estimates (Robinson, 1991) of γ_i are given by

$$\hat{\gamma}_i = \mathbf{D}\mathbf{Z}_i^{\mathrm{T}}\mathbf{V}^{-1}(\mathbf{y}_i - \mathbf{X}_i\hat{\beta}), \tag{10.25}$$

where $\mathbf{V}_i$ is obtained from equation 10.24. Once these values are obtained, we can predict the $\mathbf{y}_i$ vector for each subject as

$$\hat{y}_i = \mathbf{X}_i\beta + \mathbf{Z}_i\gamma_i \tag{10.26}$$

Plotting the predicted values would yield the individual curves. The differences between these predicted values and the observed are the ϵ_i.

We include a SAS PROC IML (1988) routine to estimate the γ values and predict $\mathbf{y}_i$ in Appendix 10.1. An alternative would have been to estimate "factor scores". Since the model used to estimate the parameters is in the form of a factor model, we could have used a factor score estimation procedure to predict $\mathbf{y}_i$. The "factor scores" in η corresponding to the random effects parameters are, in this case, the random intercepts and slopes. An interactive FORTRAN program (Molenaar, 1996) called FSCORE that will determine individual factor scores based on the LISREL factor score regressions is available on request.

An example

Consider Table 10.8. In this table we present the covariance matrix and means vector for each of the three groups in the 4 Time × 3 Group repeated measures ANOVA example. Now, we are going to fit a linear curve random coefficients model to these data according to the parameterization described above.

Table 10.8. *Covariance matrices and means vectors for the random coefficients model*

Group 1 covariance matrix			
45.48653846	24.58397436	27.97692308	14.33461538
24.58397436	38.71730769	22.27948718	12.90897436
27.97692308	22.27948718	49.73333333	27.89230769
14.33461538	12.90897436	27.89230769	41.22820513
Group 1 means vector			
19.4750	16.2750	17.1000	19.0500
Group 2 covariance matrix			
29.34358974	32.52307692	21.97435897	17.40512821
32.52307692	48.21538462	30.47435897	21.44358974
21.97435897	30.47435897	40.06089744	28.99038462
17.40512821	21.44358974	28.99038462	52.92243590
Group 2 means vector			
19.8000	16.3000	16.1250	18.4750
Group 3 covariance matrix			
1.12820513	30.66666667	24.74358974	33.71794872
30.66666667	26.29743590	18.56410256	28.00000000
24.74358974	18.56410256	24.19230769	22.11538462
33.71794872	28.00000000	22.11538462	43.16666667
Group 3 means vector			
20.5000	15.4000	15.7500	19.2500

The LISREL input file that we used is:

```
lisrel random coefficients linear adjustment data
da ni=4 no=40 ma=cm ng=3
cm fu fi='c:\repeated\biomat2.dat'
me fi='c:\repeated\biomat2.dat'
mo ny=4 ne=8 ly=fu,fi ps=sy,fi te=di,fr al=fr ty=ze
pa ps
0
0 0
0 0 0
0 0 0 0
0 0 0 0 0
0 0 0 0 0 0
0 0 0 0 0 0 1
0 0 0 0 0 0 1 1
```

```
pa al
1 1 1 1 1 1 0 0
ma ly
1 1 0 0 0 0 1 1
1 2 0 0 0 0 1 2
1 3 0 0 0 0 1 3
1 4 0 0 0 0 1 4
ma ps
0
0 0
0 0 0
0 0 0 0
0 0 0 0 0
0 0 0 0 0 0
0 0 0 0 0 0   4
0 0 0 0 0 0 −.1 1
eq te 1 1 te 2 2 te 3 3 te 4 4
st .5 te(1,1)
ou ns ad=off rs nd=4 xm fs it=200
  group2
da no=40
cm fu fi='c:\repeated\biomat2.dat'
me fi='c:\repeated\biomat2.dat'
mo ly=fu,fi ps=in te=in al=in ty=ze
ma ly
1 1 1 1 0 0 1 1
1 2 1 2 0 0 1 2
1 3 1 3 0 0 1 3
1 4 1 4 0 0 1 4
ou
  group3
da no=40
cm fu fi='c:\repeated\biomat2.dat'
me fi='c:\repeated\biomat2.dat'
mo ly=fu,fi ps=in te=in al=in ty=ze
ma ly
1 1 0 0 1 1 1 1
1 2 0 0 1 2 1 2
1 3 0 0 1 3 1 3
1 4 0 0 1 4 1 4
ou
```

Table 10.9. *LISREL output for the linear curve model for the adjustment data*

LISREL estimates (maximum likelihood) given as:

Estimate
(Standard error)
t-value

(i) Ψ

	η_7	η_8
η_7	32.5975 (8.2301) 3.9608	
η_8	−3.2213 (2.1188) −1.5204	1.2053 (0.7176) 1.6796

(ii) Θ_ϵ

Var 1	Var 2	Var 3	Var 4
18.2094	18.2094	18.2094	18.2094
(1.6835)	(1.6835)	(1.6835)	(1.6835)
10.8167	10.8167	10.8167	10.8167

(iii) $\boldsymbol{\alpha}$

η_1	η_2	η_3	η_4	η_5	η_6
18.0875	−0.0450	0.6250	−0.3700	0.4875	−0.2950
(1.2394)	(0.3525)	(1.7528)	(0.4986)	(1.7528)	(0.4986)
14.5934	−0.1276	0.3566	−0.7421	0.2781	−0.5917

Goodness-of-fit statistics
χ^2 with 32 df = 165.0121 (p = 0.0)

A partial listing of the output is included in Table 10.9. We see that the fixed parameter estimates of the model appear in the $\boldsymbol{\alpha}$ vector. α_1 and α_2 represent the respective intercept and slope of the group 1 line. α_3 and α_4 represent the differences between the intercepts and slopes for groups 1 and 2. α_5 and α_6 represent the respective differences between groups 1 and 3. As we can see from the results, there is no Time effect for group 1(α_2)

and no differences in the slopes among the groups (α_4 and α_6) implying that there is no overall Time effect. The covariance matrix of the random effects (slopes and intercepts) appears in the $\mathbf{\Psi}$ matrix, and the residual variance appears in $\mathbf{\Theta}_\epsilon$. We also see that the fit of the model is poor, indicating that the linear curve model is not a particularly good model for these data.

For purposes of comparison we analyzed the same data using SAS PROC MIXED (1995). The results appear in Table 10.10. We can see that the fixed-term parameter estimates are identical. The covariance matrix coefficients and the residual variance differ slightly.

Looking at the pattern of means we might conclude that a quadratic curve model might be more appropriate for these data. We estimated this model using the following LISREL input file:

```
lisrel random coefficients quadratic adjustment data
da ni=4 no=40 ma=cm ng=3
cm fu fi='c:\repeated\biomat2.dat'
me fi='c:\repeated\biomat2.dat'
mo ny=4 ne=12 ly=fu,fi ps=sy,fi te=di,fr al=fr ty=ze
pa ps
0
0 0
0 0 0
0 0 0 0
0 0 0 0 0
0 0 0 0 0 0
0 0 0 0 0 0 0
0 0 0 0 0 0 0 0
0 0 0 0 0 0 0 0 0
0 0 0 0 0 0 0 0 0 1
0 0 0 0 0 0 0 0 0 1 1
0 0 0 0 0 0 0 0 0 1 1 1
pa al
1 1  1 1 1 1 1 1 1 0 0  0
ma ly
1 1  1 0 0 0 0 0 0 1 1  1
1 2  4 0 0 0 0 0 0 1 2  4
1 3  9 0 0 0 0 0 0 1 3  9
1 4 16 0 0 0 0 0 0 1 4 16
ma ps
```

Table 10.10. *SAS PROC MIXED output for the linear curve random coefficients model*

(i) Covariance parameter estimates (maximum likelihood)

Cov Parm	Subject	Estimate
UN (1,1)	ID	31.56760417
UN (2,1)	ID	−3.06912500
UN (2,2)	ID	1.14655000
Residual		17.89750000

(ii) Solution for fixed effects

Effect	Group	Estimate	Std error	df	T	Pr > |T|
Int		18.08750000	1.20844791	117	14.97	0.0001
Group	1	0.48750000	1.70900343	240	0.29	0.7757
Group	2	0.62500000	1.70900343	240	0.37	0.7149
Group	3	0.00000000	—	—	—	—
Time		−0.04500000	0.34373136	117	−0.13	0.8961
Time × Group	1	−0.29500000	0.48610956	240	−0.61	0.5445
Time × Group	2	−0.37000000	0.48610956	240	−0.76	0.4473
Time × Group	3	0.00000000	—	—	—	—

(iii) Tests of fixed effects

Source	ndf	ddf Type III	F	Pr > F
Group	2	240	0.07	0.9288
Time	1	117	1.81	0.1816
Time × Group	2	240	0.32	0.7237

Cov Parm, covariance parameter; ndf, numerator degrees of freedom; ddf, denominator degrees of freedom.

0
0 0
0 0 0
0 0 0 0
0 0 0 0 0
0 0 0 0 0 0
0 0 0 0 0 0 0

```
0 0 0 0 0 0 0 0
0 0 0 0 0 0 0 0 0
0 0 0 0 0 0 0 0 0   4
0 0 0 0 0 0 0 0 0 -.1   1
0 0 0 0 0 0 0 0 0 -.1 -.1 1
eq te 1 1 te 2 2 te 3 3 te 4 4
st .5 te(1,1)
ou ns ad=off rs nd=4 xm fs it=200
 group2
da no=40
cm fu fi='c:\repeated\biomat2.dat'
me fi='c:\repeated\biomat2.dat'
mo ly=fu,fi ps=in te=in al=in ty=ze
ma ly
1 1 1  1 1 1  0 0 0 1 1 1
1 2 4  1 2 4  0 0 0 1 2 4
1 3 9  1 3 9  0 0 0 1 3 9
1 4 16 1 4 16 0 0 0 1 4 16
ou
 group3
da no=40
cm fu fi='c:\repeated\biomat2.dat'
me fi='c:\repeated\biomat2.dat'
mo ly=fu,fi ps=in te=in al=in ty=ze
ma ly
1 1 1  0 0 0 1 1 1  1 1 1
1 2 4  0 0 0 1 2 4  1 2 4
1 3 9  0 0 0 1 3 9  1 3 9
1 4 16 0 0 0 1 4 16 1 4 16
ou
```

A partial listing of the output appears in Table 10.11. We see now that the fixed effects include intercept, linear, and quadratic parameters for group 1 (α_1, α_2 and α_3), the differences between groups 1 and 2 (α_4, α_5, and α_6), and differences between groups 1 and 3 (α_7, α_8, and α_9). We now see the Time effect for group 1 (α_2 and α_3). We can also see that there are no Time differences between groups 1 and 2 (α_5 and α_6), but there are differences between groups 1 and 3 (α_8 and α_9). The fit of the model has improved considerably, but the χ^2 would still indicate a relatively poor fit.

Table 10.11. *LISREL output for the quadratic curve model for the adjustment data*

LISREL estimates (maximum likelihood) given as:

Estimate
(Standard error)
t-value

(i) $\boldsymbol{\Psi}$

	η_{10}	η_{11}	η_{12}
η_{10}	31.3221		
	(20.5255)		
	1.5260		
η_{11}	0.3741	−0.4751	
	(15.9960)	(14.4578)	
	0.0234	−0.0329	
η_{12}	−0.4887	−0.2573	0.2185
	(3.0489)	(2.8570)	(0.5826)
	−0.1603	−0.0900	0.3751

(ii) $\boldsymbol{\Theta}_\epsilon$

Var 1	Var 2	Var 3	Var 4
12.1597	12.1597	12.1597	12.1597
(1.5898)	(1.5898)	(1.5898)	(1.5898)
7.6485	7.6485	7.6485	7.6485

(iii) $\boldsymbol{\alpha}$

η_1	η_2	η_3	η_4	η_5	η_6
24.5250	−6.4825	1.2875	1.5000	−1.2450	0.1750
(1.7943)	(1.4138)	(0.2890)	(2.5375)	(1.9994)	(0.4088)
13.6684	−4.5851	4.4543	0.5911	−0.6227	0.4281

(iv) $\boldsymbol{\alpha}$

η_7	η_8	η_9	η_{10}	η_{11}	η_{12}
4.8000	−4.6075	0.8625	—	—	—
(2.5375)	(1.9994)	(0.4088)			
1.8916	−2.3044	2.1099			

Goodness-of-fit statistics
χ^2 with 26 df = 72.9665 (p = 0.24309857D-05)

Conclusion

In this chapter we have presented SEM strategies that can be used to estimate different ANOVA models. With these strategies, the data analyst can reap the benefits of structural equation modeling when estimating these models. Using SEM, one can test the overall fit of the model. One can easily test certain parameters of the model in ways that may not be available using standard statistical packages. ANOVA models can be extended to include fallible covariates and other characteristics made available by the inclusion of measurement models. In general, the SEM strategy allows the data analyst to think of ANOVA under the more general framework of model fitting and model testing. For certain research questions, this can turn out to be a significant advantage.

Appendix 10.1

SAS PROC IML for estimating the random effects
[Note that the data are read in as single vectors as follows:

```
Group1
  Subject 1 vector
  Subject 2 vector
  :
  Subject 40 vector
Group 2
  Subject 1 vector
  Subject 2 vector
  :
  Subject 40 vector
Group 3
  Subject 1 vector
  Subject 2 vector
  :
  Subject 40 vector]
```

```
OPTIONS LINESIZE=72;
CMS FILEDEF SURVEY DISK;
DATA DATAONE;
INFILE 'C:\ADJUST.VEC';
INPUT Y;
```

```
PROC IML;
START;
USE DATAONE;
READ ALL INTO Y;
DO I=1 to 120;
ZI={1 1,
    1 2,
    1 3,
    1 4};
A={1 2 3 4};
ADD=4 (I-1);
ADDTO=REPEAT(ADD,1,4);
INDEX=A+ADDTO;
YI=Y[INDEX];
IF I <= 40 THEN
  XI={1  1  0  0  0  0,
      1  2  0  0  0  0,
      1  3  0  0  0  0,
      1  4  0  0  0  0};
IF I > 40 AND I<=80 THEN
  XI={1  1  1  1  0  0,
      1  2  1  2  0  0,
      1  3  1  3  0  0,
      1  4  1  4  0  0};
IF I > 80 THEN
  XI={1  1  0  0  1  1,
      1  2  0  0  1  2,
      1  3  0  0  1  3,
      1  4  0  0  1  4};
  BETAEST={18.0875, -0.0450, 0.6250, -0.3700, 0.4875,
   -0.2950};
B={32.5975 -3.2213,
   -3.2213  1.2053 };   * from lisrel output;
WI={18.2094 0 0 0,
     0 18.2094 0 0,
     0 0 18.2094 0,
     0 0 0 18.2094};
  VI=ZI*B*ZI'+WI;
  YHAT=XI*BETAEST;
  YMINYHAT=YI-XI*BETAEST;
  GAMMA=B*ZI'*INV(VI)*YMINYHAT;
```

```
YXZHAT=YHAT+(ZI*GAMMA);
EI=INV(VI)*YMINYHAT;
*PRINT YI;
*PRINT VI;
*PRINT YHAT;
*PRINT YMINYHAT;
*PRINT EI;
PRINT I GAMMA YXZHAT;
   END;
FINISH;
RUN;
```

References

Harville, D. A. (1974). Bayesian inference for variance components using only error contrasts. *Biometrika*, **61**, 383–385.

Harville, D. A. (1976). Extensions of the Gauss–Markov theorem to include the estimation of random effects. *Annals of Statistics*, **4**, 384–395.

Harville, D. A. (1977). Maximum likelihood approaches to variance component estimation and to related problems. *Journal of the American Statistical Association*, **72**, 320–340.

Hertzog, C. & Rovine, M. J. (1985). Repeated-measures analysis of variance in developmental research: selected issues. *Child Development*, **56**, 787–809.

Huitema, B. H. (1980). *The Analysis of Covariance and Alternatives*. New York: Wiley.

Jennrich, R. I. & Schlucter, M. D. (1986). Unbalanced repeated measures models with structured covariance matrices. *Biometrics*, **42**, 805-820.

Jöreskog, K. & Sörbom, D. (1993). *LISREL User's Reference Guide*. Chicago: Scientific Software, Inc.

Jones, R. H. (1993). *Longitudinal Data with Serial Correlation: A State–Space Approach*. London: Chapman & Hall.

Kirk, R. E. (1982). *Experimental Design*. Belmont, CA: Brooks/Cole.

Laird, N. M. & Ware, J. H. (1982). Random effects models for longitudinal data. *Biometrics*, **38**, 963-974.

Long, J. S. (1983). *Confirmatory Factor Analysis*. Thousand Oaks, CA: Sage.

Molenaar, P. C. M. (1996). *FSCORE: A FORTRAN Program for Factor Scores Estimation in LISREL Models*. Technical report series #96–6. The Methodology Center. Philadelphia: Pennsylvania State University.

Pedhazur, E. J. (1997). *Multiple Regression in Behavioral Research*. Fort Worth, TX: Harcourt Brace.

Robinson, G. K. (1991). That BLUP is a good thing: the estimation of random effects. *Statistical Science*, **6**, 15–51.

Rovine, M. J. & Molenaar, P. C. M. (1998). A LISREL model for the analysis of repeated measures with a patterned covariance matrix. *Structural Equation Modeling*, **5**, 318–343.

Rovine, M. J. & Molenaar, P. C. M. (2000). A structural modeling approach to the random coefficients model. *Multivariate Behavioral Research*, **35**, 51–88.

SAS PROC IML (1988). *User's Guide*. Cary, NC: SAS Institute, Inc.

SAS PROC MIXED (1995). *Introduction to the Mixed Procedure*. Cary, NC: SAS Institute, Inc.

SAS PROC REG (1989). *SAS/STAT User's Guide, Version 6*. Cary, NC: SAS Institute.

11 Comparing groups using structural equations

James B. Grace

Abstract

This chapter presents methods for comparing structural equation models among groups. It begins by giving a general definition of multigroup analysis and by describing the objectives and utility of this type of analysis. The chapter then introduces an hypothetical example for multigroup analysis that involves the effects of fire on the relationship between soil properties and plant production in ecological communities. An overall description of the steps in a multigroup analysis is presented, which includes the initial assessment of baseline models for the separate groups, followed by tests for invariance in parameters among groups. These procedures are then illustrated using the data from the hypothetical example. Finally, a discussion of the possible uses of multigroup analyses for ecological problems is presented. It is concluded that multigroup analyses using structural equations has a wide range of applications for ecological and evolutionary studies, including both experimental and descriptive data.

Introduction

As shown in other chapters in this book, as well as in numerous references, structural equations can provide both a flexible and powerful method for analyzing multivariate relationships (Hayduk, 1987; Bollen, 1989; Hoyle, 1995; Schumacker & Lomax, 1996). Modern methods permit analyses that deal explicitly with categorical as well as continuous variables, greatly increasing the range of questions that can be addressed. This chapter considers one particular type of categorical analysis, comparisons among groups (also known as multigroup analyses).

While structural equations can be used in a great variety of ways, the standard full model described by Keesling (1972), Jöreskog (1973), and Wiley (1973) consists of a structural model that comprises the relationships among latent variables and a measurement model that comprises the relationships between latent and indicator variables. Using standard LISREL notation (Bollen, 1989), the structural relations among latent variables can be described as

$$\eta = \mathbf{B}\eta + \Gamma\xi + \zeta \tag{11.1}$$

where $\boldsymbol{\eta}$ is a vector of latent endogenous variables, $\boldsymbol{\xi}$ is a vector of latent exogenous variables, $\boldsymbol{\zeta}$ is a vector of latent errors for $\boldsymbol{\eta}$, $\mathbf{B}$ is a matrix of coefficients specifying the effects of endogenous latent variables, and $\boldsymbol{\Gamma}$ is a matrix of coefficients specifying the effects of exogenous latent variables. Two additional matrices that specify covariance structure among latent variables in the structural model (equation 11.1) are $\boldsymbol{\Phi}$, which specifies the correlations among errors of exogenous latent variables, and $\boldsymbol{\Psi}$, which specifies the correlations among errors of endogenous latent variables. The LISREL notation for the measurement model is

$$x = \Lambda_x\xi + \delta \tag{11.2}$$

and

$$y = \Lambda_y\eta + \varepsilon, \tag{11.3}$$

where $\boldsymbol{x}$ represents a vector of indicator variables for the exogenous latent variables, $\boldsymbol{y}$ is a vector of indicator variables for the endogenous latent variables, $\boldsymbol{\Lambda}_x$ is a matrix of coefficients relating $\boldsymbol{x}$ to $\boldsymbol{\xi}$, $\boldsymbol{\Lambda}_y$ is a matrix of coefficients relating $\boldsymbol{y}$ to $\boldsymbol{\eta}$, $\boldsymbol{\delta}$ is a vector of measurement error values for $\boldsymbol{x}$, and $\boldsymbol{\varepsilon}$ is a vector of measurement error values for $\boldsymbol{y}$. Two additional matrices that specify covariances among indicator variables in the measurement model are $\boldsymbol{\Theta}_\delta$, the correlations among errors for x variables, and $\boldsymbol{\Theta}_\varepsilon$, the correlations among errors for y variables.

Within the LISREL nomenclatural system, eight matrices specify the coefficients occurring in both the structural and measurement models, $\mathbf{B}$, $\boldsymbol{\Gamma}$, $\boldsymbol{\Phi}$, $\boldsymbol{\Psi}$, $\boldsymbol{\Lambda}_x$, $\boldsymbol{\Lambda}_y$, $\boldsymbol{\Theta}_\delta$, and $\boldsymbol{\Theta}_\varepsilon$. The goal of multigroup analysis is to determine which if any of the coefficients in the structural and measurement models are invariant across two or more groups. For a comparison between two groups, this translates into an evaluation of eight hypotheses,

$$\mathbf{B}^{(1)} = \mathbf{B}^{(2)}, \tag{11.4}$$

$$\boldsymbol{\Gamma}^{(1)} = \boldsymbol{\Gamma}^{(2)}, \tag{11.5}$$

$$\boldsymbol{\Phi}^{(1)} = \boldsymbol{\Phi}^{(2)}, \tag{11.6}$$

$$\boldsymbol{\Psi}^{(1)} = \boldsymbol{\Psi}^{(2)}, \tag{11.7}$$

$$\boldsymbol{\Lambda}_x^{(1)} = \boldsymbol{\Lambda}_x^{(2)}, \tag{11.8}$$

$$\mathbf{\Lambda}_y^{(1)} = \mathbf{\Lambda}_y^{(2)}, \tag{11.9}$$

$$\mathbf{\Theta}_\delta^{(1)} = \mathbf{\Theta}_\delta^{(2)}, \tag{11.10}$$

$$\mathbf{\Theta}_\varepsilon^{(1)} = \mathbf{\Theta}_\varepsilon^{(2)}. \tag{11.11}$$

This chapter considers methods for assessing these hypotheses and discusses the applicability of these methods for ecological and evolutionary studies.

Illustration of multigroup analysis

Groups to be compared using multigroup analysis can be anything from different samples to different sexes of animals or plants to different experimental treatments. A later section of this chapter will consider possible applications of multigroup analyses in ecological studies in more detail to illustrate the utility of this type of analysis. The question that drives multigroup analysis is whether two or more groups might differ in terms of the relationships among parameters. In this section, I first describe an hypothetical example and then consider the steps involved in testing for invariance among groups.

Example data

The hypothetical example presented in this chapter involves the relationships between soil conditions and plant production in unburned and burned grasslands. Figure 11.1 shows a general model of the relationships between soil conditions and plant production that serves as an *a priori* model for evaluation. This example deals with a question commonly addressed in ecological studies, "How do soil conditions affect the productivity of plant communities?". In this example it is proposed that the complex relationships among the many soil properties commonly measured can be explained by the existence of three latent soil factors.

In most cases, ecological studies of soil properties and plant communities have dealt with the complex relationships among soil parameters by selecting the best one or two predictor variables and discarding the others. Such an approach not only fails to utilize the full information available to the investigator but also fails to appreciate the theoretical implication that the correlations among soil parameters can be understood in terms of a few general soil properties. A few studies have utilized principal components analysis to reduce the number of soil parameters to a few general factors (e.g., Auclair *et al.*, 1976). For exploratory studies and those primarily interested in prediction, this approach has significant merit. What is unrecognized by

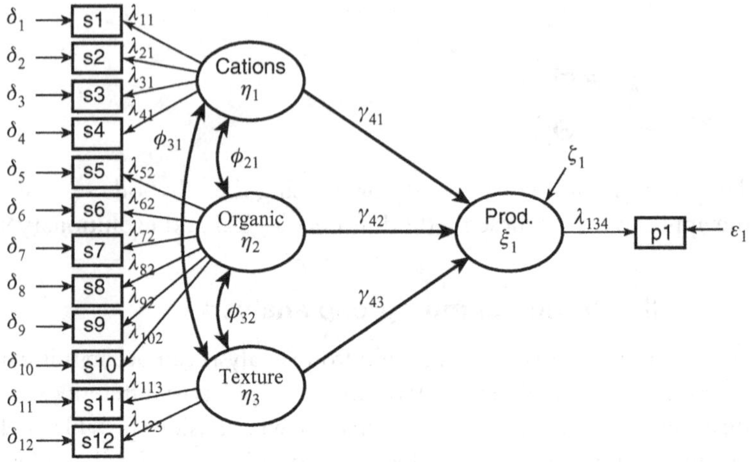

Figure 11.1. Basic structure of the common model being evaluated in this chapter. Latent variables are shown in ellipses. The three exogenous latent variables are soil cations (η_1), soil organic matter (η_2), and soil texture (η_3). The one endogenous latent variable is plant productivity (Prod.) (ξ_1), which is estimated from a single indicator, p1. Typically, the loading from ξ_1 to p1 (λ_{134}) would be fixed at some value such as 1.0, in which case the error term (ε_1) would be 0. Indicator soil variables, s1–s12, are related to latent variables by loadings (λ_{ij}) and have estimated error values (δ_i). Correlations among exogenous latent variables are indicated by φ values and dirctional path coefficients by γ values. Finally, the error estimate for the endogenous latent variable (also known as the disturbance term) is indicated by ζ_1.

many, however, is the superiority of factor analysis-type approaches (such as are employed in structural equation analysis) when the goal of a study is to develop a theoretically based understanding of general latent factors (Hair *et al.*, 1995). In the example presented in Figure 11.1, it is hypothesized that the effects of soil parameters on plant production can be explained in terms of three general properties of soils, e.g., total cation concentration, organic content, and texture.

Data in Appendix 11.1 represent a hypothetical situation where 200 samples of soil and plant biomass have been collected in an experimental research park from each of two grasslands, one that has been unburned for the preceding 10 years and another that has been burned annually during that same period. For this example, it is assumed that the samples were collected randomly and represent independent samples for each of the two grasslands. Here we have two groups, an unburned and a burned grassland. It is not possible in this study to draw general conclusions about burning because there is no replication at that level. Rather, the objective of this analysis is to determine whether the relationships between soil conditions and plant production are the same in the two grasslands.

In this hypothetical example, data are presented for 12 commonly measured soil variables and the difference between seasonal maximum and minimum standing crops (as a measure of plant production). Thus data for 13 measured parameters are presented (Appendix 11.1). The multigroup analysis (presented below) compares all estimated parameters across groups. As a result, there are two main types of question being addressed in this multigroup analysis. (1) Do the three latent soil variables reflect the same pattern of relationships among soil parameters in the two grasslands? (2) Do the two grasslands show the same relationships between latent soil variables and productivity? These questions reflect evaluations of the measurement model and the structural relations among latent variables, respectively. Evaluation of the equality of variances of the latent variables across groups is also possible, but in this example variances were set to a value of 1 for simplicity.

An overview of multigroup analysis

Historically, three approaches to multigroup analyses have been proposed. Jöreskog (1971) originally proposed the approach of first testing for complete equality among models and then progressively relaxing equality constraints until some measure of model fit, such as chi-square (χ^2) difference tests, indicated that further relaxation of constraints failed to improve fit of data to the models. This approach was found sometimes to result in inconsistent results, depending on the order used when relaxing equality constraints (Byrne, 1994) and is no longer recommended as a valid approach.

A second and commonly used strategy for multigroup analysis is one of progressively adding equality constraints. Bollen (1989, pp. 355–365) suggested the following as one possible progression:

$$\begin{array}{ll} H_{\text{form}}: & \text{same form} \\ H_{\lambda}: & \mathbf{\Lambda}^{(1)} = \mathbf{\Lambda}^{(2)} \\ H_{\lambda\Theta}: & \mathbf{\Lambda}^{(1)} = \mathbf{\Lambda}^{(2)},\ \mathbf{\Theta}^{(1)} = \mathbf{\Theta}^{(2)} \\ H_{\lambda\Theta\mathrm{B}\Gamma}: & \mathbf{\Lambda}^{(1)} = \mathbf{\Lambda}^{(2)},\ \mathbf{\Theta}^{(1)} = \mathbf{\Theta}^{(2)},\ \mathbf{B}^{(1)} = \mathbf{B}^{(2)},\ \mathbf{\Gamma}^{(1)} = \mathbf{\Gamma}^{(2)} \\ H_{\lambda\Theta\mathrm{B}\Gamma\Psi}: & \mathbf{\Lambda}^{(1)} = \mathbf{\Lambda}^{(2)},\ \mathbf{\Theta}^{(1)} = \mathbf{\Theta}^{(2)},\ \mathbf{B}^{(1)} = \mathbf{B}^{(2)},\ \mathbf{\Gamma}^{(1)} = \mathbf{\Gamma}^{(2)}, \\ & \mathbf{\Psi}^{(1)} = \mathbf{\Psi}^{(2)} \\ H_{\lambda\Theta\mathrm{B}\Gamma\Psi\Phi}: & \mathbf{\Lambda}^{(1)} = \mathbf{\Lambda}^{(2)},\ \mathbf{\Theta}^{(1)} = \mathbf{\Theta}^{(2)},\ \mathbf{B}^{(1)} = \mathbf{B}^{(2)},\ \mathbf{\Gamma}^{(1)} = \mathbf{\Gamma}^{(2)}, \\ & \mathbf{\Psi}^{(1)} = \mathbf{\Psi}^{(2)},\ \mathbf{\Phi}^{(1)} = \mathbf{\Phi}^{(2)}. \end{array}$$

Here, $\mathbf{\Lambda}$ implies both $\mathbf{\Lambda}_x$ and $\mathbf{\Lambda}_y$ and $\mathbf{\Theta}$ implies both $\mathbf{\Theta}_\delta$ and $\mathbf{\Theta}_\varepsilon$. This type of analysis begins with a multigroup test in which the data are compared with models for each group that are similar only in that they have a common form (i.e., the same number and causal order of variables). This initial analysis, which is recommended in all cases of multigroup analysis, seeks to ensure that the basic structure of the models is consistent with the data without any constraints on parameters to be equal across groups. If this analysis fails, typically it suggests that the groups are different at a fundamental level and subsequent assessments of parameter equalities are not recommended. When groups are found to fit models of the same form, subsequent analyses are used to determine which parameters are equal among groups and which ones differ. As Bollen (1989) pointed out, some flexibility exists for the order in which equality constraints are added. When primary interest is in the total covariance structure (i.e., the degree to which the specified latent variables explain the covariances among indicator variables), the above specified sequence is preferable. Here parameters of the measurement model are considered first, followed by consideration of the parameters of the structural model. Correlations among errors and latent variable variances are often of less interest, and are examined last when they are evaluated. In cases where the primary interest is in the structural model, the order of evaluation may be shifted. Model evaluation typically employs single-degree-of-freedom χ^2 tests as well as the Akaike information criterion (AIC) and other comparative fit indices that are adjusted for parsimony (Jöreskog & Sörbom, 1996).

A third approach to multigroup analysis employs Lagrange multipliers (Lee & Bentler, 1980) and is implemented in the EQS software package (Bentler & Wu, 1996; see also Chapter 14). This approach differs from the above-described methods because the Lagrange multiplier (LM) approach tests the validity of equality constraints multivariately instead of univariately. Because of its multivariate nature, it is unnecessary to evaluate a series of increasingly restrictive models in order to identify the best model. Rather, the LM approach tests all equality constraints simultaneously, which is both statistically superior and operationally preferable. In an analysis with r equality constraints to be considered, an $r \times 1$ vector of Lagrange multipliers is generated and the LM test generates multivariate test results when all constraints are considered simultaneously (Bentler, 1996, pp. 126–128). The LM test also generates univariate tests but these are not recommended in most circumstances. Because of the advantages associated with the LM approach, this is the method demonstrated in this chapter.

Illustration of a multigroup analysis

The simulation feature of the EQS program (Bentler & Wu, 1996) was used to generate data representing unburned and burned grasslands. The covariance matrices produced from these data as well as the program code for the analysis are presented in Appendix 11.1. As indicated above, the first step was to perform separate analyses on each group to ensure that the models fit the basic structure shown in Figure 11.1. The data for both groups fit the basic model well. For the unburned grassland, the resultant χ^2 value was 62.7 with 61 degrees of freedom (df) ($p = 0.414$) and a Bentler–Bonett nonnormed fit of 0.996. For the burned grassland, the resultant χ^2 was 63.3 with 62 df ($p = 0.430$) and a Bentler–Bonett nonnormed fit of 0.997. Both the low χ^2 and high fit values indicate a good concordance between data and models.

To perform a multigroup analysis using EQS, program statements for the two data sets were "stacked" by placing one above the other, specifying that there were two groups in the analysis, and specifying a number of constraints across groups (Appendix 11.1). Using this procedure, the EQS program first provides an assessment of the degree to which both groups collectively fit the common model and then performs an LM test of the potential for improving fit by relaxing equality constraints. For the initial multigroup analysis, all parameters were constrained to be equal across groups. The fit of the data to a common model was indicated to be reasonably good by a χ^2 of 154.8 with 138 df ($p = 0.156$). However, the results of the LM analysis (Table 11.1) indicated opportunities for improvement. This analysis found that significant improvements could be made if three equality constraints were relaxed: (1) the pathway from Organic to Prod., (2) the correlation between Organic and Texture, and (3) the pathway between Texture and Prod. No other constraints were indicated to warrant deletion.

Because the LM analysis indicated that some parameters differed between groups, it was necessary to redo the multigroup analysis with those equality constraints eliminated. When this less restricted model was analyzed, the χ^2 obtained was 126.8 with 138 df ($p = 0.742$) and a Bentler–Bonett nonnormed fit value of 1.01. Consistent with the earlier analysis, the LM test revealed no additional differences between groups. Thus, this model was accepted as the final model. The parameter estimates for the two groups obtained in this final analysis are presented in Figure 11.2.

As the results of the final analysis show (Figure 11.2), the two groups could be judged to be equal in most parameters. This does not mean

Table 11.1. *Lagrange multiplier test results for multigroup analysis with all parameters constrained to be equal across groups*

Constraints are described such that V1 refers to the number 1 indicator variable and F1 refers to the first latent factor in the model. Thus F4, F2 refers to the path from Organic to Prod. while V12, F3 refers to the loading of s12 on TEXTURE (Figure 11.2). "Cum. χ^2" refers to the cumulative χ^2 improvement based on the multivariate assessment of the Lagrange multipliers. "χ^2 increment" shows the incremental improvement of model fit if each individual constraint was relaxed. "Probability" values less than 0.05 indicate those constraints that should be relaxed across groups.

Constrained to be equal across groups	Cum. χ^2	χ^2 increment	Probability (p)
F4, F2	13.07	13.07	0.000
F3, F2	23.15	10.08	0.001
F4, F3	27.39	4.24	0.039
V12, F3	27.71	0.32	0.572
F3, F1	27.74	0.03	0.860
V4, F1	27.76	0.02	0.886
F4, F1	27.76	0.01	0.937
V3, F1	27.77	0.00	0.947
V5, F2	27.77	0.00	0.956
V9, F2	27.77	0.00	0.971
V11, F3	27.78	0.01	0.978
V2, F1	27.78	0.00	0.979
V1, F1	27.78	0.00	0.981
V7, F2	27.78	0.00	0.986
V6, F2	27.78	0.00	0.985
V10, F2	27.78	0.00	0.988
V8, F2	27.78	0.00	0.990
F2, F1	27.78	0.00	0.995

that the values were identical for the two groups, only that they were not significantly different. For these parameters, the two groups can be said to fit a common model. Parameters that did differ between groups are indicated in Figure 11.2 with asterisks. In this example, the correlation between productivity and soil organic matter was reasonably strong in the unburned grassland and nonsignificant in the burned one. Another difference was the absence of a relationship between soil texture and productivity in the unburned grassland but not in the burned one. Finally, no correlation between

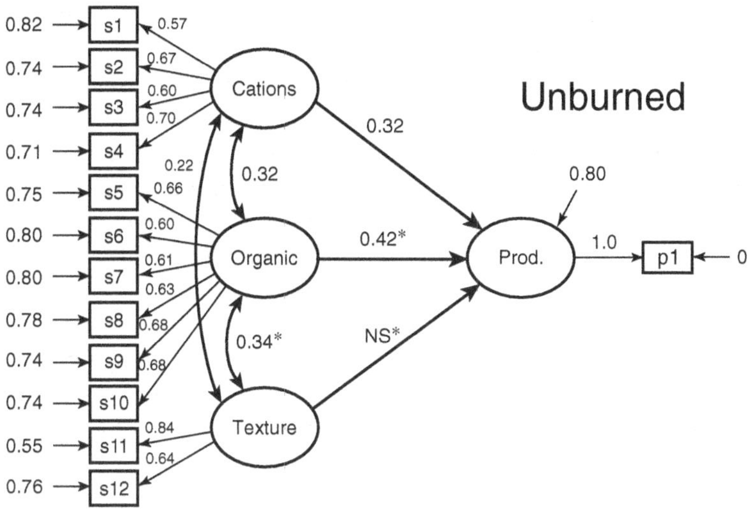

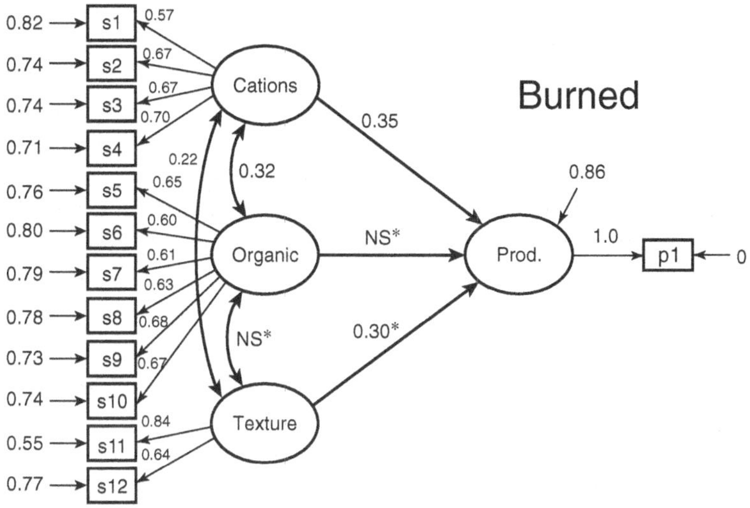

Figure 11.2. Results from the final multigroup analysis for unburned and burned grasslands. Coefficients presented are standardized. See Figure 11.1 for a definition of the elements in these models. Coefficients that differed significantly between groups are shown by asterisks. NS, nonsignificant.

soil organic and soil texture was found in the burned grassland, though one did exist for the unburned grassland.

A number of mechanistic interpretations are possible for the pattern of results shown in Figure 11.2. For this hypothetical example, however, the pattern results from the presumption that over a significant period of time burning would result in a lower accumulation of organic matter in the soil (Scifres & Hamilton, 1993). It was further presumed that variations in plant growth within the unburned grassland were controlled primarily by variations in the soil organic matter, while, in the burned grassland, variations in productivity were associated with variations in soil texture because of a greater importance of drought stress. Because this is an hypothetical example, there is little point in elaborating greatly on the underlying mechanisms except to say that the intended differences between groups were clearly reflected in the results.

Applications of multigroup analysis

To date, I am aware of only three published examples of multigroup analysis in the fields of ecology and evolutionary biology. The first was conducted as part of an examination of how morphological traits relate to survival for house sparrows (*Passer domesticus*) by Pugesek & Tomer (1996). In this study, multigroup analysis was performed to determine the degree to which male and female sparrows had the same multivariate relationships. A second application of multigroup analysis was published by Grace & Pugesek (1998) and involved the re-examination of data comparing the population dynamics of two mice species. In this analysis, the use of multigroup methods permitted an assessment of model fit when models for the individual groups were saturated (i.e., had no degrees of freedom). A third application of multigroup analysis was employed by Grace & Jutila (1999), who examined the factors controlling plant species diversity in grazed and ungrazed meadows. Here, group comparisons using structural equations permitted a clear assessment of grazing effects in the presence of a complex interplay of additional factors.

As the use of structural equations increases in ecological and evolutionary studies, reliance on multigroup comparisons can be expected to be a standard part of the methodology. In addition to comparing multivariate relationships among different groups, such as male versus female animals and plants or among separate populations, multigroup analysis is well suited for use with experimental data. Compared to many other scientific disciplines, ecological experiments are easily confounded due to the large number of factors that are difficult to control under field conditions. The ecological literature is replete with examples of studies that were heavily influenced by covariates (e.g., Tilman, 1983; Grace, 1984). Standard approaches such as analysis of variance (ANCOVA) and multivariate analysis of variance (MANCOVA) provide only limited capabilities for dealing with covariates in the analysis of experimental data. Among the weaknesses of such methods are an inability to deal with problems of multicollinearity, the absence of a way of partitioning measurement error, and the lack of capabilities for evaluating networks of relationships. In contrast, multigroup methods for structural equations analysis permit the handling of all these complexities. As methods develop further for the application of structural equation analysis to hierarchically structured designs (McArdle & Hamagami, 1996), these methods will be able to cover a broader array of experimental data.

As mentioned above, one additional advantage of multigroup analysis is the ability to work with saturated models. When the models for individual groups are saturated (i.e., all pathways need to be estimated), there are no degrees of freedom available for assessing model fit. In such a case, equality constraints across groups can provide degrees of freedom and permit evaluations of model fit.

Finally, a discussion of multigroup comparisons would not be complete without mentioning the comparison of means across groups (also known as "modeling mean structures"). Through an extension of multigroup analysis, it is possible to compare means across groups in addition to variances and covariances (Byrne, 1994; Bentler & Wu, 1996). Aside from comparing the means of observed variables, the analysis of means using structural equations can also be applied to latent variables. From a regression perspective, standard multigroup analysis permits a comparison of variances and covariances among groups while means analysis also permits comparisons of the means and intercepts.

Overall, then, multigroup analysis can be seen to be an important extension of the use of structural equations. It can be applied to comparisons among any categorical groups or samples as long as the measured parameters are the same among groups and the models for each group have the same basic structure. Multigroup analysis not only permits comparisons among groups or samples but greatly enhances its applicability for experimental data. Its utility for the study of ecological and evolutionary biology would seem to be great indeed.

Appendix 11.1. Data and program commands used in analyses

Table 11.A1. *Covariance matrices generated by the simulation procedure of EQS to represent 13 measured variables for unburned and burned grasslands. The first 12 variables are soil properties while the last is plant production*

Unburned grassland												
1.535												
0.665	2.113											
0.706	0.872	1.946										
0.684	0.987	0.937	1.983									
0.174	0.373	0.355	0.369	1.654								
0.199	0.346	0.065	0.187	0.512	1.712							
0.221	0.270	0.303	0.291	0.813	0.640	1.733						
0.140	0.240	0.232	0.109	0.686	0.734	0.660	1.599					
0.282	0.223	0.226	0.094	0.773	0.696	0.614	0.617	1.632				
0.238	0.425	0.209	0.092	0.642	0.735	0.645	0.655	0.848	1.667			
0.186	0.289	0.148	0.382	0.474	0.267	0.115	0.194	0.349	0.388	2.024		
0.064	0.142	0.097	0.200	0.329	0.290	0.194	0.224	0.368	0.388	1.046	1.977	
0.506	0.780	0.693	0.646	0.778	0.615	0.579	0.708	0.663	0.795	0.495	0.309	2.604
Burned grassland												
1.535												
0.665	2.113											
0.706	0.872	1.946										
0.684	0.987	0.937	1.983									
0.174	0.373	0.355	0.369	1.654								
0.199	0.346	0.065	0.187	0.512	1.712							
0.221	0.270	0.303	0.291	0.813	0.640	1.733						
0.140	0.240	0.232	0.109	0.686	0.734	0.660	1.599					
0.282	0.223	0.226	0.094	0.773	0.696	0.614	0.617	1.632				
0.238	0.425	0.209	0.092	0.642	0.735	0.645	0.655	0.848	1.667			
0.166	0.237	0.106	0.401	0.134	−0.113	−0.261	−0.186	−0.028	−0.045	2.159		
0.044	0.089	0.055	0.219	−0.011	−0.089	−0.182	−0.155	−0.009	−0.045	1.170	2.089	
0.394	0.630	0.515	0.595	0.302	0.054	−0.002	0.155	0.136	0.192	0.678	0.481	2.153

Table 11.A2. *EQS program statements for the initial multigroup analysis that generated the Lagrange Multiplier test results shown in Table 11.1 in the main body of the text. In order to generate the final results shown in Figure 11.2, three constraints (indicated below) were removed from the program*

```
/TITLE
mg1001: grazed grassland, group 1
/SPECIFICATIONS
  DATA='C:\EQS\MULTIGRP\MG1001.DAT';
  VARIABLES=13; CASES=200; GROUPS=2;
  METHODS=ML;
  MATRIX=RAW;
/EQUATIONS
V1 = + *F1 + E1;
V2 = + *F1 + E2;
V3 = + *F1 + E3;
V4 = + *F1 + E4;
V5 = + *F2 + E5;
V6 = + *F2 + E6;
V7 = + *F2 + E7;
V8 = + *F2 + E8;
V9 = + *F2 + E9;
V10 = + *F2 + E10;
V11 = + *F3 + E11;
V12 = + *F3 + E12;
V13 = + 1.0F4 + E13;
F4 = + *F1 + *F2 + *F3 + D4;
/VARIANCES
F1 = 1;
F2 = 1;
F3 = 1;
E1 = *;
E2 = *;
E3 = *;
E4 = *;
E5 = *;
E6 = *;
E7 = *;
E8 = *;
E9 = *;
E10 = *;
E11 = *;
E12 = *;
E13 = 0;
```

Table 11.A2. (*cont.*)

```
D4 = *;
/COVARIANCES
F2, F1 = *;
F3, F1 = *;
F3, F2 = *;
/END
/TITLE
mg2001: burned grassland, group 2
/SPECIFICATIONS
  DATA='C:\EQS\MULTIGRP\MG2001.DAT';
  VARIABLES=13; CASES=200;
  METHODS=ML;
  MATRIX=RAW;
/EQUATIONS
V1 = + *F1 + E1;
V2 = + *F1 + E2;
V3 = + *F1 + E3;
V4 = + *F1 + E4;
V5 = + *F2 + E5;
V6 = + *F2 + E6;
V7 = + *F2 + E7;
V8 = + *F2 + E8;
V9 = + *F2 + E9;
V10 = + *F2 + E10;
V11 = + *F3 + E11;
V12 = + *F3 + E12;
V13 = + 1.0F4 + E13;
F4 = + *F1 + *F2 + *F3 + D4;
/VARIANCES
F1 = 1;
F2 = 1;
F3 = 1;
E1 = *;
E2 = *;
E3 = *;
E4 = *;
E5 = *;
E6 = *;
E7 = *;
E8 = *;
E9 = *;
E10 = *;
```

Table 11.A2. (*cont.*)

```
E11 = *;
E12 = *;
E13 = 0;
D4 = *;
/COVARIANCES
F2, F1 = *;
F3, F1 = *;
F3, F2 = *;
/CONSTRAINTS
(1,V1,F1)=(2,V1,F1);
(1,V2,F1)=(2,V2,F1);
(1,V3,F1)=(2,V3,F1);
(1,V4,F1)=(2,V4,F1);
(1,V5,F2)=(2,V5,F2);
(1,V6,F2)=(2,V6,F2);
(1,V7,F2)=(2,V7,F2);
(1,V8,F2)=(2,V8,F2);
(1,V9,F2)=(2,V9,F2);
(1,V10,F2)=(2,V10,F2);
(1,V11,F3)=(2,V11,F3);
(1,V12,F3)=(2,V12,F3);
(1,F4,F1)=(2,F4,F1);
(1,F4,F2)=(2,F4,F2); !THIS CONSTRAINT ELIMINATED IN FINAL
  MODEL
(1,F4,F3)=(2,F4,F3); !THIS CONSTRAINT ELIMINATED IN FINAL
  MODEL
(1,F2,F1)=(2,F2,F1);
(1,F3,F1)=(2,F3,F1);
(1,F3,F2)=(2,F3,F2); !THIS CONSTRAINT ELIMINATED IN FINAL
 MODEL
/LMTEST;
/END
```

References

Auclair, A. N. D., Bouchard, A. & Pajaczkowski, J. (1976). Productivity relations in a *Carex*-dominated ecosystem. *Oecologia*, **26**, 9–31.

Bentler, P. M. (1996). *EQS: Structural Equations Program Manual*. Encino, CA: Multivariate Software Inc.

Bentler, P. M. & Wu, E. J. C. (1996). *EQS for Windows: User's Guide*. Encino, CA: Multivariate Software Inc.

Bollen, K. A. (1989). *Structural Equations with Latent Variables*. New York: Wiley.
Byrne, B. M. (1994). *Structural Equation Modeling with EQS and EQS/Windows*. Thousand Oaks, CA: Sage.
Grace, J. B. (1984). Effects of tubificid worms on the germination and establishment of *Typha*. *Ecology*, **65**, 1689–1693.
Grace, J. B. & Jutila, H. (1999). The relationship between species density and community biomass in grazed and ungrazed coastal meadows. *Oikos*, **85**, 398–408.
Grace, J. B. & Pugesek, B. H. (1998). On the use of path analysis and related procedures for the investigation of ecological problems. *American Naturalist*, **152**, 151–159.
Hair, J. F., Jr, Anderson, R. E., Tatham, R. L. & Black, W. C. (1995). *Multivariate Data Analysis*, 4th edition. New York: Prentice-Hall.
Hayduk, L. A. (1987). *Structural Equation Modeling with LISREL: Essentials and Advances*. Baltimore, MD: Johns Hopkins University Press.
Hoyle, R. H. (ed.) (1995). *Structural Equation Modeling*. London: Sage.
Jöreskog, K. G. (1971). Simultaneous factor analysis in several populations. *Psychometrika*, **57**, 409–426.
Jöreskog, K. G. (1973). A general method for estimating a linear structural equation system. pp. 85–112, In A. S. Goldberger & O. D. Duncan, eds., *Structural Equation Models in the Social Sciences*. Academic Press, New York.
Jöreskog, K. G. & Sörbom, D. (1996). *LISREL 8: User's Reference Guide*. Chicago: Scientific Software International.
Keesling, J. W. (1972). Maximum likelihood approaches to causal analysis. PhD thesis. Department of Education, University of Chicago.
Lee, S.-Y. & Bentler, P. M. (1980). Some asymptotic properties of constrained generalized least squares estimation in covariance structure models. *South African Statistical Journal*, **14**, 121–136.
McArdle, J. J. & Hamagami, F. (1996). Multilevel models from a multiple group structural equation perspective. In G. A. Marcoulides & R. E. Schumacker (eds.) *Advanced Structural Equation Modeling*, pp. 89–124. Mahwah, NJ: Lawrence Erlbaum Associates.
Pugesek, B. H. & Tomer, A. (1996). The Bumpus house sparrow data: a reanalysis using structural equation models. *Evolutionary Ecology*, **10**, 387–404.
Schumacker, R. E. & Lomax, R. G. (1996). *A Beginner's Guide to Structural Equation Modeling*. Mahwah, NJ: Lawrence Erlbaum Associates.
Scifres, C. J. & Hamilton, W. T. (1993). *Prescribed Burning for Brushland Management*. College Station, TX: Texas A&M University Press.
Tilman, D. (1983). Plant succession and gopher disturbance along an experimental gradient. *Oecologia*, **60**, 285–292.
Wiley, D. E. (1973). The identification problem for structural equation models with unmeasured variables. In A. S. Goldberger & O. D. Duncan (eds.), *Structural Equation Models in the Social Sciences*, pp. 69–83. New York: Academic Press.

12 Modeling means in latent variable models of natural selection

Bruce H. Pugesek

Abstract

This chapter describes theory and methods for measuring mean changes in phenotypes when the phenotypic variable is considered as a latent construct with multiple indicators. Data for exposition of the methods are generated using EQS simulation techniques. Two datasets are used, one simulating the population prior to a selection event, the pre-selection group, and another simulating the population after the selection event, the post-selection group. The example includes an environmental construct that may also influence phenotypic variable means. The example demonstrates how structural equation modeling (SEM) may be used to statistically control for such environmental influences to arrive at a more accurate estimate of the effect of the selection event. Data are analyzed with LISREL version 8.30. The analysis includes a measurement model and stacked model of covariances to test model assumptions, and a final means model to estimate the mean phenotypic response to selection.

Introduction

Considerable interest has been generated in recent years in developing methods for the measurement of the effects of natural and sexual selection on phenotypic traits. Methods devised by Lande, Arnold and Wade (Lande, 1979; Lande & Arnold, 1983; Arnold & Wade, 1984a,b) provided the analytical tools that sparked a resurgence in such selection studies (see recent issues of *Evolution*). Their model employs multiple regression techniques to estimate the direct and indirect effects of selection on phenotypic traits.

The expected multivariate response to selection is given by Lande (1979) as

$$\Delta\bar{z} = \mathbf{G}\mathbf{P}^{-1}s = \mathbf{G}\beta$$

where $\Delta\bar{z}$ is the $q \times 1$ vector in mean phenotype of the characters, **G** is the matrix of additive genetic variances and covariances between characters, **P** is the phenotypic variance–covariance matrix, s is a column vector of selection differentials for each phenotypic character, and $\boldsymbol{\beta} = \mathbf{P}^{-1}\boldsymbol{s}$ is a vector of selection gradients. The selection differentials represent the within-generation change of the phenotypic means produced by selection. Each element β_i of the selection gradient $\boldsymbol{\beta}$ represents the effects of selection acting directly on character i, with effects of selection on other characters removed.

The elements of $\boldsymbol{\beta}$ are formally equivalent to the partial regression of relative fitness on the characters and are estimated by multiple regression (Lande & Arnold, 1983). The equation above shows that the change in phenotypic mean resulting from selection is dependent both on the elements of the selection gradient $\boldsymbol{\beta}$ that measure the force of selection and on the genetic variability of each character as represented in **G**.

The approach outlined above assumes that selection takes place at the level of the measured phenotypic characteristics rather than at the more abstract level, which can be modeled as a common factor of several measured phenotypic characters. An example will be body size, which may be considered at a more abstract level of several measurements such as weight, length, girth, etc. Such an approach has been advocated as appropriate in studies of evolution (e.g., Crespi, 1990; Kingsolver & Schemske, 1991) because selection may often act directly on the abstract construct.

It is well known that error variance produced by either measurement error or by lack of construct validity is a common source of bias in the estimation of partial regression coefficients. Even small amounts of error variance can produce severe bias and distortion to partial regression coefficients, especially when independent variables are intercorrelated (Pugesek & Tomer, 1995).

Under the circumstances outlined above, more accurate estimates of relations between measurement level variables and fitness may be obtained using latent constructs and SEM. Pugesek & Tomer (1995) advocated reformulation of the Lande–Arnold equation to incorporate latent variables. In this reinterpretation **P**, the phenotypic variance–covariance matrix, includes as elements variances and covariances conceived at the more abstract level of latent variables. Consequently, the selection gradient is expressed by the vector of structural coefficients for the paths relating the latent variable to fitness.

Evolutionary biologists may also be interested in change of mean phenotypic characters as measured by latent variables. This chapter provides a LISREL formulation for analysis of changes in means of latent variables.

A LISREL formulation of the means model

Modeling means requires the addition of four new matrices, $\boldsymbol{\tau}_x$, $\boldsymbol{\tau}_y$, $\boldsymbol{\alpha}$, and $\boldsymbol{\kappa}$ to the traditional LISREL model. They are incorporated into the LISREL submodels as follows:

Submodel 1: $\boldsymbol{x} = \boldsymbol{\tau}_x + \boldsymbol{\Lambda}_x\boldsymbol{\xi} + \boldsymbol{\delta}$
Submodel 2: $\boldsymbol{y} = \boldsymbol{\tau}_y + \boldsymbol{\Lambda}_y\boldsymbol{\eta} + \boldsymbol{\varepsilon}$
Submodel 3: $\boldsymbol{\eta} = \boldsymbol{\alpha} + \mathbf{B}\boldsymbol{\eta} + \boldsymbol{\Gamma}\boldsymbol{\xi} + \boldsymbol{\zeta}$

where $\boldsymbol{\tau}_x$, $\boldsymbol{\tau}_y$, and $\boldsymbol{\alpha}$ are vectors of constant intercept terms. It is assumed, as in the traditional LISREL model, that ς is uncorrelated with $\boldsymbol{\xi}$, $\boldsymbol{\varepsilon}$ is uncorrelated with $\boldsymbol{\eta}$, and that $\boldsymbol{\delta}$ is uncorrelated with $\boldsymbol{\xi}$. As in the traditional LISREL model, it is further assumed that $\mathrm{E}(\boldsymbol{\zeta}) = 0$, $\mathrm{E}(\boldsymbol{\varepsilon}) = 0$, and $\mathrm{E}(\boldsymbol{\delta}) = 0$ (E is the expected value operator). However, it is not assumed that $\mathrm{E}(\boldsymbol{\xi}) = 0$ and $\mathrm{E}(\boldsymbol{\eta}) = 0$.

The expectancies of endogenous variables, $\mathrm{E}(\boldsymbol{\xi})$, are estimated and recorded in the column vector $\boldsymbol{\kappa}$. The expectancies of the endogenous variables, $\mathrm{E}(\boldsymbol{\eta})$, do not require estimation because they are determined by the structural coefficients in the model and by the estimates of the exogenous concept means. This can be shown as follows. The equation for submodel 3 may be rearranged by moving $\mathbf{B}\boldsymbol{\eta}$ to the left side of the equation. Next, by factoring out $\boldsymbol{\eta}$ and then premultiplying both sides of the equation by $(\mathbf{I} - \mathbf{B})$ we obtain:

$$\boldsymbol{\eta} = (\mathbf{I} - \mathbf{B})^{-1}(\boldsymbol{\alpha} + \boldsymbol{\Gamma}\boldsymbol{\xi} + \boldsymbol{\zeta})$$

Next we take the expectations of both sides of the equation. Note that $\mathrm{E}(\boldsymbol{\zeta}) = 0$ and that all matrices other than $\boldsymbol{\eta}$ and $\boldsymbol{\xi}$ contain constants that can be factored out of the expectation operator. Thus, means of the $\boldsymbol{\eta}$ concepts can be obtained provided that we have a set of estimates for the $\mathbf{B}$ and $\boldsymbol{\Gamma}$ slopes, the $\boldsymbol{\alpha}$ intercepts, and the means of the exogenous concepts $\boldsymbol{\kappa}$.

$$\mathrm{E}(\boldsymbol{\eta}) = (\mathbf{I} - \mathbf{B})^{-1}(\boldsymbol{\alpha} + \boldsymbol{\Gamma}\boldsymbol{\kappa})$$

The model-implied means of the observed variables are obtained in a similar fashion. We take the expectations of submodels 1 and 2, and then insert equivalent representations of the means of the exogenous concepts, $\mathrm{E}(\boldsymbol{\xi}) = \boldsymbol{\kappa}$, and endogenous concepts, into the appropriate portions of the equations.

$$\mathrm{E}(\boldsymbol{x}) = \boldsymbol{\tau}_x + \boldsymbol{\Lambda}_x\mathrm{E}(\boldsymbol{\xi})$$
$$\mathrm{E}(\boldsymbol{x}) = \boldsymbol{\tau}_x + \boldsymbol{\Lambda}_x\boldsymbol{\kappa}$$

and

$$E(\boldsymbol{y}) = \boldsymbol{\tau}_y + \boldsymbol{\Lambda}_y E(\boldsymbol{\eta})$$
$$E(\boldsymbol{y}) = \boldsymbol{\tau}_y + \boldsymbol{\Lambda}_y(\boldsymbol{I} - \mathbf{B})^{-1}(\boldsymbol{\alpha} + \boldsymbol{\Gamma}\boldsymbol{\kappa}).$$

The fit function employed for the LISREL means model is the general fit function for multisample models and is defined as

$$F = \sum_{g=1}^{G} \frac{N_g}{N} F_g,$$

where

$$F_g = \left(\boldsymbol{s}^{(g)} - \boldsymbol{\sigma}^{(g)}\right)' \mathbf{W}_{(g)}^{-1} \left(\boldsymbol{s}^{(g)} - \boldsymbol{\sigma}^{(g)}\right) + \left(\bar{\boldsymbol{z}}^{(g)} - \boldsymbol{\mu}^{(g)}\right)' \mathbf{V}_{(g)}^{-1} \left(\bar{\boldsymbol{z}}^{(g)} - \boldsymbol{\mu}^{(g)}\right).$$

The fit of the multigroup model is the size-weighted average of the fits obtained for each group individually. Note that the first portion of the equation for F_g is identical to the general function for fitting covariance structures (Jöreskog & Sörbom, 1996a) where $\boldsymbol{s}$ is a vector of elements from the sample variance–covariance matrix, $\boldsymbol{\sigma}$ is corresponding vector of elements of $\boldsymbol{\Sigma}$ and $\mathbf{W}^{-1}$ is a weight matrix. The second portion of the equation involves the sample mean vector $\boldsymbol{z}^{(g)}$, the population mean vector $\boldsymbol{\mu}^{(g)}$, and a weight matrix $\mathbf{V}_{(g)}$ defined as:

$$\mathbf{V}_{(g)} = \mathbf{S}^{(g)} \quad \text{for ULS, GLS, WLS, DWLS}$$

$$\mathbf{V}_{(g)} = \boldsymbol{\Sigma}^{(g)} \quad \text{for ML,}$$

where ULS is unweighted least squares; GLS is generalized least squares; WLS is generally weighted least squares; DWLS is diagonally weighted least squares; ML, maximum likelihood. Note that if $\boldsymbol{\tau}_y$, $\boldsymbol{\tau}_x$, $\boldsymbol{\alpha}$, and $\boldsymbol{\kappa}$ are default, the second term in the equation for F_g is a constant, no mean structures are required, and the problem reduces to the multisample formulation.

Means of latent variables with multiple indicators are unidentified in a single group. However, group differences in the means of latent variables can be estimated provided that the latent variables are on the same scale in all variables (Sörbom, 1974). Mean structures are defined using submodel 1 as

$$\boldsymbol{x}^{(g)} = \boldsymbol{\tau}_x + \boldsymbol{\Lambda}_x \boldsymbol{\xi}^{(g)} + \boldsymbol{\delta}^{(g)}, \quad g = 1, 2, \ldots, G,$$

with the mean of $\boldsymbol{\xi}^{(g)} = \boldsymbol{\kappa}^{(g)}$. The origins and units of measurement of the $\boldsymbol{\xi}$-factors are defined by setting $\boldsymbol{\kappa}^{(g)} = 0$ and fixing one nonzero value in each column of $\boldsymbol{\Lambda}_x$. The parameters to be estimated are:

$\boldsymbol{\tau}_x$, $\boldsymbol{\Lambda}_x$	assumed invariant over groups
$\boldsymbol{\kappa}^{(1)}, \boldsymbol{\kappa}^{(2)}, \ldots, \boldsymbol{\kappa}^{(G)}$	mean vectors of $\boldsymbol{\xi}$
$\boldsymbol{\Phi}^{(1)}, \boldsymbol{\Phi}^{(2)}, \ldots, \boldsymbol{\Phi}^{(G)}$	covariance matrices of $\boldsymbol{\xi}$
$\boldsymbol{\Theta}_\delta^{(1)}, \boldsymbol{\Theta}_\delta^{(2)}, \ldots, \boldsymbol{\Theta}_\delta^{(G)}$	error covariance matrices of $\boldsymbol{\xi}$ variables

In most cases error covariance matrices, $\boldsymbol{\Theta}_\delta$, are diagonal; however, they may be set to be invariant over groups.

A simulated study of phenotypic selection

I created a simulation study to illustrate how means models may be analyzed in studies of phenotypic selection. Two samples of six variables were generated, the first to represent the population before a selection event (pre-selection group) and the second to represent the population after the selection event (post-selection group). Each sample consisted of one variable constructed as an exogenous latent variable with three indicators and one endogenous latent variable with three indicators (Figure 12.1). The exogenous latent variable is some environmental or biotic construct. The endogenous variable is a complex assembly of phenotypic traits, such as body size, upon which selection may act directly. It is assumed that the population is sufficiently large that the two samples are independent.

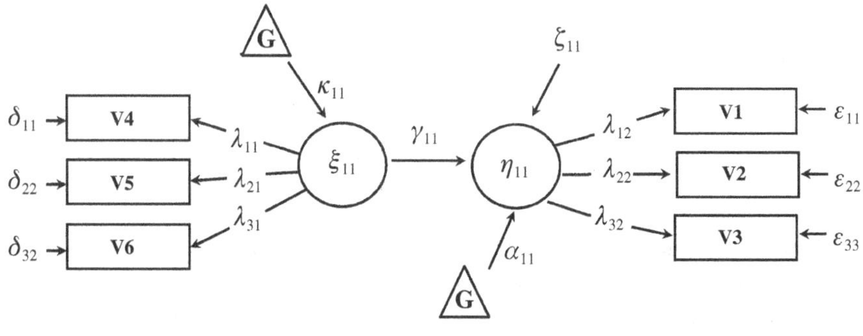

Figure 12.1. A path diagram of the simulation model consisting of one exogenous and one endogenous latent variable (circles), each with three indicators (rectangles). Triangles denote constants, in this case latent variable mean differences between pre-selection and post-selection groups. Paths and error terms are listed in LISREL notation.

Data were generated for each of the two samples with EQS (Bentler & Wu, 1995; see Appendix 12.1). The main EQS programming steps were essentially the same as those described by Pugesek for Shiras Moose (see Chapter 3) and will not be reiterated here. Two added features of this EQS simulation, however, require some elaboration.

First, the simulation presented here has the added feature of including constants in the equation lines that are specified by a numerical value followed by the variable name V999. This has the effect of creating a mean value for the variable specified in the equation. The two samples created for the simulation differ only in that a mean value for the latent variable F1 is added to the equation in the post-selection group sample and the two simulations have different seed numbers. Therefore, the two groups have similar variance–covariance structure and mean values for indicators of F2. The small differences that are evident are attributable to sampling variation. Means of indicators variables of F1 are higher in the post-selection group as compared with the pre-selection group as a consequence of adding a constant to the equation for F1.

Second, the simulation in Chapter 3 used a covariance matrix as signified by the Analysis=COV statement in the *specification* section of the program. In this simulation, variances, covariances, and means are necessary to generate data. Therefore, a moments matrix is used and is specified by the statement Analysis=MOM. Results were output as an ASCII data set.

Analysis of the data was performed with PRELIS2 (Jöreskog & Sörbom, 1996b) and LISREL 8.30 (Jöreskog & Sörbom, 1996a). PRELIS2 generated the moments matrices necessary for analysis by LISREL8.

Data analysis

The measurement model and tests of assumptions

The analysis begins with a measurement model designed to test factorial structure within groups (Figure 12.2a,b). This model consists of two factors, each with three indicators. Variables V1 to V3 represent the endogenous latent variable F1, and V4 to V6 represent the exogenous latent variable F2 from Figure 12.1. A correlation is modeled between F1 and F2 to account for covariance between latent variables. The measurement model fits well in both the pre- and post-selection event data sets, $\chi^2_{df=8} = 5.68$, $p = 0.68$, *GFI* = 1.0, RMSEA = 0.0 and $\chi^2_{df=8} = 3.26$, $p = 0.92$, GFI = 1.0, RMSEA = 0.0, respectively (where GFI is the goodness-of-fit

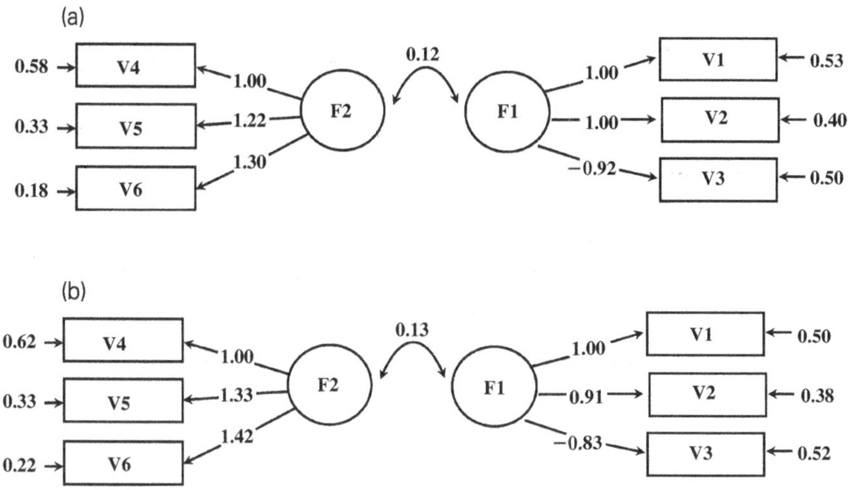

Figure 12.2. A path diagram of results of the measurement model for the pre-selection (a) and post-selection (b) groups. Curved double-headed arrows linking latent variables denote a modeled correlation.

index, RMSEA is the root mean square error of approximation, and df is degrees of freedom). No significant modification indices occur indicating good factorial structure.

Next a structural model considering only covariances is run as a stacked model to test the assumption of between group factorial invariance. It is not assumed that the path from the F1 to F2 is invariant between groups so this parameter is set to be free. The multigroup model fits the data well, $\chi^2_{\mathrm{df}=20} = 10.50$, $p = 0.96$, GFI = 1.0, RMSEA = 0.0. There are no significant modification indices for factor loadings or error terms. Therefore, the assumption of factorial invariance is met. One additional model is run to determine whether groups differ with respect to the strength of the dependence of F2 on F1. Setting the path from F1 to F2 as invariant does not significantly decrease model fit, and thereby indicates no between group differences, $\chi^2_{\mathrm{df}=21} = 10.55$, $p = 0.97$, GFI = 1.0, RMSEA = 0.0. This model is retained, since it is more parsimonious and conserves one degree of freedom (Figure 12.3).

The means model

Having determined that the latent variables adequately represent patterns of covariation among measurement level indicators, and that these relations are consistent between groups, it is now possible to examine mean differences

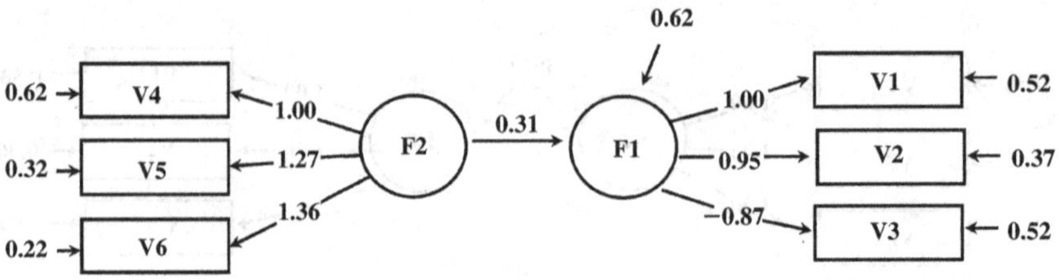

Figure 12.3. A path diagram of the multigroup model of covariances in the pre-selection and post-selection samples.

among latent variables. Since the model contains only one exogenous and one endogenous variable, the structural model is:

$$\eta^{(g)} = \alpha^{(g)} + \gamma^{(g)}\xi^{(g)} + \zeta^{(g)}.$$

As previously mentioned, it is not possible to estimate values for α or κ for single groups because the model would be unidentified. However, we can compare means between groups by setting slopes to be invariant between groups, setting factor means of one group to zero, and estimating the difference in means between groups. The following LISREL program analyzes the simulation data.

```
1. PRE-SELECTION GROUP
2. DA NI=6 NOBS=400 NG=2
3. CM FI=TE1.COV
4. !SD FI=TE1.SD
5. ME FI=TE1.M
6. LA
7. V1 V2 V3 V4 V5 V6/
8. SE
9. V1 V2 V3 V4 V5 V6/
```

Line 1 provides a title for the first model. The DA line (line 2), as in all analyses, specifies the number of input variables and the number of observations. Here an NG statement is added to notify the program that this is a multigroup model that consists of two groups. Lines 3 through 5 provide paths to the covariance, standard deviations, and means matrices for the pre-selection data set generated by PRELIS. Lines 6 and 7 specify the order of variables as they appear in the data set. Lines 8 and 9 specify the order of variables as they appear in the model. Endogenous variables, V1 to V3, are listed first, followed by exogenous variables V4 to V6.

```
10. MO NX=3 NK=1 NY=3 NE=1 LX=FR TX=FR LY=FR TY=FR GA=FR
    PH=FR PS=FR TD=DI,FR TE=DI,FR AL=FR KA=FR
11. LE
12. F1
13. LK
14. F2
15. FI LX 1 LY 1
16. VA 1 LX 1 LY 1
17. OU ALL
```

Lines 10 through 17 complete the specification of the model for the pre-selection data. The MO line specifies one exogenous (NK=1) and one endogenous variable (NE=1), each with three indicators (NX=3, NY=3). Elements of the $\boldsymbol{\xi}$, η and γ matrices are set to be free (LX=FR, LY=FR, GA=FR). Matrices $\boldsymbol{\Phi}$ (PH) and $\boldsymbol{\Psi}$ (PS) are set to be free. Each has one element to be estimated, the variance of the exogenous variable and the error term for the endogenous variable respectively. $\boldsymbol{\Theta}_\delta$ and $\boldsymbol{\Theta}_\varepsilon$ are designated as diagonal matrices with elements on the diagonal free. Also included on the MO line are matrices for $\boldsymbol{\alpha}$ (AL) and $\boldsymbol{\kappa}$ (KA), both set to be free. Lines 11 through 14 list variable names for the endogenous and exogenous variable. Line 15 fixes (FI) the value of one element of each latent variable. Line 16 specifies the value of each fixed element from line 15 as 1.0. The OU line (line 17) requests the output desired in the LISREL output file.

```
18. POST-SELECTION GROUP
19. DA NOBS=400
20. CM FI=TE2.COV
21. !SD FI=TE2.SD
22. ME FI=TE2.M
23. LA
24. V1 V2 V3 V4 V5 V6/
25. SE
26. V1 V2 V3 V4 V5 V6/
27. MO NX=3 NK=1 NY=3 NE=1 LX=IN TX=IN LY=IN TY=IN GA=FR
    PH=FR PS=FR TD=SP TE=SP AL=FI KA=FI
28. LE
29. F1
30. LK
31. F2
32. OU ALL
```

Lines 18 to 30 specify the model for the post-selection group. Program code takes essentially the same form as the previous model, calls up the appropriate data set, specifies the number of variables and their order of entry in to the model, etc. The notable exception is the MO line (line 27). Here matrices LX, LY, TX and TY are specified as IN so that factor loadings and intercepts respectively are forced to be invariant. α and κ are fixed (FI), the default condition being set to 0.0. GA is set to be free, and thus, γ will be estimated independently for each group. The condition SP for TD and TE specifies Θ_δ and Θ_ε in the post-selection model to have the same pattern of fixed and free elements as does the pre-selection model.

The model for each group contains specifications for the all matrices necessary from the general LISREL model as well as four additional matrices for intercepts of x and y variables, and means of exogenous and endogenous variables (τ_x, τ_y, α, and κ). A matrix for β is not included because there are no paths to model between endogenous latent variables. The LISREL program will often include matrices by default, thereby making it unnecessary to list each matrix in the MO line. For example, in the model presented here, PH and PS specifications could be omitted because the default condition would be to set elements of these matrices to be free. Until the practitioner is experienced, however, it is a good idea to specify all necessary model matrices.

The model described above fit the data well, $\chi^2_{df=24} = 18.08$, $p = 0.80$, GFI = 1.0, RMSEA = 0.0. There appeared to be little difference between γ, the path linking the exogenous and endogenous variables, estimated independently for the two groups. Values are 0.29 (standard error (SE) $\pm$ 0.08) in the pre-selection group and 0.31 (SE $\pm$ 0.07) in the post-selection group. A second model is fit, identical to the first, with the exception that γ was modeled as invariant between groups. This model also fit the data well $\chi^2_{df=25} = 18.13$, $p = 0.84$, GFI = 1.0, RMSEA = 0.0. The difference between models, $\chi^2_{df=1} = 0.05$, is not significant, therefore, the second model is retained. Inspection of the results of the second model indicates that ϕ and ψ may also be modeled as invariant across groups. The final model with invariant parameters γ, ϕ, and ψ is retained as the most parsimonious model ($\chi^2_{df=27} = 18.52$, $p = 0.89$, GFI = 1.0, RMSEA = 0.0) (Tables 12.1 and 12.2).

There is little difference between groups for means of the exogenous ξ variables. κ, the difference in means between exogenous variables is estimated as 0.02 (SE $\pm$ 0.05) ($T = 0.32$, NS). The mean difference between endogenous variables, α, is -0.69 (SE $\pm$ 0.07) and significant ($T = -10.44$). Recalling that the latent variable means are set to zero in the post-selection group, the negative value for the difference in means

Table 12.1. *Maximum likelihood estimates, standard errors, and t-values for model parameters that are freely estimated in both groups*

Parameter	Pre-selection group	Post-selection group
θ^{δ}_{11}	0.58	0.62
	(0.05)	(0.05)
	12.63	12.59
θ^{δ}_{22}	0.33	0.32
	(0.04)	(0.04)
	8.61	8.28
θ^{δ}_{33}	0.18	0.23
	(0.04)	(0.04)
	4.99	5.81
$\theta^{\varepsilon}_{11}$	0.52	0.53
	(0.05)	(0.05)
	9.79	9.96
$\theta^{\varepsilon}_{22}$	0.39	0.35
	(0.05)	(0.04)
	8.47	7.99
$\theta^{\varepsilon}_{33}$	0.52	0.53
	(0.05)	(0.05)
	10.93	11.06
α	−0.69	0
	(0.07)	
	10.44	
κ	0.02	0
	(0.05)	
	0.32	

indicates that the mean of the pre-selection group is 0.69 less than that of the post-selection group. Controlling for variation in α due to κ only slightly modified the estimate of α. The estimate of α without statistical control, −0.68, differs by only −0.01. The adjustment is slight because κ is essentially the same in both groups, differing only as a result of sampling variation.

Discussion

The analysis provides several useful pieces of information. The factorial structure of the exogenous and endogenous latent variables does not change

Table 12.2. *Maximum likelihood estimates, standard errors, and t-values of parameters that are modeled as invariant across groups. Note that λ_{11} of the ξ and η variables are fixed to 1.00; therefore, estimates are not calculated*

Parameter	Combined groups
$\tau_1^{(x)}$	0.94
	(0.04)
	21.72
$\tau_2^{(x)}$	5.98
	(0.05)
	127.20
$\tau_3^{(x)}$	2.95
	(0.05)
	61.23
$\tau_1^{(y)}$	1.62
	(0.05)
	31.78
$\tau_2^{(y)}$	1.65
	(0.05)
	34.41
$\tau_3^{(y)}$	0.41
	(0.05)
	8.92
$\lambda_{21}^{(x)}$	1.27
	(0.07)
	18.01
$\lambda_{31}^{(x)}$	1.36
	(0.08)
	17.76
$\lambda_{21}^{(y)}$	1.00
	(0.05)
	19.41
$\lambda_{31}^{(y)}$	−0.86
	(0.05)
	−18.36
γ_{11}	0.30
	(0.05)
	5.53
ϕ_{11}	0.60
	(0.05)
	11.14
ψ_{11}	0.41
	(0.04)
	9.39

as a consequence of the selection event. In other words, relations among indicator variables remain the same. Had the dependent latent variable been a measure of body size, for example, we could conclude that the relative proportions of body size measures remained the same after the selection event. In addition, the effect of the exogenous latent variable on the endogenous latent variable is the same before and after the selection event. κ, the difference in exogenous latent variable means, is not significant. We can conclude, therefore, that the selection event has no material effect on the exogenous latent variable or its relationship with the endogenous latent variable.

α can be interpreted as the difference in endogenous latent variable means resulting from the episode of selection holding constant the effect of the exogenous latent variable. The post-selection group has a significantly higher mean as compared with the pre-selection group. Returning to the example of body size we conclude that the selection event produced a shift in phenotypes towards larger relative body size.

The SEM methodology presented here allows researchers the opportunity to explore selection events at the level of latent variable. Questions regarding the selection event including "Do the nature of latent variables change?", "Do relationships between latent variables change?", and "Do latent variable means change in response to selection?" are all attainable within the framework of rigorous hypothesis tests. SEM methodology has the further advantage of controlling for environmental factors that may bias estimates of phenotypic responses to a selection event (Rausher, 1992).

The simulation capabilities of SEM packages should not be overlooked by the evolutionary ecologist interested in theoretical studies of selection (e.g., Shaw & Shaw, 1992; Podolsky *et al.*, 1997). Both PRELIS (Jöreskog & Sörbom,1996b), the data handling companion to LISREL (Jöreskog & Sörbom, 1996a), and EQS (Bentler & Wu, 1995) have well-developed simulation capabilities that can generate large datasets and moments matrices for large numbers of variables and models. These packages are capable of generating large numbers of moments matrices, covariance, correlation, etc. from pre-specified probability distributions. In addition, features include both boot strapping and Monte Carlo methods.

Appendix 12.1. EQS data simulation program

EQS simulation for pre-selection group data

```
/TITLE Modeling Phenotypic Means - Group 1
/SPECIFICATION
```

```
CAS=400; VAR=6; ME=ML; ANALYSIS=MOM;
/EQUATIONS
F1=.25*F2+D1;
V1=1*V999+.75*F1+E1;
V2=1*V999+.8*F1+E2;
V3=1*V999-.7*F1+E3;
V4=1*V999+.6*F2+E4;
V5=6*V999+.8*F2+E5;
V6=3*V999+.9*F2+E6;
/VARIANCES
E1=.56; E2=.36; E3=.51; E4=.64; E5=.36; E6=.19;
D1=.93;
/SIMULATION
SEED=1295756891;
DATA='te1';
SAVE=SEPARATE;
POP=MODEL;
/OUTPUT
LISTING; PARAMETER ESTIMATES; STANDARD ERRORS; CO;
/END
```

EQS simulation for post-selection group data

```
/TITLE Modeling Phenotypic Means - Group 2
/SPECIFICATION
 CAS=400; VAR=6; ME=ML; ANALYSIS=MOM;
/EQUATIONS
F1=.75*V999+.25*F2+D1;
V1=1*V999+.75*F1+E1;
V2=1*V999+.8*F1+E2;
V3=1*V999-.7*F1+E3;
V4=1*V999+.6*F2+E4;
V5=6*V999+.8*F2+E5;
V6=3*V999+.9*F2+E6;
/VARIANCES
E1=.56; E2=.36; E3=.51; E4=.64; E5=.36; E6=.19;
D1=.93;
/SIMULATION
SEED=1294756891;
DATA='te2';
```

```
SAVE=SEPARATE;
POP=MODEL;
/OUTPUT
LISTING; PARAMETER ESTIMATES; STANDARD ERRORS; CO;
/END
```

References

Arnold, S. J. & Wade, M. J. (1984a). On the measurement of natural and sexual selection: theory. *Evolution*, **38**, 709–719.

Arnold, S. J. & Wade, M. J. (1984b). On the measurement of natural and and sexual selection: applications. *Evolution*, **38**, 720–734.

Bentler, P. M. & Wu, E. J. C. (1995). *EQS for Windows: Users Guide.* Encino, CA: Multivariate Software.

Crespi, B. J. (1990). Measuring the effect of natural selection on phenotypic interaction systems. *American Naturalist*, **135**, 32–47.

Jöreskog, L. & Sörbom, D. (1996a). *LISREL 8: User's Reference Guide.* Chicago, IL: Scientific Software International.

Jöreskog, L. & Sörbom, D. (1996b). *PRELIS 2: User's Reference Guide.* Chicago, IL: Scientific Software International.

Kingsolver, J. G. & Schemske, D. W. (1991). Path analyses of selection. *Tree*, **6**, 276–280.

Lande, R. (1979). Quantitative genetic analysis of multivariate evolution, applied to brain body size allometry. *Evolution*, **33**, 402–416.

Lande, R. & Arnold, S. (1983). The measurement of selection on correlated characters. *Evolution*, **37**, 1210–1226.

Podolsky, R.H., Shaw, R.G. & Shaw, F. (1997). Population structure of morphological traits in *Clarkia dudleyana.* II. Constancy of within-population genetic variance. *Evolution*, **51**, 1785–1796.

Pugesek, B. H. & Tomer A. (1995). Determination of selection gradients using multiple regression versus structural equation models (SEM). *Biometrical Journal*, **37**, 449–462.

Rausher, M. D. (1992). The measurement of selection of quantitative traits: biases due to enviornmental covariance between traits and fitness. *Evolution*, **46**, 616–626.

Shaw, R. G. & Shaw, F. (1992). Quercus: programs for quantitative genetic analysis using maximum likelihood. Published electronically on the Internet, available directly from the authors or via anonymous ftp from ftp.bio.indiana.edu: directory path biology/quantgen/quercus.

Sörbom, D. (1974). A general method for studying differences in factor means and factor structures between groups. *British Journal of Mathematical Statistical Psychology*, **72**, 229–239.

13 Modeling manifest variables in longitudinal designs – a two-stage approach

Bret E. Fuller, Alexander von Eye, Phillip K. Wood, and Bobby D. Keeland

Abstract

Increasingly, the analysis of biological data involves collection of data over time, especially when the phenomena of interest concern identification of patterns of differential growth or response to long-term environmental stressors. When different patterns over time can be identified, they may indicate the presence of different survival strategies in a population of interest, species-specific responses to stressors, or may be markers for as yet unidentified subspecies in the population. In this chapter, two techniques for the analysis of biological data over time are described. The first technique is a two-stage analysis of data over time in which polynomial regressions are conducted to identify distinct growth patterns in the data. These growth curve patterns are then used in subsequent structural models to identify the correlates of growth curves. The second technique, the method of "Tuckerized" growth curves, proposes identification of growth curves by an eigenvalue decomposition. This solution is then rotated to make the solution conceptually interpretable within the context of overall growth over time. Factor scores from this rotated solution are then used to investigate whether such growth patterns differ as a function of environmental or species classification. Both techniques are compared and contrasted using longitudinal data on tree circumference taken from three species of trees located in the wetlands of two North American states, South Carolina and Louisiana. Rather than being competing models of growth over time, these two techniques represent complementary analyses, which differ in terms of their assumptions of categorization and growth curve processes over time.

Introduction

The application of latent variables to longitudinal data has generated increasing interest (e.g., von Eye & Clogg, 1994, 1996; Rose *et al.*, 2000) because such models permit specification of differential patterns of performance or growth over time, make less restrictive assumptions about the data, and are designed to document phenomena fundamentally different from those investigated in traditional analyses of variance and related linear models. Although the pace and scope of the development of applications of latent variable models to a variety of cases now includes models of statistical interaction (e.g., Marcoulides & Schumacher, 1998), and integrative models of both categorical and continuous variables (Bartholomew, 1987; Muthén, 1998), this chapter will focus specifically on the use of two different latent variable models for estimation of growth curves. When the latent variable approach is used to assess differential growth over time, such models can generate statements about both the general and individual overall levels of performance and underlying growth parameters. An additional advantage of the latent variable approach over the usual analytical strategy is that growth parameters and relationships between constructs are not attenuated by measurement error, meaning that change parameters represent inherent change over time and not some confound of actual change patterns and variability in performance due to measurement error (James *et al.*, 1982; Kenny & Judd, 1984; Huba & Harlow, 1987; Bollen, 1989; Willett & Sayer, 1996; DeShon, 1998).

An additional advantage of recent latent variable models of differential growth and change is that they can be used to specify and evaluate whether organisms demonstrate substantially different change and growth patterns over time. Using latent variable models of growth, for example, it is possible to determine whether more than one trajectory of performance over time exists within a particular population. At first blush, this point may seem obvious to biologists. For example, if a researcher is interested in studying trees of different species over time, it seems obvious that different trajectories of growth are in fact present. Such differences in growth could occur as a function of tree species or tree size at the start of the study, or position in the canopy. It may be, however, that distinct differences in such trajectories may exist even within a homogeneous assay over time, which may reflect the presence of multiple survival, reproductive, or growth strategies. Conversely, even if the population is composed of multiple species, biologists may differ in their evaluations of whether one or more than one general growth trend may be present in the data. The biological sciences in particular have a long history of debate concerning classification of organisms (Appel, 1987). Although recent advances in structural equation modeling software have made this approach more accessible to many, it should be noted that this approach has a long history in both developmental psychology and the psychology of individual differences. Estes (1956) observed that simply computing the average performance at each time point could mask important patterns of growth or learning. Statistical models of growth curves began with the early work of Rao (1958) and Tucker (1958, 1966) on longitudinal factor analysis and "Tuckerized" growth curves.

That some organisms demonstrate more variability than others in response to the environment has led biologists to speculate that such variability may have particular reproductive or survival value in some settings. Under a variety of environmental conditions, some subspecies of organisms manifest a wide variety of phenotypes, while others demonstrate very few changes in characteristics. This differential performance of the genotype under different environments is termed the "range of reaction" for the genotype (Dobzhansky *et al.*, 1977). For example, in a classic study of the range of reaction, Clausen *et al.* (1940) used plants of the same species (*Potentilla glandulosa*) native to alpine, mid-elevation, and sea level. Plants from each type of environment were cut into three parts, and these cuttings planted in gardens at sea level, mid-altitude, and alpine zones. The study found markedly different ranges of reaction for these three subspecies of the same plant. Plants native to sea level grow to about the same sizes at both mid-elevation and sea level, but died at alpine elevation. Cuttings from the alpine elevation were uniformly small and grew at all elevations, but grew largest under mid-elevation conditions. Cuttings from the mid-elevation site produced the largest range of reaction. These cuttings thrived at mid-elevation, and survived at other elevations, but were much smaller than cuttings native to the elevation. The conclusions of the study were that the adaptability of the three subspecies (or ecotypes) of the same plant was markedly different and that the cuttings of the plants were not only different in appearance but also adaptively fittest within their native habitats.

Similar research strategies in animal research have explored the role of the genetic substrate in problem-solving behavior. Henderson (1970), for example, has found marked differences in the variability of performance of inbred and cross-strains of mice to "problem solve" in a food finding task which necessitated climbing a wire-mesh ladder, running along a narrow ramp or other such "real-world" situations. Such variability has been termed "heterosis" (Parsons, 1972, p. 95) and is frequently observed in enriched laboratory conditions (Henderson, 1970, 1981). Heterosis is characterized by both an overall better performance on the task and a wide variation of performance on the task. Although this variability has led some researchers to advocate transforming the data so that additive relationships obtain and heteroscedasticity is reduced (e.g., Henderson, 1968, 1970, 1973, 1981), such variability may actually represent a remnant component of the natural genetic variability of organisms in its own right. Further, although such transformations may reduce heteroscedasticity in the cells of a research design, such transformations also make any inferences regarding statistical interaction difficult, because tests of interactions are not scale invariant. Tests of whether differential change occurs as a function of time, for example, could be made to appear or disappear under various choices of transformation. Finally, when organisms are observed in the wild, environmental changes may differentially affect some organisms and not others, resulting in differential patterns of variability over time, which may or may not be identifiable as species-specific variation.

Although more traditional methods such as repeated measures analysis of variance (ANOVA) and other linear models have been typically applied to problems of this nature and have sometimes been advanced as the uniformly preferred approach, there are several limitations to these approaches. Repeated measures ANOVA takes into account only mean structures whereas structural equation modeling (SEM) techniques can also take into account the covariance structure, means, and variances of the growth parameters and errors (Meredith & Tisak, 1990). The combination of time-variant variables and time-invariant variables allows for a powerful way of modeling individual changes in time and provides many advantages over simpler ANOVA models (Duncan & Duncan, 1995; Duncan *et al.*, 1999). Further, Raykov (1997) noted that SEM has advantages over repeated measures ANOVA because the assumptions such as covariance matrix homogeneity across levels of between-subject factors and sphericity are more flexible in SEM analysis. He further suggested that an advantage of SEM is that it allows greater flexibility when testing models because SEM analysis can accommodate any error structures as long as the model is identified. This is far less restrictive than the error structures that are required for repeated measures ANOVA (Willett & Sayer, 1994). Recently, Hancock *et al.* (2000) have extended these arguments to the case of MANOVA as well and have presented Monte Carlo simulations to support the conclusion that the statistical power is especially affected by such violations of the model. McArdle (1989, 1991), McArdle & Epstein (1987), Meredith & Tisak (1984, 1990) and Muthén (1989, 1997) showed the technical developments of using SEM for longitudinal data analysis.

Development of growth curve methodology

Bryk & Raudenbush (1992) are among the pioneers of the development of hierarchical growth modeling, which is a multilevel regression model (often called a random coefficients model) for hierarchical or nested data. They developed an algorithm based on Bayes' estimation that allowed for estimation of the level 1 and level 2 parameters simultaneously. When applied to longitudinal data this methodology is very similar to latent growth models. Simply put, most multilevel data is nested, e.g., students nested within schools or trees nested within a particular watershed. The level 1 variables are the students or trees, and the level 2 variables are schools or watersheds, which are identical for groups of individuals attending a particular school or living in a particular watershed. When this approach is applied to longitudinal data, the data are considered nested by grouping occasions within individuals. Although these methods are highly similar, we choose to discuss the applications of SEM in this text. We refer the reader to an excellent comparison of these two approaches by MacCallum *et al.* (1997).

Another approach to the assessment of intra-individual change over time is group-based modeling of developmental trajectories (e.g., Nagin, 1999). Under these models, the researcher believes that there exist separate groups that differ in their performance or growth over time, but that the researcher does not know how many groups there are *a priori*, nor does the researcher know which individuals belong in which groups. This family of models, which can include both continuous and categorical variables, is distinct from the continuous models presented here. In group-based models, individuals are thought to be members of a discrete, unique trajectory over time. No individual differences in degree are thought to exist under this model. Raudenbush (2001) contrasted the continuous growth models described in this discussion with group-based models by citing examples of two psychological constructs, vocabulary growth in children and depression in a population. A continuous latent curve model better represents children's vocabulary scores because the processes are generally monotonic and vary regularly within the population. In biological data, assays of size, height, or weight would seem best modeled by a latent curve approach if the individuals possess these characteristics to varying degrees. Raudenbush cited the example of depression as a developmental process that is better modeled by a group-based approach. A group-based model for depression is more appropriate than a latent curve approach because it is not reasonable to assume, for example, that everyone is increasing (or decreasing) in depression. Further, some individuals will always be high in depression, while others will always be low and others may become depressed (or recover) over a given period. In biological research, a group-based approach may be more appropriate when a biologist is interested in understanding reproductive or territorial behaviors. Transit or breeding-nest location, for example, cannot be thought of as increasing in degree over time, but may be part of a coherent strategy over the life course of an organism.

The crux of most of these models is that individuals (i.e., people, animals, plants) are allowed to vary or change in some systematic way over the course of time. These methods allow for other variables to predict individual developmental differences for the individual units of analysis (Duncan & Duncan, 1995). Latent curve analysis or LCA is a technique using SEM with a very structured set of fixed parameters. Rather than allowing the individual paths to be freely estimated, LCA sets these paths in such a way as to structure the meaning of the variance of each factor in a particular way. This technique uses information obtained from both the covariance structure of the variables in the data matrix and the means, variances, and covariances for the growth parameters (Curran, 2000). Willet & Sayer (1996), Duncan & Duncan (1996), Raykov (1997), and Curran *et al.* (1997) have provided good examples of the use of this approach in applied research in psychology.

In the standard latent growth model, a set of factors is extracted from a set of repeated measures that correspond to the intercept, linear, and exponential growth parameters. It is only possible to extract $K-1$ factors where K is equal to the number of time variant indicators. The paths from the intercept factor to the longitudinal indicators are all set to 1. This intercept has different interpretations based on how the time scores for the other parameters are set. The scores for the linear component would be fixed to increasing values such as 0, 1, 2, 3, and 4 for five time points. In this case, the intercept would be defined as the mean of the time point where the linear time point is zero (i.e., the first point). The selection of the location of the intercept is conceptually important depending on the research question because the interpretation of this intercept depends upon its location. For instance, if the research question is the differential growth of the same species of tree under different water and light conditions, setting the intercept at the first time point would not be wise considering that the factors influencing growth have not occurred yet and the trees are all the same size. In this case, it would make sense to set the time point at maybe the last time point as an outcome measure. In this case the growth parameters would be $-4, -3, -2, -1, 0$. It should be noted in this case, however, that such changes in scale, although not dramatically affecting the recovered quadratic curves can also be specified by squaring all of the time scores (i.e., 0, 1, 4, 9, 16) and by adjusting the order and sign of the time point in the previous example (i.e., $-16, -9, -4, -1$, and 0). The factors called growth parameters can then be predicted by other factors such as water level (a continuous factor) or species (a categorical factor). Muthén & Muthén (1998) have pioneered the techniques and algorithms for the latter approach.

An important aspect of the unrestricted growth model is the ability to model the covariance between the intercept and slope (i.e., linear growth curve) parameters. This can be useful when it is of interest to determine whether a relationship exists between slope and intercept. For example, if a higher starting value for some individuals predicted a higher rate of growth (higher slope) than those individuals who start at a lower value, this would suggest the "rich get richer" scenario. A significant and positive covariance would be indicative of this relationship. In psychology, this is often found when higher achieving students have greater gains in mathematics scores than lower achieving students. In biology, such patterns could occur when one set of organisms within a population demonstrates a trajectory leading to a substantially larger phenotype than other organisms, or could achieve a larger phenotype due to such properties as hybrid vigor.

The conditional growth curve model includes variables that explain the variance associated with the intercept and linear growth parameters. Here it may be determined whether other variables influence the rate of growth or the mean level of the intercept (see Figure 13.2). These models can be calculated with any standard SEM program. However, programs such as EQS (Bentler, 1995), LISREL (Jöreskog & Sörbom, 1993), Mx (Neale, 1994) and Mplus (Muthén & Muthén, 1998) are superior due to the flexibility of modeling and the number of options available to the user.

MacCallum *et al.* (1997) reviewed a set of methods in which two sets of time-varying variables can be modeled to determine the relationships of change over both variables. In this case, rather than having time-invariant variables to explain the magnitude of intercepts and growth curves, two time-varying variables are modeled together to show the relationship between growth in one variable (i.e., growth of human bones) and changes in another variable (i.e., levels of human growth hormone in the body). This adds to the flexibility of these models by showing how two variables interact over time. Our data example will demonstrate both a conditional growth model and a multivariate latent growth curve model.

LCA has the same statistical advantages as regular SEM models. It allows testing of model fit to determine the adequacy of a hypothesized model, individual growth trajectories to be averaged into one mean trajectory, and for time-varying covariates (Duncan & Duncan, 1995). However, SEM requires very large sample sizes, multivariate normality in all variables in the model, and that the spacing of time points be identical. However, for violations of the latter requirement, the programs HLM (Raudenbush *et al.*, 2000) and Mplus (Muthén & Muthén, 1998) have incorporated ways of adjusting for unequal spacing of time points.

A limitation to the conditional growth model is that the covariance between the slope and intercept cannot be assessed directly except by a covariance between the error terms for the parameters. As will be shown below, an advantage of manifest growth curve modeling is the way in which the model is set up to allow the growth parameters to be exogenous, thereby allowing the researcher to be able to assess this covariance directly even in a conditional growth model. Further limitations are that there is no easy way in LCA to specify a factor beyond anything higher than a quadratic growth curve (i.e., S-shaped curves are cubic because of two curves in the line of best fit). At most the quadratic and cubic terms used can represent only an exponential growth curve rather than specifying the actual shape of the curve such as a parabolic curve (quadratic) or an S-curve (cubic). This is because the time scores that are used in the structural matrix are set to be integers around the zero point, at which you will set the intercept values.

The present chapter presents two methods for this purpose. Both methods consist of two steps. In the first step parameters are estimated that describe the series of observations at the level of individuals. The second step incorporates these parameters within a general structural model.

Polynomial extraction

The present chapter presents methods for this purpose. Specifically, a method is presented that consists of two steps. The first step involves estimating parameters that describe a series of observations at the level of individuals. The second step involves devising a structural model for these parameters. Consider the full LISREL model

$$\eta = \mathbf{B}\eta + \boldsymbol{\Gamma}\boldsymbol{\xi} + \boldsymbol{\zeta}, \qquad (13.1)$$

where $\boldsymbol{\eta}^{\mathrm{T}} = (\eta_1, \ldots, \eta_m)$ are the latent factors on the dependent variable side, $\mathbf{B}$ is the covariance matrix of the $\boldsymbol{\eta}$ factors, $\boldsymbol{\xi}^{\mathrm{T}} = (\xi_1, \ldots, \xi_n)$ are the latent factors on the independent variable side, $\boldsymbol{\Gamma}$ is the matrix of the covariances of $\boldsymbol{\eta}$ and $\boldsymbol{\xi}$ factors, and $\boldsymbol{\zeta}^{\mathrm{T}} = (\zeta_1, \ldots, \zeta_m)$ is a vector of residuals (see Jöreskog & Sörbom, 1993; Hershberger *et al.*, Chapter 1,). The measurement model for the dependent variables is

$$y = \boldsymbol{\Lambda}_y \boldsymbol{\eta} + \boldsymbol{\varepsilon}, \qquad (13.2)$$

and the measurement model for the independent variables is

$$x = \boldsymbol{\Lambda}_x \boldsymbol{\xi} + \boldsymbol{\delta}, \qquad (13.3)$$

where $\boldsymbol{y}^{\mathrm{T}} = (y_1, \ldots, y_p)$ and $\boldsymbol{x}^{\mathrm{T}} = (x_1, \ldots, x_q)$ are the observed dependent and independent variables, respectively, and $\boldsymbol{\varepsilon}^{\mathrm{T}} = (\varepsilon_1, \ldots, \varepsilon_p)$ and $\boldsymbol{\delta}^{\mathrm{T}} = (\delta_1, \ldots, \delta_q)$ are residual vectors (also called error variables). In the following sections we describe how, in order to model longitudinal information, polynomials can be inserted into $\boldsymbol{y}^{\mathrm{T}}$ and then predicted from time-constant or time-varying variables $\boldsymbol{x}^{\mathrm{T}}$.

Approximation of time series using orthogonal polynomials

There exist many methods that can be used to approximate or smooth time series. Among the most flexible functions are polynomials. Polynomials are used in repeated measures analysis of variance (Neter *et al.*, 1996) and in regression analysis (von Eye & Schuster, 1998). Their main benefits include the following: that

1. Virtually any series can be satisfactorily approximated using polynomials; it should be noted however, that cyclical series and series that approach an asymptote often require complex or special polynomials.
2. The polynomial coefficients often have a natural interpretation.
3. Polynomial coefficients typically are estimated using least squares methods; thus, the resulting estimates have the same desirable characteristics as all least squares estimates, but they also share the problematic characteristics such as outlier dependency.
4. Standard statistical software can be used to estimate polynomial parameters.

Most useful are *orthogonal polynomials* (see Abramowitz & Stegun, 1972; von Eye & Schuster, 1998). Consider a function that is defined for all real-valued $\boldsymbol{x}$ and does not involve any divisions. This function is called a *polynomial* and has the form

$$y = b_0 + b_1\boldsymbol{x} + b_2\boldsymbol{x}^2 + \cdots = \sum_{j=0}^{J} b_j \boldsymbol{x}^j, \tag{13.4}$$

where the real-valued b_j are the polynomial parameters. The $\boldsymbol{x}$ values are vectors that contain the so-called polynomial coefficients. These coefficients describe polynomials in standard from. The parameters in the equation are linear. The $\boldsymbol{x}$ values are raised to the power that corresponds to the order (also called degree) of a polynomial. For example, a third-order polynomial contains a term x^3. Polynomial coefficients can be found in most textbooks of analysis of variance (e.g., Kirk, 1995).

In this chapter we focus on *orthogonal polynomials*. Consider the form

$$y = b_0 + b_1\xi^T + b_2\xi_2^T + \cdots = \sum_{j=0}^{J} b_j\xi_j^T, \tag{13.5}$$

where the ξ_j^T are polynomials of the j-th order themselves. Equation 13.5 describes a *system of polynomials*. The equation describes a system of *orthogonal* polynomials if any pair of different polynomials in the system meets the condition that

$$\sum_i y_{ij} y_{ki} = 0, \tag{13.6}$$

for $j \neq k$, where j and k index polynomials of different order, and i indexes polynomial coefficients. In other words, the orthogonality condition given in equation 13.6 states that for any two orthogonal polynomials from a system to be orthogonal, the cross-product must be zero.

Consider the following example. A time-series involves five observations. A researcher asks if this series can be described using a first-order and a second-order polynomial from a system of orthogonal polynomials. The coefficients for the linear polynomial are $\boldsymbol{x}_1^T = (-4, -2, 0, 2, 4)$. The coefficients for the quadratic polynomial are $\boldsymbol{x}_2^T = (2, -1, -2, -1, 2)$. The cross-product (inner product) of these two polynomials' coefficient vectors is $-4 \times 2 + -2 \times (-1) + 0 \times (-2) + 2 \times (-1) + 4 \times 2 = 0$.

In addition to being mutually orthogonal, the coefficients in a vector $\boldsymbol{x}$ are typically set up so that they sum to zero. It should be noted that this second characteristic can always be produced by centering variables. Centering, however, does not always yield orthogonal vectors.

The ξ_j^T in equation 13.5 are polynomials themselves. To illustrate, consider a second-order, i.e., quadratic, polynomial. By replacing the ξ_j^T, this polynomial can be expressed as

$$y = \beta_0 + \beta_1(\alpha_{10} + \alpha_{11}x) + \beta_2(\alpha_{20} + \alpha_{21}x + \alpha_{22}x^2), \tag{13.7}$$

where the β_j are the weights (parameters) for the polynomials of degree j, and the α_{il} are the parameters within the polynomials (for $l = 0, 1, 2$).

When smoothing time-series using polynomials, orthogonal polynomials have a number of important advantages over nonorthogonal polynomials. First, the coefficients of orthogonal polynomials are standardized. As a consequence, polynomial parameter estimates will be comparable across samples or observation points in time, as long as the number of observation points and the scale units of the dependent measure remain the same. Second, and most importantly, the parameter estimates are independent of the degree of the polynomial. Thus, if researchers decide that they need polynomials of a higher degree, all they need to do is estimate the parameters for the higher degree polynomial. The lower-order parameter estimates will not change and, therefore, do not need to be estimated. Accordingly, dropping any of the polynomials will not affect the remaining parameter estimates.

It is important to note that there exist a large number of systems of orthogonal polynomials. The system introduced and used here is widely used in repeated measures analysis of variance. It has, however, disadvantages when the series assumes certain forms (see above). Other polynomials may be better suited under certain conditions. For example, Tshebysheff polynomials (these are orthogonal trigonometric functions) are well suited to smooth cyclical series (for more detail, see Abramowitz & Stegun, 1972).

Data example

In psychology, the concept of *cognitive complexity* has met with uneven interest over the last 30 years. The number of articles published to this topic is, in five-year intervals between 1971 and 1995: 1971 to 1975, 137; 1976 to 1980, 206; 1981 to 1985, 198; 1986 to 1990, 115; 1991 to 1995, 125 (see von Eye, 1999).

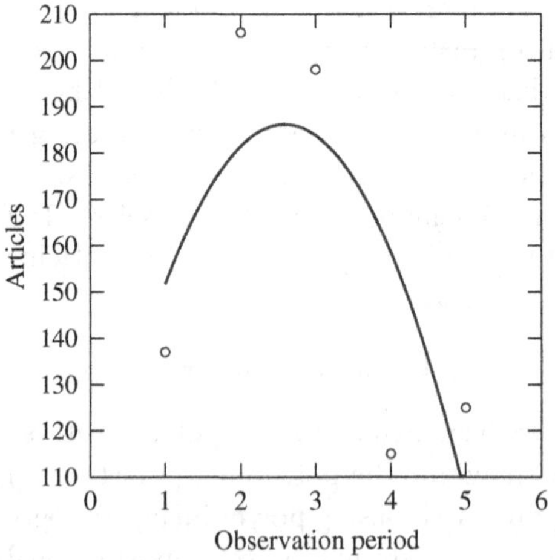

Figure 13.1. Quadratic approximation polynomial for number of publications.

Table 13.1. *Regression analysis for the estimation of orthogonal polynomial coefficients*

Effect	Coefficient	Std Error	Std Coef	Tolerance	T	p (2-tail)
Constant	156.200					
P1	−5.750	3.137	−0.427	1.000	−1.833	0.318
P2	−13.786	5.303	−0.605	1.000	−2.600	0.234
P3	17.000	6.274	0.631	1.000	2.710	0.225

Dep Var: Articles N: 5, multiple R: 0.973, multiple R^2: 0.946, Adjusted multiple R^2: 0.783, standard error of estimate: 19.841

The raw data, depicted by circles, and the least squares-estimated second-order polynomial for these data appear in Figure 13.1. To estimate the polynomial we apply a standard multiple regression model of the form $y = \mathbf{X}\beta + \varepsilon$. The y-values are given by the observed time-series. The column vectors in $\mathbf{X}$ contain the coefficients of the polynomials. For the present example we obtain the following regression equation:

$$\begin{bmatrix} 137 \\ 206 \\ 198 \\ 115 \\ 125 \end{bmatrix} = \begin{bmatrix} 1 & -4 & 2 & -1 \\ 1 & -2 & -1 & 2 \\ 1 & 0 & -2 & 0 \\ 1 & 2 & -1 & -2 \\ 1 & 4 & 2 & 1 \end{bmatrix} \begin{bmatrix} b_0 \\ b_1 \\ b_2 \\ b_3 \end{bmatrix} + \begin{bmatrix} e_1 \\ e_2 \\ e_3 \\ e_4 \\ e_5 \end{bmatrix}$$

Estimating the parameters using SYSTAT yields the regression values shown in Table 13.1. The results in Table 13.1 suggest that a third-order polynomial allows us to explain almost 95% of the variance of the time series. Yet not one of the polynomials by itself makes a significant contribution. This does not come as a surprise because there are only five measures in the series. The tolerance scores in the third column from the right indicate the proportion to which each predictor remains *un*explained by the respective other predictors. The tolerance values in Table 13.1 show that the variance of each of the predictors, i.e., the polynomial coefficients, remains 100% unexplained by the respective other predictors, because they are orthogonal. This is also indicated by the variance–covariance matrix of estimated regression coefficients (not reproduced here). All coefficients in the off-diagonal cells of the correlation matrix are zero, indicating that estimates of regression weights are independent of each other.

The polynomial coefficients have a natural interpretation. The zero-order parameter is the arithmetic average of the dependent measure. This is always the case when predictors are centered. In the present example this is easily confirmed. The constant 156.2 is the average number of publications over the observation interval. The first-order parameter is identical to the standard first-order regression coefficient. It indicates the number of steps on the dependent measure that go hand in hand with a one unit step on the independent measure, given all the other predictors in the equation. In the present example, the estimate of $b_1 = -5.75$ suggests that, with each step on the x-scale from -4 to 4, the number of publications decreases by 5.75. The second-order parameter describes the strength of the acceleration ($b_2 > 0$) or deceleration ($b_2 < 0$) of the linear trend. In the present example we find that the linear trend is decelerated by $b_2 = -13.79$. This indicates that the number of publications on cognitive complexity decreases at an accelerated pace as the distance from the mean of the observation intervals increases. This is counterbalanced by a positive cubic trend, as can be seen in Figure 13.1.

In addition to the above-mentioned characteristics, the polynomials estimated using the methods described here have a number of characteristics that make the present approach particularly interesting. Two of these characteristics will be reviewed briefly here. First, in practically all instances in which scientific parsimony guides data analysis, the number of parameters estimated will be smaller than the number of data points in a time-series. Thus the transformed time-series will be more parsimonious than the observed one. When the sample is of limited size, this can be a major plus. Second, the smoothed series may describe the time-varying (or constant) behavior even better than the original data. The raw data typically come with measurement errors that can alter trends. The smoothed curves typically indicate less erratic development over time.

One constraint inherent in the approximation method reviewed here is that the observation points must be equidistant in time. If they are not equidistant, they must be at the same distances for each respondent. If this cannot be achieved, methods of orthogonal polynomial approximation must be used that allow for nonequidistant points on the x-axis (for an application of such methods see, for instance, von Eye & Nesselroade, 1992).

When estimating polynomial parameters for individuals for later use in a structural model we set the following criteria:

1. The significance of the regression model or of single parameters is of subordinate importance because the time-series we smooth are typically short. However, even if a series is longer than the one

used in the example, and significance testing may be more meaningful, the multiple R^2 is more important for assessing the polynomial goodness-of-fit.

2. The polynomial must be of the same type and degree for each case.
3. The multiple R^2 must be at the same level or below the reliability of the measure used in the time-series.

Using polynomial parameters in structural equations modeling

The present approach starts modeling, just as in any other structural equations modeling approach, from a rectangular data matrix, **X**, of size $N \times q$. **X** is transformed in some matrix of covariation, typically a variance–covariance matrix. In contrast to modeling with **X** containing raw data, the present approach proposes using polynomial parameter estimates instead. Thus the input data matrix still has the usual form. However, the approximation of raw data by orthogonal polynomials preceded the modeling step. Thus **X** no longer contains raw data but the parameter estimates of the polynomials.

Specifically, we propose the following two-step procedure for modeling time series.

1. Estimate polynomial parameters for each of the repeatedly observed p-dependent and the q-independent variables; create a data matrix, **X**, with N rows and $\sum_i p_i c_i^p + \sum_j q_j c_j^q$ columns, where p_i is the i-th dependent variable, c_i^p is the number of polynomial parameters estimated for the i-th dependent variable, and q_j and c_j^q represent the dependent variables.
2. Proceed with structural modeling using the thus-created data matrix.

To illustrate, consider a situation where two independent variables, A and B, are observed six times. The researchers approximate the resulting series of measures using third-order orthogonal polynomials, which reduces the number of data points from 12 to 8 per person. The resulting data matrix, **X**, for this example is

$$\begin{bmatrix} a_{01} & a_{11} & a_{21} & a_{31} & b_{01} & b_{11} & b_{21} & b_{31} \\ \vdots & \vdots & \vdots & \vdots & \vdots & \vdots & \vdots & \vdots \\ a_{0N} & a_{1N} & a_{2N} & a_{3N} & b_{0N} & b_{1N} & b_{2N} & b_{3N} \end{bmatrix}, \tag{13.8}$$

where a_{01} is the zero-order polynomial parameter estimate for the first person for the first dependent measure, and so forth. The variance–covariance matrix, $\mathbf{\Sigma}$, which is modeled through structural equations, describes the variances and covariances of the polynomial parameters rather than the original, repeatedly observed variables. Note that the covariances can be unequal to zero although the polynomials are orthogonal because, in the population, the time series can have a common structure.

The supermatrix analyzed by programs such as LISREL (Jöreskog & Sörbom, 1993) is

$$\mathbf{\Sigma} = \begin{bmatrix} \mathbf{\Sigma}_{yy} & \mathbf{\Sigma}_{yx} \\ \mathbf{\Sigma}_{xy} & \mathbf{\Sigma}_{xx} \end{bmatrix}, \qquad (13.9)$$

where $\mathbf{\Sigma}_{yy}$ contains terms concerning the covariation of the dependent manifest variables, $\mathbf{\Sigma}_{yx}$ and $\mathbf{\Sigma}_{xy}$ contain terms concerned with the covariation of the dependent and the independent variables, and $\mathbf{\Sigma}_{xx}$ contains terms concerned with the covariation of the independent variables. More specifically, Σ_{yy} explains the covariation among the y-variables in terms of their factor structure, the relationships among the factors on the y-side, and the relationships among the factors on the y-side and the factors on the x-side. In more technical terms, this submatrix is

$$\mathbf{\Sigma}_{yy} = \left[\mathbf{\Lambda}_y[\mathbf{I} - \mathbf{B}]^{-1}(\mathbf{\Gamma\Phi\Gamma}^{\mathrm{T}} + \mathbf{\Psi})(\mathbf{I} - \mathbf{B}^{\mathrm{T}})^{-1}\,\mathbf{\Lambda}_{y^{\mathrm{T}}} + \mathbf{\Theta}_{\varepsilon}\right], \qquad (13.10)$$

where $\mathbf{\Lambda}_y$ is the matrix of loadings of the y-variables on their factors, $\mathbf{I}$ is an identity matrix, $\mathbf{B}$ is the matrix of asymmetric relationships among the factors on the y-side, $\mathbf{\Gamma}$ is the matrix of relationships between the factors on the y-side and the factors on the x-side, $\mathbf{\Psi}$ is the matrix of covariations between the factors on the y-side, and $\mathbf{\Theta}_{\varepsilon}$ is the variance–covariance matrix of the residuals of the variables on the y-side.

Most important from the present perspective is that the variables on the y-side are no longer the original variables. Rather, they are the parameter estimates of the polynomials that describe the time series. Thus we do not explain the covariation among repeatedly observed variables. Instead, we explain variances and covariances of polynomial parameters. So, for instance, we explain the covariation between the linear slopes of two variables. In different words, rather than modeling the instantiations of a series of measures, we model the characteristics of the series. If zero-order polynomial parameters are part of a model, the means structure of a time-series will also be described.

This applies accordingly if the x-variables are time varying, too. The submatrix for the x-variables is

$$\mathbf{\Sigma}_{xx} = [\mathbf{\Lambda}_x \, \mathbf{\Phi} \, \mathbf{\Lambda}_{x^{\mathrm{T}}} + \mathbf{\Theta}_{\delta}], \tag{13.11}$$

where $\mathbf{\Lambda}_x$ is the matrix of the loadings of the independent variables on their factors and $\mathbf{\Theta}_{\delta}$ is the variance–covariance matrix of the residuals on the x-side. As for the y-variables, the loadings in equation 13.1 are not the loadings of the original x-variables on their factors but rather the loadings of the polynomial parameter estimates. Thus equation 13.1 describes the covariation of the polynomials that depict the time-series of the independent variables.

Finally, the covariation of the variables on the x- and the y-sides is

$$\mathbf{\Sigma}_{xy} = \left[\mathbf{\Lambda}_x \, \mathbf{\Phi}\mathbf{\Gamma}^{\mathrm{T}}(\mathbf{I} - \mathbf{B}^{\mathrm{T}})^{-1} \, \mathbf{\Lambda}_y^{\mathrm{T}}\right]. \tag{13.12}$$

Equations 13.9 through 13.12 are the same as those found in standard introductory textbooks of structural equations modeling (e.g., Schumacker & Lomax, 1996). However, again, the covariation to be explained is between parameter estimates of polynomials. As a matter of course, time-constant variables and predictors that were measured only once can be included together with the polynomial parameters. The following section presents a data example.

Data example using manifest growth curve modeling

A data example using biological data will be presented here using the methods described above. The data used in this example have been previously published (Keeland & Sharitz, 1995, 1997; Young *et al.*, 1995; Keeland *et al.*, 1997) and were designed and collected by Bobby D. Keeland. The circumference of the trunk of each tree was measured at the beginning of the spring and monitored and recorded every week through the growth season for 34 weeks. These measurements were collected on three species of tree in the wetlands of South Carolina and Louisiana. Tree species were swamp tupelo (*Nyssa sylvatica*), water tupelo *(Nyssa aquatica) and* baldcress cypress *(Taxodium distichum)* and will be referred to henceforth as NS, NA, and TD, respectively. Also measured was the type of flood plain in which the tree grew and whether the tree grew at a canopy or sub-canopy level. The data have some unique characteristics. First, they have an S-shaped growth curve, due to the fact that growth during the spring is slower at first as the trees begin to leaf out, then rapid growth is observed during the summer months followed by reduced growth nearing the fall when the trees begin to lose their leaves. This pattern of data is ideal for growth modeling using this methodology. Growth may be perceived as linear if regression type methods are used, but the quadratic (one bend in the line) and cubic (two bends in the line) are more accurate in describing this S-shaped curve (note that third-order polynomials may not be able to describe an S-shaped curve well if there exist asymptotes; in this case, higher-order polynomials or other types of functions may be needed).

Second, the data have 34 time points of measurement. For our purposes this is too many. Although some techniques exist to analyze this number of time points (e.g., spectral analysis), for our purposes the data were reduced to eight time points by aggregating between two and four contiguous time points (i.e., one-month intervals). MacCallum *et al.* (1997) recommended that the number of time points be moderate when the models under consideration are simple. No more than five or six time points should be used if only a linear function is to be extracted from the model. For more extensive models when the shape of the curve is to be estimated, like a growth curve with an S-shape, considerably more time points should be added. For extracting three parameters, eight time points should be sufficient.

Third, there is a two-level structure to our data. The time-varying data consist of the growth measurements (level 1 variables). These data are subjected to a regression analysis with orthogonal polynomials in order to extract the growth parameters. The growth parameters will consist of the intercept, the mean circumference of the trees at time 1, and three growth patterns – linear (straight line growth), quadratic (exponential growth), and cubic (S-shaped growth). This regression used the aggregated weekly growth measurements as the dependent variable (since there were eight measurements, this would be calculated as an *n* of eight) with the orthogonal contrasts for an eight time point series with three polynomials as the independent variables (linear, quadratic, and cubic). In the unstandardized solution the intercept is calculated automatically by adding a vector of 1s in the matrix of orthogonal contrasts. When choosing weights it is important to remember to choose contrasts that have a shape similar to the pattern you are interested in testing. Although it makes no difference in the magnitude of the regression weights, the pattern of positive and negative values in the contrasts will affect the signs of the regression weights. (Please see Figure 13.3 to refer to the signs and shapes of the three curves used in this analysis.) The weights above refer to positive curves. Reversing the signs will extract negative curves.

Table 13.2. *Orthogonal contrasts for eight groups deriving linear, quadratic, and cubic functions*

Intercept	1	1	1	1	1	1	1	1
Linear	−7	−5	−3	−1	1	3	5	7
Quadratic	7	1	−3	−5	−5	−3	1	7
Cubic	7	−5	−7	−3	3	7	5	−7

This regression is run for each tree (subject or unit of analysis), thus producing four parameters for each subject. In the second level of data are the time-invariant variables (level 2 variables), which are those that do not change over time. In this example, the time-invariant variables consist of the flood plain, species, and canopy of the tree. One of these variables is a dichotomous variable (canopy) and is coded 0 = subcanopy and 1 = canopy. With independent variables of this kind, the t-value of the path essentially becomes a mean difference of the endogenous variable by means of the level of the exogenous variable with the other exogenous variables controlled for. Another variable (species) is a trichotomous (three-group categorical) variable. For variables of this kind it is necessary to dummy-code the variable. Standard regression textbooks address dummy- and effect-coding and it is up to the researcher to decide which coding procedure to use for each case. In this case the dummy-coding scheme compared one species of tree with the other two in the series by using a standard dummy-coding strategy: the last variable, flood plain, is continuous. Even though this can be considered categorical, the flood plain patterns have a logical order of increased water exposure: very shallow infrequent flooding; shallow flooding, deep frequent flooding, shallow permanent flooding, and deep permanent flooding. Thus the order of this variable is as above with the anchors being 1 = shallow infrequent flooding to 5 = deep permanent flooding. By means of a structural equation model, the level 2 variables (i.e., flood, canopy, and the two species dummy-coded variables) are regressed upon the four level 1 variables. The model was solved through a maximum likelihood solution. Please see Figure 13.2 for the diagram of this structural model. Several covariances were added to the disturbance terms and the exogenous variables in order to improve model fit. It was decided that, in order to improve model fit, all covariances between each of the disturbance terms were specified except for one. This exception was needed to ensure model identification. The same strategy intercorrelated the

Table 13.3. *Dummy-codes used for categorical variables*

Species	Dummy code 1	Dummy code 2
Baldcress Cypress (TD)	0	0
Water Tupelo (NA)	1	0
Swamp Tupelo (NS)	0	1

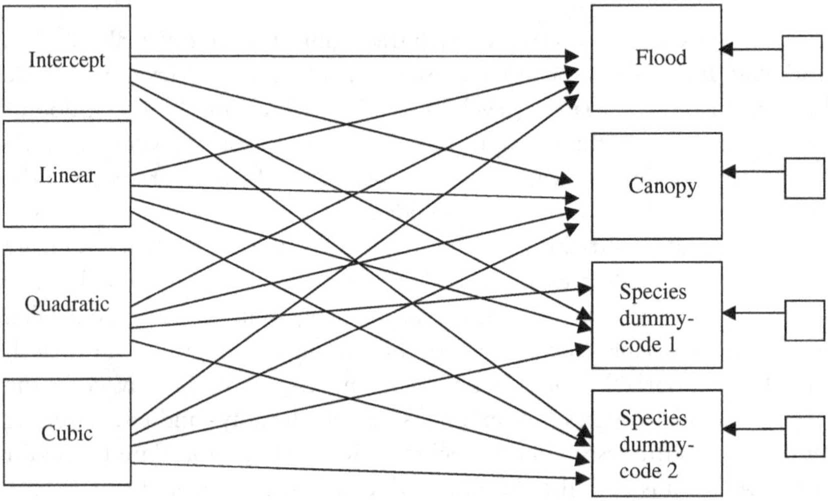

Figure 13.2. Manifest structural equations model depicting the intercept, linear, quadratic, and cubic derived parameters predicting the dependent continuous and categorical variables.

exogenous parameters (i.e., intercept, linear, quadratic, and cubic growth parameters), leaving one unspecified so as to ensure identifiability.

The model showed acceptable model fit by some indicators including the CFI, NFI, and GFI; however, two often used indicators showed poor fit, the NNFI is very low and the chi-square (χ^2) is significant (see Table 13.4 for explanations of abbreviations). Often with large sample sizes, the χ^2 is significant and is often not a good indicator of model fit, but the NNFI result is puzzling considering that it is usually higher, for this kind of analysis than the NFI. It may be that this indicator is not accurate rather than lower. In any event, this model appears to fit well and its residual matrix showed very few high values.

The paths in the model are shown in the next three tables. Table 13.5 shows the values from the $\mathbf{\Gamma}$ matrix, which consists of the paths between

Table 13.4. *Fit functions*

Fit function	Value
Goodness-of-fit index (GFI)	0.9670
Chi-square (χ^2)	87.93
Degrees of freedom (df)	1
Probability	<0.0001
Bentler's comparative fit index (CFI)	0.9629
Nonnormed fit index (NNFI)	0.2415
Normed fit index (NFI)	0.9728

Table 13.5. *Standardized path coefficients:* $\mathbf{\Gamma}$ *matrix*

	Intercept	Linear	Quadratic	Cubic
Flood	0.0733	0.7280*	−0.2825*	−0.7865*
Canopy	0.6622*	−0.0826	−0.1075*	0.1031
Species 1	−0.2549*	0.4103*	−0.0145	−0.0785
Species 2	0.2648*	−0.2212	0.1541*	−0.0362

Paths significant at $p < 0.05$ are marked with an asterisk.

the exogenous variables and the endogenous variables, excluding disturbances. Standardized path coefficients are displayed along with an asterisk to show which paths are significant at the $p < 0.05$ level. We will begin our interpretation of this table with the continuous variable flood in order to show how a continuous variable is interpreted with polynomials. This table shows that flooding depth and frequency is a major determinant of the rate of growth in these trees. The intercept indicates that the initial growth rate of the trees does not depend upon the level of water in which they grow. However, the amount of water in the flood plain definitely does have an effect on growth. The linear function indicates that the slope is positive and that increasing water levels in a certain flood plain will produce greater growth in these trees. The positive linear function is the ascending one in the first graph in Figure 13.3. The negative linear function is the descending one. There are other patterns independent of this linear function that also account for the pattern of these data. The quadratic term is significant and negative. This implies that the growth function begins gaining momentum exponentially. Towards the end of the growth cycle the curve asymptotes, thus indicating a slowing of the rate of growth as the season closes. A positive

(a)

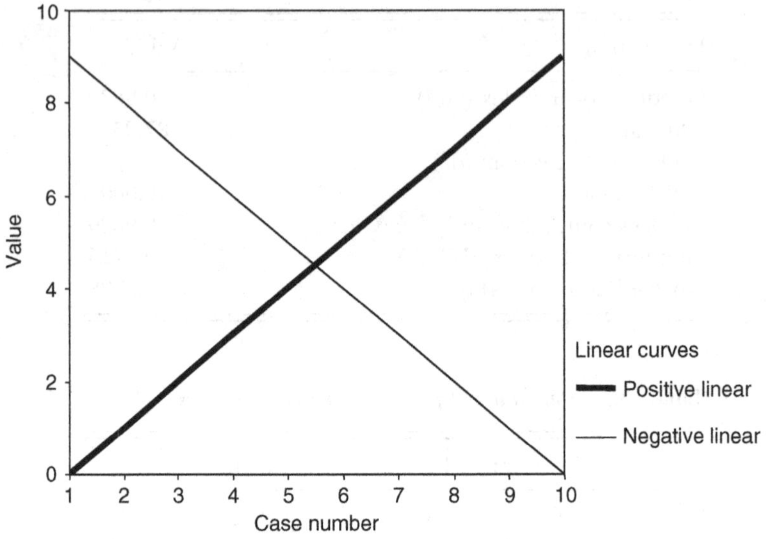

(b)

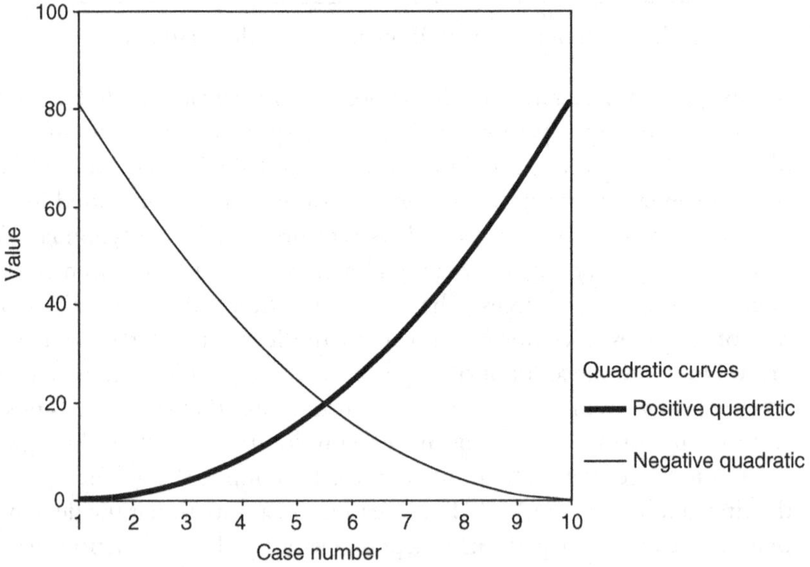

Figure 13.3. Depictions of the line of best fit for (a) liner (b) quadratic and (c) cubic.

(c)

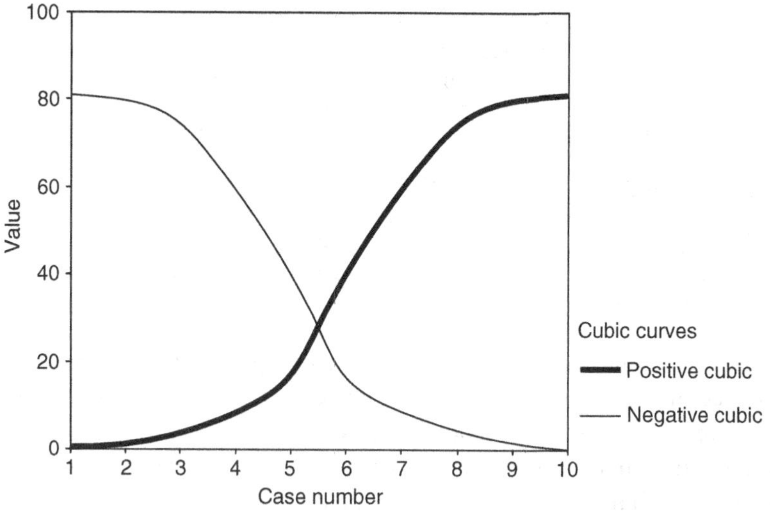

Figure 13.3. *(cont.)*

quadratic function would imply that the growth was gradual at first, gaining momentum towards the end. Please refer to Figure 13.3c for a graphic of the positive (ascending) and negative (descending) curves.

Even though this growth function may not adequately explain all the variance in the pattern of growth, it contributes considerable variance to the pattern of data observed. In addition to this, the cubic function is also significant and positive. A negative cubic function is the classic S-shaped curve, which indicates that growth is slow in the beginning and then speeds up and then slows down towards the end. The positive function is the reverse of this. Remember that each consecutive polynomial adds a bend to the function, implying a change in the rate of growth over time. This significant negative cubic function indicates that, depending on flooding, the rate of growth is influenced by increases in the amount of water available for the trees. However, the cubic function also indicates that growth is slower in the spring and fall, and faster in the summer, thus modifying the effect by which the water level influences the growth of the tree. There is also a large negative correlation between the linear and cubic slopes. This is natural, as the positive linear function is very similar to the one described by the negative cubic function.

The dichotomous variable indicates that canopy level shows a significant positive intercept and a negative quadratic slope. This indicates that

Table 13.6. *Means of the growth parameters*

	Canopy trees ($N = 288$)		Sub-canopy trees ($N = 420$)	
Parameter	Mean	Std	Mean	Std.
Intercept	46.12	10.10	23.45	9.31
Linear	0.65	0.55	0.44	0.49
Quadratic	−0.08	0.08	−0.05	0.08
Cubic	−0.13	0.12	−0.08	0.10

the mean circumference for those trees that are a part of the canopy, and have increased sun exposure, is larger at the time of initial measurement than those that were not a part of the canopy. This of course makes sense, since larger and taller trees make up the canopy. Interpretation of the growth curves is a little trickier. The significant negative quadratic growth curve indicates the shape of the curve only. It does not provide information regarding whether canopy or subcanopy trees grew at a greater rate. To do this, one must calculate the means of the parameters for each level of the dependent variable and run a test on the means to determine whether significant differences in the slopes exist for the means for the growth parameters, (see Table 13.6).

Analyses of variance procedures reveal that, out of four pairs of parameter means, all of the mean differences are significant however, only the quadratic function can be interpreted because it was the only parameter that was significant in the structural equation model above for the canopy variable. Therefore, it is concluded that the shape of the growth curve is negative quadratic and that the canopy trees grew at a greater rate than subcanopy trees. In summary, there are three interpretations to draw from these results: (1) trunk size is bigger for canopy trees; (2) the shape of the growth curve for the canopy variable is curvilinear; and (3) the curvilinear growth is steeper in the canopy trees than in the sub-canopy trees.

Interpretation of the dummy-coding variables is similar. With these contrasts the user can create a set of contrasts that will test mean differences in intercept and slope parameters. Referring back to the dummy-codes listed above, the reader can see how these hypotheses are tested. The first intercept parameter specifies that the second species of tree, NA, is compared with the other two species. The data indicate that the NA had a mean size circumference smaller than that of the other two species of tree (i.e., negative weight). This also indicates that the linear function was significant, large in magnitude and positive, indicating that there was greater growth for these

Table 13.7. *Growth parameters*

	Dummy-code 1: 0		Dummy-code 1: 1	
	NS and TD ($N = 388$)		NA ($N = 320$)	
Parameter	Mean	Std	Mean	Std
Intercept	36.4223549	14.0322382	28.1304559	14.2815907
Linear	0.3702742	0.4371010	0.7149817	0.5600959
Quadratic	−0.0468949	0.0735107	−0.0859955	0.0840062
Cubic	−0.0699042	0.0882498	−0.1360964	0.1176977

	Dummy-code 2: 0		Dummy-code 2: 1	
	NA and TD ($N = 526$)		NS ($N = 182$)	
Parameter	Mean	Std	Mean	Std
Intercept	31.2057688	14.2712225	36.9196989	15.2299975
Linear	0.6121790	0.5487433	0.2772218	0.3453830
Quadratic	−0.0764217	0.0857653	−0.0303075	0.0504846
Cubic	−0.1167710	0.1127803	−0.0508357	0.0719050

NS, Swamp Tupelo; TD, Baldcress Cypress; NA, Water Tupelo.

trees than for the other two. Although we can assess which species of tree was smaller by examining the intercept, remember that the linear growth function gives only the significance of the path and the valence, but the size of each of the linear growth curves by species must be assessed from an examination of the means of the parameters. These means indicate, as we already know, that the NA have smaller trunk circumferences. Further, an examination of the means of the linear growth function show that the linear growth was higher for the NA than for the other two species of tree. Thus this species of tree is faster growing that the other two (for these growth parameters, see upper part of Table 13.7).

The other dummy code compared the NS with the other two trees and shows that these trees grow bigger as compared with the other two and have a shallower but significant positive quadratic growth curve. This indicates that growth is faster in the initial part of the year and tapers off towards the end of the growing season. An examination of the means of the unstandardized regression coefficients show that the linear regression

Table 13.8. *Variances and covariances:* **Θ** *matrix*

	Intercept	Linear	Quadratic	Cubic
Intercept	216.87	0.24696	−0.21686 (n.s.)	0.26379
Linear	1.90922	0.27560	−0.69005	0.97323
Quadratic	−0.25789 (n.s.)	−0.41840	0.00652	−0.65276
Cubic	−0.41840	0.05503	−0.00568	0.01160

n.s., not significant.

coefficient is higher for the other two trees and the NS actually grows at a slower rate than the other two trees (see lower part of Table 13.7). Dummy-coding strategies should be selected on the basis of the researcher's hypotheses regarding the data. This approach offers considerable flexibility in the selection of dummy-coding strategies.

One advantage of this way of conducting growth curve modeling is that the direct covariances between the exogenous growth and intercept parameters can be evaluated. These are shown in Table 13.8. All the values shown are significant at the 0.05 level. The abbreviation n.s. stands for "not specified", indicating that this covariance was not included in the model in order to achieve identification. The general purpose software package SAS provides these estimates as a part of the printout. In Table 13.8 standardized covariances (i.e., correlations) are displayed in the upper triangular, unstandardized covariances are shown in the lower triangular and variances are displayed across the diagonal. Examining this table we can see that linear and cubic functions are highly intercorrelated with a negative valence. This makes sense when you remember that the positive linear and negative cubic trends basically travel from lower left to upper right (cf. to Figure 13.3). This very high correlation is what is most likely responsible for the cubic trend being positive in the structural equation model and negative in the correlation matrix of the predictors and flood. Being able to directly assess these correlations is helpful.

It is also helpful to examine the matrix between the disturbances in the model. This indicates whether there is any covariance between the endogenous variables that is not being accounted for by the exogenous variables in the model. Establishing these covariances is a good way to achieve better model fit by lessening the elements of the residual matrix that are high because of unspecified covariance. The correlations look relatively small except for the −0.478 correlation between D3 and D4. This is the correlation between flood and species dummy-code 2 (see Table 13.9). This

Table 13.9. Ψ *matrix*

	D1	D2	D3	D4
D1	0.10456	0.12675	0.03353	0.16239
D2	0.01770	1.7885	−0.25668	n.s.
D3	−0.00417	0.04265	0.18651	−0.47797
D4	−0.07022	n.s.	−0.24582	0.14788

n.s., not specified.

is most likely due to the fact that some species of tree live in certain flood plains in which may account for the high correlation between these two disturbances. Note: all values are significant. The abbreviation n.s. stands for "not specified", indicating that this covariance was not included in the model in order to achieve identification. SAS provides these estimates as a part of the printout. Correlations are displayed in the upper triangular, covariances in the lower triangular and variances on the diagonal.

"Tuckerized" growth curves

Tucker (1958) proposed another model for the analysis of growth curves based on a direct factorization of the sum of squares and cross-products matrix. As such, this technique is similar to the usual exploratory factor analysis/eigenvalue decomposition approaches readily available in many statistics packages, except that eigenvalues and eigenvectors are calculated on the basis of the sum of squares and cross-products of the variables rather than the usual correlation matrix or variance–covariance matrix of the data. Such an approach has some advantages relative to the polynomial decomposition outlined above. First, no assumption is made regarding the timing of assessments and no assumption is made about the mathematical relationships between assessments within a reference curve. For this reason, the Tucker method has been described by some as "nonparametric". In a manner similar to the polynomial decomposition, however, any given individual in the study can demonstrate more than one pattern of change over time. This may be particularly useful to biologists who wish to identify the presence of determinants of growth which onset at a fixed point in time, as may occur when a genetic component responsible for growth "switches on" in some members of the population under study. One disadvantage of the approach, however, is that the resulting eigenvalues and eigenvectors of the data yield an indeterminate solution, and, like any exploratory factor analytical approach, rotation of the eigenvectors of the solution will yield any number of solutions, all of which fit the data equally well. Tucker originally proposed three criteria that would guide the analyst to a growth curve solution that was maximally interpretable. First, all eigenvalues for the solution should be positive numbers, reflecting the characteristic that each eigenvalue reflects a positive increment to performance or size. Second, when examined over time, eigenvalues should increase or at least stay the same across the period considered. Third, each curve should be maximally correlated with time; curves that are largely flat are not thought to describe growth processes.

When evaluating this approach relative to biological data, it seems reasonable to believe the assumption that all eigenvectors should be positive, indicating no loss over time. Obviously, if the organisms under study are hosts to a parasite or undergoing some other stress resulting in the organism getting smaller, then a curve containing negative values for some eigenvectors may be appropriate. The second assumption, that the curves increase over time, would seem to make sense for biological growth data as well. The third assumption, that each curve be maximally correlated with time does not appear to be as reasonable, given that individuals may grow asymptotically at any given time in the study.

For the tree data considered above, a Tuckerized growth curve analysis is based on the sum of squares and cross-product matrix (SSCP) $\mathbf{X}^{\mathrm{T}}\mathbf{X}$, where the matrix $\mathbf{X}$ is a matrix with columns representing the times of measurement in the data and the rows corresponding to the individual organisms under study. This was done simply by premultiplying the matrix $\mathbf{X}$ by its transpose within SAS's PROC IML. This SSCP matrix is then factored, and the eigenvalues of this matrix considered. These are shown in Table 13.10. These values are then inspected in order to determine the number of curves needed to retain in the final solution. Given that the SSCP matrix is not centered to a mean of zero, as covariance and correlation matrices are, we can expect eigenvectors from the SSCP matrix to contain several large values. For this reason, it seems reasonable to employ a scree test (Cattell, 1966) to determine the number of curves to retain. For these data, it seems reasonable to retain four curves to describe the data, given that the drop between the fourth and fifth eigenvalues is a factor of 3, while those beyond that drop at most by a factor of 2 (between the sixth and seventh eigenvalues). Although it is possible that as many as six curves are present in the data, we decided to examine the four-curve solution because the resulting curves appeared to correspond to four easily interpretable processes and the rotations of the curves seemed relatively straightforward. Another way to explore the dimensionality of the SSCP matrix is by reference to mean

Table 13.10. *Eigenvalues of the SSCP matrix*

Factors	Eigenvalues	Factors	Eigenvalues	Factors	Eigenvalues
1	129 578 591	12	14.60396	23	3.4329988
2	86 843.185	13	11.813353	24	3.3808333
3	3083.07	14	8.9873795	25	3.1830318
4	301.49515	15	8.8538559	26	2.8821814
5	115.78105	16	7.5554835	27	2.4770578
6	91.30477	17	7.0355831	28	2.29794
7	44.379725	18	6.1387246	29	1.9946476
8	37.303769	19	5.7539972	30	1.5121401
9	32.338563	20	5.4089204	31	0.1035624
10	19.189979	21	4.253703	32	4.656E-10
11	17.714757	22	3.8644768	33	0

Table 13.11. *Revised eigenvalues of the SSCP matrix*

Curves	MSR	Curves	MSR	Curves	MSR
1	4212.364	12	3.1908802	23	1.8703906
2	647.29343	13	2.8318531	24	2.0507935
3	112.10038	14	2.314398	25	2.2067226
4	17.527317	15	2.477949	26	2.3553617
5	8.6813899	16	2.2833905	27	2.4689238
6	8.9716705	17	2.3082982	28	3.1316503
7	5.0892863	18	2.1792912	29	4.8721377
8	5.0024113	19	2.227243	30	43.332742
9	5.1198075	20	2.3076495	31	21227778
10	3.3725229	21	1.9635962	32	0.0502674
11	3.4857263	22	1.9372538	33	0

square residuals (MSR) proposed by Tucker (1966). These MSRs are theoretically distributed as F-statistics, but the resulting values are often liberal in practice, and as such the MSR values obtained are only heuristic as a result. Again, consonant with the decision above, it seems that there is a clear break between the fourth and fifth MSRs, indicating the presence of four curves in the data. The break between the sixth and seventh MSR seems less pronounced than it was for the eigenvalues in the preceding table (see Table 13.11). The eigenvectors, which correspond to the first four characteristic roots, are then rotated according to Tucker's criteria outlined above.

Table 13.12. *Means and standard deviations (SD) of "Tuckerized" curves by species*

		Curve 1		Curve 2		Curve 3		Curve 4	
Species	*N*	Mean	SD	Mean	SD	Mean	SD	Mean	SD
NA	97	0.59A	0.90	0.53	1.04	0.13A	0.90	0.57	0.84
NS	107	0.20	0.53	−0.22	0.21	0.35AB	0.64	0.86	0.44
TD	94	0.71A	0.98	0.19	1.12	0.43B	1.32	0.79	0.63

NS, Swamp Tupelo; TD, Baldcrees Cypress; NA, Water Tupelo.

For these data, SAS macros described in Wood (1992) were used to rotate to a conceptually meaningful solution.

The resulting growth curves are shown in Figure 13.4a–d and appear relatively straightforward interpretations. The first curve characterizes a period of rapid growth, in that it appears to capture a single period of growth from weeks 7 to 17. The second growth curve appears to describe a more gradual period of growth extending from the seventh to the twenty-fifth week. The third curve also appears to describe a pattern of gradual growth over the course of time, except that the onset of this growth is later than that indicated by the second curve, appearing to start at week 10. The final curve appears to document a growth pattern similar to that of the rapid growth curve in that it shows increases beginning in week 5. However, this curve also seems to capture short periods of no growth, producing a step-wise pattern in the curve. These "steps" appear in weeks 14 to 16 and 30 to 32 in the data.

As with the two-stage procedure outlined above, it is then possible to use these patterns in the data as the object for a subsequent analysis of the time-invariant differences in the trees under study. Using the SAS macros from Wood (1992), factor scores were then generated for each tree under study and these factor scores were then used as dependent variables in general linear models. Means of these factor scores are by species and canopy status as shown in Tables 13.12 and 13.13. For curve 1, species main effect ($F(2, 392) = 10.02$, $p < 0.0001$). It would appear that this curve does not describe the growth over time for NS as it does for NA and TD. Although no main effect for canopy was found ($F(1, 392) = 0.08$, $p = 0.78$), a significant species by canopy interaction was found ($F(3, 392) = 21.88$, $p < 0.0001$), which seems to indicate that, though this curve is more often typical of canopy trees of species NS and TD, it is typical only of sub-canopy NA trees.

(a)

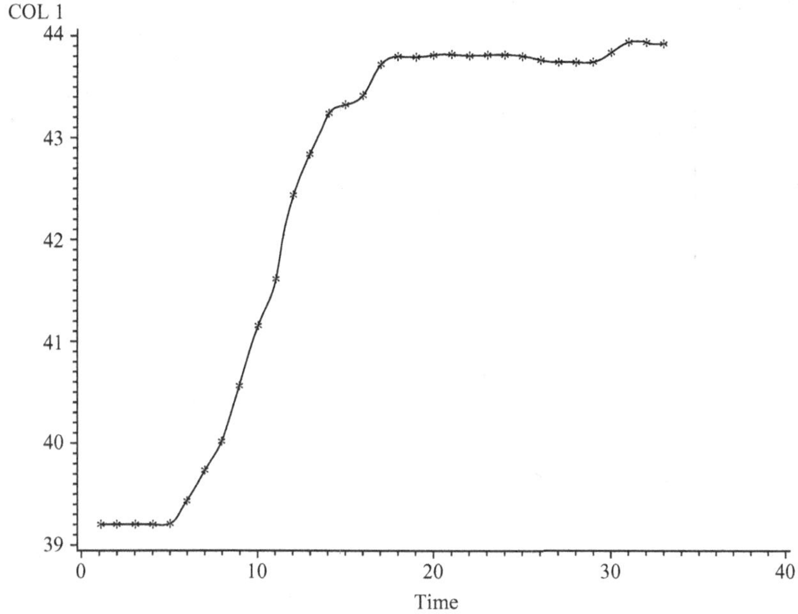

(b)

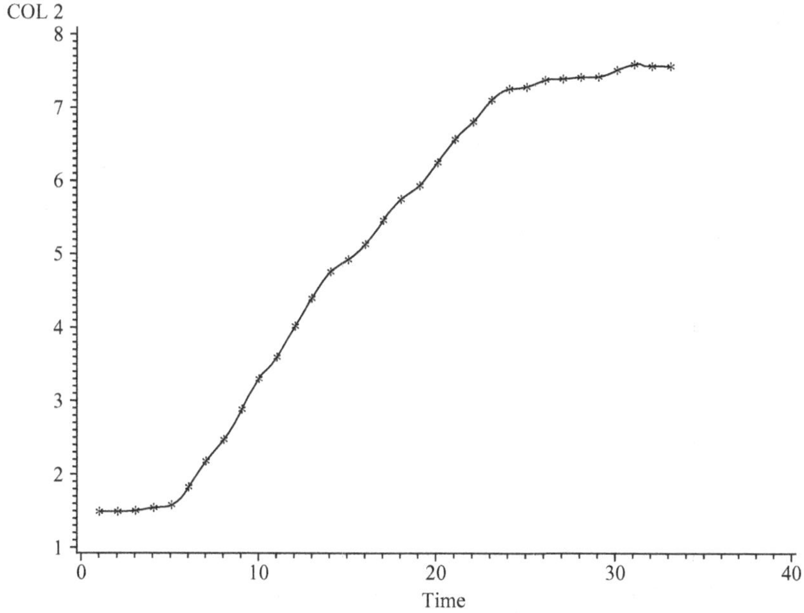

Figure 13.4. Growth curve describing rapid growth; (b) growth curve describing gradual growth; (c) stepwise growth pattern, delayed onset; (d) stepwise growth pattern.

(c)

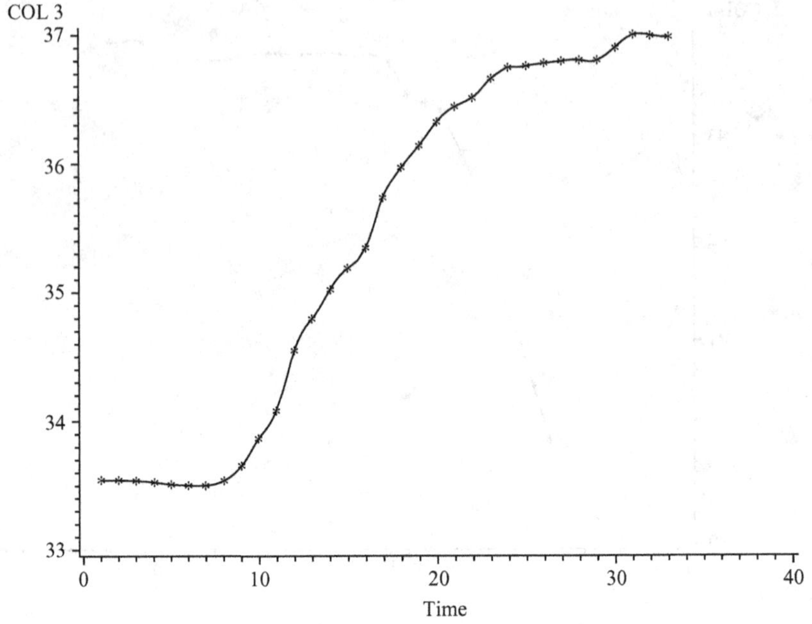

(d)

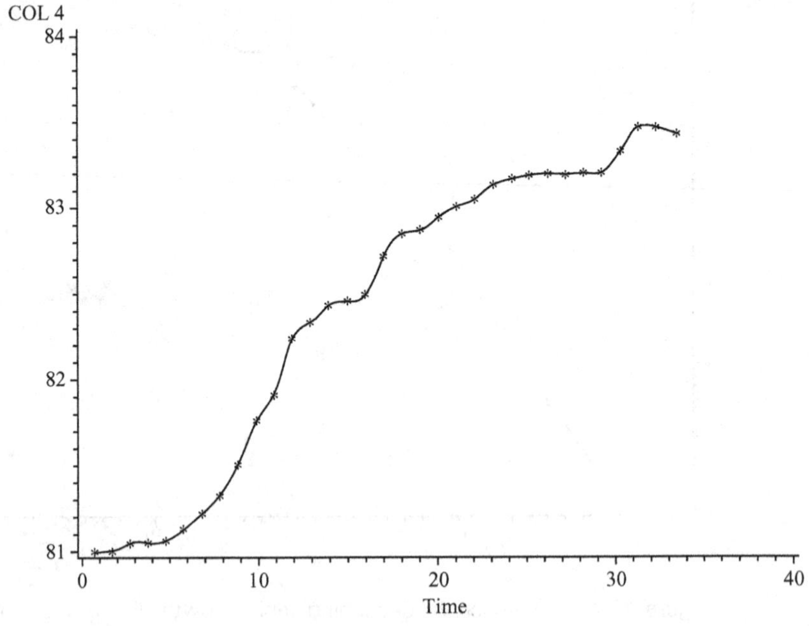

Figure 13.4. *(cont.)*

Table 13.13. *Means and standard deviations of "Tuckerized" curves by species and canopy status*

		Subcanopy		Canopy	
Species	Curve	Mean	SD	Mean	SD
NA (sub-canopy $N = 133$; canopy $N = 64$)	1	0.822	0.936	0.117	0.576
	2	0.901	1.026	−0.248	0.535
	3	0.078	0.956	0.248	0.755
	4	0.117	0.475	1.517	0.623
NS (sub-canopy $N = 57$; canopy $N = 50$)	1	0.021	0.306	0.409	0.650
	2	−0.157	0.160	−0.286	0.238
	3	0.136	0.543	0.598	0.656
	4	0.777	0.381	0.948	0.493
TD (subcanopy $N = 50$; canopy $N = 44$)	1	0.525	0.875	0.913	1.060
	2	0.394	1.164	−0.035	1.026
	3	0.094	1.196	0.812	1.371
	4	0.682	0.401	0.907	0.808

NS, Swamp Tupelo; TD, Baldcress Cypress; NA, Water Tupelo.

For the second curve, which represents a more gradual rate of growth, main effects were found for species, canopy, and the species-by-canopy interaction ($F(2, 392) = 14.38$, $F(1, 392) = 40.49$ and $F(2.392) = 13.67$, respectively, all p values less than 0.0001). *A posteriori* Waller–Duncan multiple range tests found that all three species were different from each other, with NA being larger than TD, which was in turn larger than NS. The canopy main effect, overall, found that canopy trees fell along this curve more than did sub-canopy trees, which is reasonable given that the canopy trees are larger.

The third curve, which represented a delayed, but rapid pattern of growth also yielded a main effect for species ($F(2, 392) = 3.36$, $p = 0.04$), a main effect for canopy ($F(1, 392) = 20.05$, $p = 0.0001$), but no species by canopy interaction ($F(2, 392) = 2.69$, $p = 0.07$). As can be seen from Table 13.13, this type of curve is more typical of TD and NS trees than it is of NA trees. As with the other recovered curves, the main effect for canopy merely means that, on the whole, canopy trees have higher factor scores than sub-canopy trees.

For the last curve, which described a slower, gently curving pattern of growth, no main effect for species was found ($F(2, 392) = 0.44$,

Table 13.14. *Correlations between factor scores*

	Curve 1	Curve 2	Curve 3	Curve 4
Curve 1	1.00000	−0.15409	−0.16313	−0.58774
Curve 2		1.00000	−0.06959	−0.25073
Curve 3			1.00000	−0.26544

$p = 0.64$), but significant effects were found for canopy ($F(2.392) = 110.89$, $p < 0.0001$) as well as a species × canopy interaction ($F(2.392) = 59.85$, $p < 0.0001$). Conceptually, this interaction appears to be due to the fact that this curve is more typically found in NS and TD trees, but not in sub-canopy NA trees.

It is also helpful to explore the degree to which these growth curves are correlated over time. Table 13.14 shows the correlations between factor scores for the four curves across all trees. These correlations are largely unremarkable, except for the negative correlations between curve 4 and the others, especially with curve 1. The low correlations are not surprising, given that the extraction and rotation of the curves was assumed to be orthogonal. The negative correlations with curve 4 appear to make sense, given that curve 4 tends to identify NS, TD, and NA canopy trees, while curve 1 tends to be associated with NA sub-canopy trees.

Discussion

At this point, a summary of the proposed techniques is appropriate, as well as a reminder of what these techniques do and do not do, from a conceptual standpoint. First, the two statistical approaches proposed here are similar in that they attempt to decompose general patterns of growth over time. As such, these techniques are more complex than models that relate general change over time as a linear function, either by means of difference scores, or by assessing growth as a linear function over time. Both techniques are also similar in that they propose that analysis of the data proceeds in two stages. In the first stage, the general form of individual differences in growth is determined and, in the second stage, estimates of these parameters are then used as the objects of a traditional linear model designed to determine which interindividual differences are systematically associated with these growth parameters.

Earlier discussion of these models has also stressed the assumptions regarding underlying change mechanisms that are assumed to be at work in the population under investigation. It is assumed that the organisms under investigation possess the growth parameters of interest to varying degrees and do not adopt a particular strategy in an all-or-none fashion. As mentioned above, in cases where the population under study represents a mixture of discrete types of organism or distinct reproductive strategies, latent class mixture models may be more appropriate for the data.

Given that reviewers of research articles often (with good reason) express a desire to forego statistical elegance in favor of more familiar statistical techniques, it is also appropriate to discuss the relative hazards of employing traditional regression and analysis of variance models for data that demonstrate the intraindividual changes over time described here. First of all, if a researcher simply enters the data from such as a study as a repeated measures factor, the analysis would be somewhat misleading because of the failure of the data to meet several of the assumptions of the general linear model. Individual differences in trajectory of growth over time would result in increasing variability over time. The result is that the heteroscedasticity within cells would result in an overly inflated error term associated with tests of main effects and interactions in the model. Although it is well known that F-statistics for such models are relatively robust to such violations of multivariate normality, it is also well known that statistical power suffers dramatically in such situations (Conover & Iman, 1977). Second, a statistical approach such as the general linear model does not allow for the identification of multiple strategies of growth unless the researcher can specify the form of such strategies prior to analyzing the data.

Extraction of linear and quadratic trends within the general linear model is not new, and the reader may be puzzled as to why the first approach advanced in this chapter is necessary or even whether it is different from the more common linear and quadratic trend extractions in multivariate analysis of variance (MANOVA). Further, alternative approaches to measuring growth that involve the parametric specification of the mathematical function underlying the curve, such as logistic approaches, may also seem like attractive alternatives. The default approach to extracting such curves in the data or specifying such mathematical model functions is that they assume that all individuals under study partake of the specified parameters such as linear and quadratic trends and that any deviation from the general form is due only to measurement error (albeit a measurement error which would seem more reasonable for survival data in the case of logistic regression). The approaches in this chapter are different in that they allow the organisms under study to partake of some, all, or even none of the distinct growth functions identified.

The models proposed here are not without their weaknesses, however, and these must be kept in mind in evaluating the adequacy of any proposed model resulting from them. Because of their two-stage approach, the assumption that the correct number and shape of curves has been captured can affect the validity of the second stage of analysis. Generally speaking, failure to include additional growth parameters actually present in the data will result in the heteroscedasticity mentioned above and result in systematic bias against finding statistically significant relationships in the data. The relative drawbacks associated with extracting too many parameters are relatively less grave, but also would generally result in reduced statistical power, given that the inclusion of such parameters would result in a decrease in the error degrees of freedom associated with tests of hypotheses.

The two models presented in this chapter also involve trade-offs when considered against each other. First, the polynomial model presented at the beginning of the chapter has a major advantage in that it involves the analysis of relatively few grouped time points for the data and, as such, would seem to make analysis more tractable. The gain in reducing the time variable to a manageable level seems considerable. It is possible, however, that arbitrary aggregation of the data across time may mask important developmental processes. In addition, the researcher may not be confident of how much the results would change under slightly different aggregations of the time variable. As such, a good familiarity with the descriptive statistics of the data and consideration of the Tuckerized approach may be appropriate ancillary analyses when such questions seem relevant. The Tuckerized approach would seem to have several advantages in this regard in that it makes no assumptions regarding the sampling of time in measurement occasions in the study, and requires no assumption regarding the polynomial nature of the data (for an application, see Pugesek & Wood, 1992). The chief limitation of this approach, however, is that the extraction of curves is necessarily arbitrary and, lacking a clear mapping of the rotation of the growth curves onto conceptual processes, one could argue that other researchers might come up with curves that are somewhat different in shape, even though mathematically equivalent.

On a methodological note, it is appropriate to note that, as Meredith & Tisak (1990) demonstrated, more advanced models are possible which blend the growth curve models outlined here with parametric assumptions regarding the data. Such models hold the promise of making more realistic assumptions about the nature of the processes under investigation and also doing a more satisfactory job of modeling the measurement error involved in studies of growth of organisms. More elegant models that simultaneously extract the growth curves of interest and then relate these to other study variables are possible in such programs as Mplus (Muthén & Muthén, 1998). The means by which such models can be compared and contrasted with latent mixture models of cluster analytical approaches remains to be explicated.

In summary, the approaches outlined here for individual differences in growth offer biologists tools for investigating the effects of environmental influences that operate to a matter of degree (such as periodic flooding), and which may assist in the identification of multiple survival strategies on the part of some organisms. As such, they provide a useful adjunct to commonly used statistical techniques. As John Tukey observed, however, "All statistical models are false, but some are useful." Users of these techniques must employ them with a mindfulness of their limitations and the alternative explanations that a reasonable skeptic might raise regarding the research conclusions made using the techniques.

References

Abramowitz, M. & Stegun, I. (1972). *Handbook of Mathematical Functions*. New York: Dover.

Appel, T. A. (1987). *The Cuvier–Geoffroy Debate: French Biology in the Decades before Darwin*. New York, Oxford University Press.

Bartholomew, D. J. (1987). *Latent Variable Models and Factor Analysis*. New York: Oxford University Press.

Bentler, P. M. (1995). *EQS Structural Equations Program Manual*. Encino, CA: Multivariate Software, Inc.

Bollen, K. A. (1989). *Structural Equations with Latent Variables*. New York: Wiley.

Bryk, A. & Raudenbush, S. (1992). *Hierarchical Linear Models*. Thousand Oaks, CA: Sage.

Cattell, R. B. (1966). The scree test for the number of factors. *Multivariate Behavioral Research*, **1**, 245–276.

Clausen, J., Keck, D. D. & Heisey, W. M. (1940). *Experimental Studies on the Nature of Species. Effects of Varied Environments on Western North American Plants*. Publication 520. Washington, DC: Carnegie Institution of Washingtion,

Conover, W. J. & Iman, R. L. (1977). On the power of the t-test and some ranks tests when outliers may be present. *Canadian Journal of Statistics*, **5**, 187–193.

Curran, P. J. (2000). A latent curve framework for studying developmental trajectories of adolescent substance use. In J. Rose, L. Chassin, C. Presson & J. Sherman (eds.), *Multivariate Applications in Substance Use Research*, pp. 1–42. Hillsdale, NJ: Lawrence Erlbaum Associates.

Curran, P. J., Stice, E. & Chassin, L. (1997). The relation between adolescent and peer alcohol use: a longitudinal random coefficients model. *Journal of Consulting and Clinical Psychology*, **65**, 130–140.

DeShon, R. P. (1998). A cautionary note on measurement error corrections in structural equation models. *Psychological Methods*, **3**, 412–423.

Dobzhansky, T., Ayala, F. J., Stebbins, G. L. & Valentine, J. W. (1977). *Evolution*. San Francisco: W. H. Freeman.

Duncan, S. C. & Duncan, T. E. (1996). A multivariate latent growth curve analysis of adolescent substance abuse. *Structural Equation Modeling*, **3**, 323–347.

Duncan, T. E. & Duncan, S. C. (1995). Modeling the processes of development via latent variable growth curve methodology. *Structural Equation Modeling*, **2**, 187–213.

Duncan, T. E., Duncan, S. C., Strycker, L. A., Li, F. & Alpert, A. (1999). An introduction to latent variable growth curve modeling: concepts, issues and applications. Mahwah, NJ: Lawrence Erlbaum Associates.

Estes, W. K. (1956). The problem of inference from curves based on group data. *Psychological Bulletin*, **53**, 134–140.

Hancock, G. R., Lawrence, F. R. & Nevitt, J. (2000). Type I error and power of latent mean methods in factorially invariant and noninvariant latent variable systems. *Structural Equation Modeling*, **7**, 534–556.

Henderson, N. D. (1968). The confounding effects of genetic variables in early experience research: can we ignore them? *Developmental Psychobiology*, **1**, 146–152.

Henderson, N. D. (1970). Genetic influences on the behavior of mice can be obscured by laboratory rearing. *Journal of Comparative and Physiological Psychology*, **72**, 505–511.

Henderson, N. D. (1973). Brain weight changes resulting from enriched rearing conditions: a diallel analysis. *Developmental Psychobiology*. **6**, 367–376.

Henderson, N. D. (1981). Genetic influences on locomotor activity in 11-day-old housemice. *Behavior Genetics*, **11**, 209–225.

Huba, G. J. & Harlow, L. L. (1987). Robust structural equation models: implications for developmental psychology. *Child Development*, **58**, 147–166.

James, L. R., Mulaik, S. A. & Brett, J. M. (1982). *Causal Analysis: Assumptions, Models, and Data. Studying Organizations: Innovations: Innovations in methodology*, Vol. 1. Newbury Park, CT: Sage.

Jöreskog, K. & Sörbom, D. (1993). *LISREL 8 User's Reference Guide*. Chicago: Scientific Software Inc.

Keeland, B. D. & Sharitz R. R. (1995). Seasonal growth patterns of *Nyssa sylvatica* var. *biflora, Nyssa aquatica* and *Taxodium distichum* as affected by hydrologica regime. *Canadian Journal of Forestry Research*, **25**, 1084–1096.

Keeland, B. D. & Sharitz R. R. (1997). The effects of water-level fluctuations on weekly tree growth in a Southeastern USA swamp. *American Journal of Botany*, **84**, 131–139.

Keeland, B. D., Conner, W. H. & Sharitz, R. R. (1997). A comparison of wetland tree growth response to hydrologic regime in Louisiana and South Carolina. *Forest Ecology and Management*, **90**, 237–250.

Kenny, D. A. & Judd, C. M. (1984). Estimating the nonlinear and interactive effects of latent variables. *Psychological Bulletin*, **96**, 201–210.

Kirk, R. E. (1995). *Experimental Design: Procedures for the Behavioral Sciences.* Monterey: Brooks/Cole.

MacCallum, R. C., Kim, C., Malarkey, W. B. & Kiecolt-Glaser, J. K. (1997). Studying multivariate change using multilevel models and latent curve models. *Multivariate Behavioral Research*, **32**, 215–225.

Marcoulides, G. A. & Schumaker, R. E. (1998). *Advanced Structural Equation Modeling.* Mahwah, NJ: Lawrence Erlbaum Associates.

McArdle, J. J. (1989). A structural modeling experiment with multiple growth functions. In P. Ackerman, R. Fanfer & R. Cudek (eds.), *Learning and Individual Differences: Abilities, Motivation, and Methodology*, pp. 71–117. Mahwah, NJ: Lawrence Erlbaum Associates.

McArdle, J. J. (1991). Structural models of developmental theory in psychology. *Annals of Theoretical Psychology*, **7**, 139–159.

McArdle J. J. & Epstein, D. (1987). Latent growth curves within developmental structural equation models. *Child Development*, **58**, 110–133.

Meredith W. & Tisak, J. (1984). "Tuckerizing" curves. Paper presented at the Annual Meeting of the Psychometric Society, Santa Barbara, CA.

Meredith W. & Tisak, J. (1990). Latent curve analysis. *Psychometrika*, **55**, 107–122.

Muthén, B. (1989). Latent variable modeling in heterogeneous populations. *Psychometrika*, **54**, 557–585.

Muthén, B. (1997). Latent variable modeling with longitudinal and multilevel data. In A. Raferty (ed.), *Sociological Methodology*, pp. 453–480. Boston: Blackwell.

Muthén, B. (1998). Second-generation structural equation modeling with a combination of categorical and continuous latent variables: new opportunities for latent class/latent growth modeling. In L.M. Collins & A. Sayer (eds.), *New Methods for the Analysis of Change.* Washington, DC: American Physciological Association.

Muthén, L. & Muthén, B. (1998). *Mplus: The Comprehensive Modeling Program for Applied Researchers: User's Guide.* Los Angeles: Muthén & Muthén.

Nagin, D. S. (1999). Analyzing developmental trajectories: A semiparametric, group-based approach. *Psychological Methods*, **4**, 139–157.

Neale, M.C. (1994). *Mx: Statistical Modeling*, 2nd edition. Box 710 MCV, Richmond, VA 23298: Department of Psychiatry, Virginia Commonwealth University.

Neter, J., Kutner, M. H., Nachtsheim, C. J. & Wasserman, W. (1996). *Applied Linear Statistical Models*, 4th edition. Chicago, IL: Irwin.

Parsons, P. A. (1972). Genetic determination of behavior (mice and men). In L. Ehrman, G. S. Omenn & E. Caspari (eds.), *Genetics, Environment, and Behavior*, pp. 75–98. New York: Academic Press.

Pugesek, B. H., & Wood, P. (1992). Alternate reproductive strategies in the California gull. *Evolutionary Ecology*, **6**, 279–295.

Rao, C. R. (1958). Some statistical methods for the comparison of growth curves. *Biometrics*, **14**, 1–17.

Raudenbush, S. W. (2001). Toward a coherent framework for comparing trajectories of individual change. In A. Sayers & L. Collins (eds.), *New Methods for the Analysis of change*, pp. 33–64. Washington, DC: American Psychological Association.

Raudenbush, S. E., Bryk, A. S. & Congdon, R. T. (2000). *HLM 5.04 Hierarchical Linear Modeling Software*. Lincolnwood, IL: Scientific Software International, Inc.

Raykov, T. (1997). Growth curve analysis of ability means and variances in measures of fluid intelligence of older adults. *Structural Equation Modeling*, **4**, 283–319.

Rose, J. S., Chassin, L., Presson, C. C. & Sherman, S. J. (2000). *Multivariate Applications in Substance Use Research*. Mahwah, NJ: Lawrence Erlbaum Associates.

Schumacker, R. E. & Lomax, R. G. (1996). *A Beginner's Guide to Structural Equation Modeling*. Mahwah, NJ: Lawrence Erlbaum Associates.

Schumacker, R. E. & Marcoulides, G. A. (1998). *Interaction and nonlinear effects in Structural Equation Modeling.* Mahwah, NJ: Lawrence Erlbaum Associates.

Tucker, L. R. (1958). Determination of parameters of a functional relation by factor analysis. *Psychometrika*, **50**, 203–228.

Tucker, R. (1966) Some mathematical notes on three-mode factor analysis. *Psychometrika*, **31**, 279–311.

von Eye, A. (1999). Kognitive Komplexität – Messung und Validität [Cognitive complexity – assessment and validity]. *Zeitschrift für Differentielle und Diagnostische Psychologie*, **20**, 81–96.

von Eye, A. & Clogg, C. C. (eds.) (1994). *Latent Variables Analysis. Applications for Developmental Research*. Thousand Oaks, CA: Sage.

von Eye, A. & Clogg, C. C. (eds.) (1996). *Categorical Variables in Developmental Research*. San Diego: Academic Press.

von Eye, A. & Nesselroade, J. R. (1992). Types of change: application of configural frequency analysis in repeated measurement designs. *Experimental Aging Research*, **18**, 169–183.

von Eye, A. & Schuster, C. (1998). *Regression Analysis for the Social Sciences*. San Diego: Academic Press.

Willett, J. B. & Sayer, A. G. (1994). Using covariance structure analysis to detect correlates and predictors of change. *Psychological Bulletin*, **116**, 363–381.

Willett, J. B. & Sayer, A. G. (1996). Cross-domain analyses of change over time: combining growth modeling and covariance structure analysis. In G. A. Marcoulides & R. E. Schumacker (eds.), *Advanced Structural Equation Modeling: Issues and Techniques*, pp. 125–157. Mahwah, NJ: Lawrence Erlbaum Associates.

Wood, P. K. (1992). Generation and objective rotation of generalized learning curves using matrix language. *Multivariate Behavioral Research*, **27**, 21–29.

Young, P. J., Keeland, B. D. & Sharitz, R. R. (1995). Growth response of bald cypress (*Taxodium distichum* (l.) Rich.) to an altered hydrologic regime. *American Midland Naturalist*, **133**, 206–212.

Section 3 Computing

14 A comparison of the SEM software packages Amos, EQS, and LISREL

Alexander von Eye and Bret E. Fuller

Abstract

This chapter presents a comparison of three of the most capable and popular software packages for structural equation modeling (SEM), Amos, EQS, and LISREL. The programs are compared in regard to available input formats, output formats, and many technical options, including options concerning the methods of estimation. Results suggests that the three packages typically provide equivalent results. However, they differ, for instance, in how they handle admissibility problems. The programs also differ in documentation. All three programs contain options that are not even mentioned in the manuals. All three programs are very capable. Thus user preferences may reside in program characteristics unrelated to SEM.

Introduction

Structural equation modeling (SEM) is a methodology that allows one to specify and test models that describe the relationships among manifest, i.e., observed, and latent, i.e., unobserved, variables. SEM has revolutionized data analysis in the social and behavioral sciences. The method is complex enough that the attempt to estimate models using pocket calculators is, given the 1998 state-of-the art calculator, hopeless. Relatively powerful desk-top computers, workstations or main frame computers are needed, along with the appropriate software, to estimate models. This chapter is concerned with software. More specifically, we compare in this chapter three of the most capable and popular software packages for SEM. The three packages are, in alphabetical order, Amos 3.6 (Arbuckle, 1996), EQS 5.7 (Bentler & Wu, 1995), and LISREL 8.20 (Jöreskog & Sörbom, 1996a)[1]. The packages can be purchased from the following addresses:

[1] The authors decided to discuss the programs in the same order throughout the manuscript. The order is alphabetical and does not indicate a preference for one program over another.

- Amos
  - SmallWaters Corp.
  - 1507 E 53rd Street, no. 452
  - Chicago, IL 60615, USA
  - website: www.smallwaters.com
- EQS
  - Multivariate Software Inc.
  - 4924 Balboa Blvd, no. 368
  - Encino, CA 91316, USA
  - website: www.mvsoft.com
- LISREL
  - Scientific Software International
  - 7383 Lincoln Avenue – Suite 100
  - Chicago, IL 60646–1704, USA
  - website: www.ssicentral.com

Prices of packages vary depending on number of licences purchased and whether or not any discounts apply. As usual, educational discounts reduce the cost of purchase considerably. There are a good number of other SEM programs. Examples include, but are not restricted to, RAMONA (see Wilkinson, 1996), which implements McArdle & McDonald's (1984) reticular action model and is now part of the general purpose statistical software package SYSTAT (Wilkinson, 1996), COSAN (Fraser & McDonald, 1988), and Mx (Neale *et al.*, 1999). We do not include these programs, for the three that are included here are currently the most popular. We want to keep the comparison within certain limits. The programs that are not included are also very capable. The fact that they are not included does not reflect our judgment concerning their capability and quality.

In the following sections we first provide a brief overview of SEM (for more detail, see, for instance, the manuals that come with the programs; Bollen, 1989; Schumacker & Lomax, 1996; Hershberger *et al.*, Chapter 1; for more advanced treatments, see von Eye & Clogg, 1994; Arminger *et al.*, 1995; Marcoulides & Schumacker, 1996). Subsequent sections cover the software comparison and provide sample input and output files.

Structural equation modeling

In this section we provide a brief overview of the general structural equation model. We use the notation introduced by Jöreskog & Sörbom (1996a,b). Occasionally, to illustrate similarities between notations, reference is made to the Bentler–Weeks model (Bentler & Weeks, 1980, 1985). A structural model can be written in the form

$$\eta = \mathbf{B}\eta + \boldsymbol{\Gamma}\xi + \zeta, \tag{14.1}$$

where η denotes the $[m \times 1]$ vector of the m latent dependent variables; ξ is the $[n \times 1]$ vector of the latent independent variables; $\mathbf{B}$ is the $[m \times m]$ matrix of structure coefficients that describe the relationships among the latent dependent variables; $\boldsymbol{\Gamma}$ is the $[m \times n]$ matrix of structure coefficients that describe the relationships between the latent independent and the latent dependent variables; and ζ is a vector that contains the prediction errors (residuals) for the manifest (observed) variables and disturbances for the latent variables (factors). In addition to the elements contained in equation (14.1), there is a matrix $\boldsymbol{\Phi}$ that contains the variances and covariances among the latent independent variables, and a matrix $\boldsymbol{\Psi}$ that contains the variances and covariances among the latent dependent variables. Additional variables contain the residual variances and covariances of the manifest dependent and independent variables (see below), and the residual covariances of the dependent with the independent variables.

The Bentler–Weeks model describes the structural model equation as

$$\eta = \beta\eta + \boldsymbol{\Gamma}\xi, \tag{14.2}$$

where η is the same column vector for dependent variables as before, ξ is the same column vector for independent variables as before, and $\boldsymbol{\Gamma}$ contains the coefficients that describe the strength of regression of the dependent variables η upon the independent variables, ξ. The matrix β contains the regression coefficients that describe the relationships among the dependent variables. $\boldsymbol{\Phi}$ is the matrix of covariances of the independent variables. Thus the models in equations (14.1) and (14.2) and the parameter matrices of these models are essentially the same.

Researchers often discriminate between the measurement models on the dependent and the independent variables sides. Measurement models are of importance because they allow researchers to evaluate the relationships between the postulated factors and their indicators. The paths from factors to indicators are supposed to be significant and have the right sign. The measurement model for the dependent variables is, in Jöreskog & Sörbom's notation,

$$y = \boldsymbol{\Lambda}_y\eta + \epsilon, \tag{14.3}$$

where $\boldsymbol{y}$ denotes the $[p \times 1]$ vector of observed dependent variables; $\boldsymbol{\eta}$ is the $[m \times 1]$ vector of latent dependent variables; and $\boldsymbol{\Lambda}_y$ is the $[p \times m]$ matrix of the coefficients that describe the relationships of observed with latent dependent variables. The coefficients in $\boldsymbol{\Lambda}_y$ are also known as the *loadings* of the observed variables on the latent factors $\boldsymbol{\eta}$. $\boldsymbol{\epsilon}$ contains the $[p \times 1]$ measurement errors on the dependent variable side. Accordingly, the measurement model for the independent variables is

$$\boldsymbol{x} = \boldsymbol{\Lambda}_x \boldsymbol{\xi} + \boldsymbol{\delta}, \tag{14.4}$$

where the $[q \times 1]$ vector $\boldsymbol{x}$ contains the observed independent variables; $\boldsymbol{\xi}$ is the $[n \times 1]$ vector of latent independent variables; $\boldsymbol{\Lambda}_x$ is the $[q \times m]$ matrix of the factor loadings of the independent variables; and $\boldsymbol{\delta}$ is the $[q \times 1]$ vector of measurement errors on the independent variable side.

There are two additional matrices that need to be considered. First, there is the $[p \times p]$ variance–covariance matrix $\boldsymbol{\theta}_\epsilon$ of the measurement errors of the observed dependent variables y. Second, there is the $[q \times q]$ variance–covariance matrix $\boldsymbol{\theta}_\delta$ of the errors of the observed independent variables x. These matrices are of utmost importance. They allow researchers to take relationships among variables into account that have not been captured by the factors. Applications of this option are vital in repeated observations studies or, in more general terms, whenever variables systematically covary because of shared characteristics that are not included in the definition of a factor.

A comparison of Amos, EQS, and LISREL

In this section we compare Amos, EQS, and LISREL. This comparison considers such objective criteria as speed and number of options, and such subjective criteria as ease of input and readability of output. It is not the goal of this comparison to create a rank order of the three software packages. Although the authors have their preferences (which differ in some respects), it is important to recognize that readers may have their own system of weighing the characteristics of software. Among those may be the criterion that a user may have made positive experiences with one of the packages and, therefore, sees no reason to switch. Thus we only list but do not weigh our criteria in the next section. The following sections present sample applications and their results. The following comparison focuses on Windows operation systems (OS). Sample runs were performed on both Windows 95 and Windows NT 4.0. There exist versions of these programs for other systems such as the Mac OS and OS/2 Warp. For information concerning Amos, EQS, and LISREL for these systems, please contact the distributors.

Criteria and results of comparison

Most of the criteria used here are objective. In addition, these criteria overlap widely with the criteria used in earlier software comparisons of similar nature (e.g., Bollen & Ting, 1991). There is a number of new criteria that result from the development of both the mathematical and statistical features of SEM on the one hand, and the programs on the other. The versions of the programs that we compared are the Windows versions of Amos 3.6, EQS 5.7, and Interactive LISREL 8.20. The criteria that we employ can be classified as six groups: (1) system requirements; (2) documentation; (3) data management and entry; (4) modeling options; (5) output options; and (6) ease of use. Only the last criterion is subjective.

System requirements (Table 14.1)

The following requirements are taken from the manuals that come with the programs. Additional information is inserted on the basis of the experiences made when installing the programs and when preparing the examples for this chapter.

A few comments concerning the required processors and hard disk space are needed. When processors such as the Intel 80386 are used, math coprocessors are supported by all three programs. In fact, the latest versions of these SEM programs are so numerically intensive that a user of an 80386 processor must sometimes be very patient before a program completes its run, even with a math coprocessor. This applies in particular to LISREL 8.20. We do not recommend runs with many variables on machines with processors slower than a Pentium. As far as the hard disk space is concerned, the requirements can be reduced by, for instance, moving the data files to some external storage device. The gain, however, is only marginal. Processing speed can be increased in particular by (1) faster processors, (2) multiple processors, and (3) more RAM.

As far as printers are concerned, each of these programs provide very impressive graphical input and/or output of path models. Some of the output comes in color (EQS, LISREL). The LISREL path models can be sent directly to a color printer. Thus, researchers interested in using the graphics for publication or presentation purposes may consider using a high resolution color printer.

Table 14.1. *System requirements*

Requirement	Amos 3.6	EQS 5.7	LISREL 8.20
Hardware			
Processor	Intel 80386 or higher (or compatible)	Intel 80386 or higher (or compatible)	Intel 80386 or higher (or compatible)
RAM	4 minimum (varies depending on OS, Windows NT requires 16 MB)	4 minimum (varies depending on OS)	4 minimum (varies depending on OS)
Mouse or equivalent	Needed	Needed	Needed
Graphics adapter/monitor	EGA or VGA or better	EGA or VGA or better	EGA or VGA or better
Hard disk space for programs and examples	6.4 MB (4.2 MB as indicated by the Windows Explorer)	3 MB (5.44 MB as indicated by the Windows Explorer)	6.06 MB (as indicated by the Windows Explorer)
CD/floppy disk reader (for installation)	Floppy disk	Floppy disk	CD
Software			
OS	Windows 3.1 or higher	Windows 3.1 or higher	Windows 3.1 or higher
Drivers	—	Requires printer driver installed	—

The *installation* of the three programs was smooth and problem-free[2]. The usual SETUP command started a fully automated installation sequence for all three programs. There is no need to dwell on installation any further.

Documentation

Each of the three SEM software distributors provide documentation in three forms. The first are the manuals, presented in soft-cover book form. The second form are the help files integrated in the programs. The third form is the documentation available on the web sites of the program authors and distributors. Website addresses were given above. In addition, there are independent user groups that share information on the web. The following comparison focuses on the manuals that come with the programs.

Overall, the manuals that come with the three programs are very impressive. They provide a large amount of useful information. Most of the program features are covered. In addition, the statistical model is described in a way that the manuals can be integrated in courses on SEM. These descriptions go far beyond what one typically expects from a statistical software package manual. It should be emphasized that all manuals are incomplete in the sense that each of the three programs contains commands that are not documented or that are available (1) in form of flyers that also come with the program (EQS), (2) from user groups, (3) in Readme files that come with the latest releases of the program CDs or floppy disks, or (4) on the distributors' websites. The reason for this lack of completeness of manuals is that the programs are updated more often than the manuals. In addition, it is well known that most statistical software packages are not fully documented. Why this is the case, is largely obscure. However, some of the nondocumented commands are first test versions of commands that may be incorporated in later releases. Other commands are no longer needed but were not deleted from the program and are still functional.

Documentation for Amos

The Amos documentation consists of one 600 page volume entitled *Amos Users' Guide Version 3.6*, authored by J. L. Arbuckle (1996). The first two chapters are the introduction and the installation instructions. The next chapter provides a tutorial to Amos' graphical approach to model specification. The fourth chapter gives a tutorial to Amos' option to specify models

[2] It should be noted, however, that certain restrictions apply when using Amos under Windows NT.

via command text files. Chapters 5 and 6 give compilations of commands, called reference guides, for the graphical and the text-oriented ways to input information. The following chapter which covers more than two-thirds of the volume provides examples. From a practical perspective a very important example is no. 17 on missing data. Unlike EQS and LISREL/PRELIS (see below), Amos does not provide an option for missing data estimation and imputation. Instead, the program calculates the full information maximum likelihood estimates in the presence of missing data. This approach is based on the assumption that data are missing at random. More will be discussed on this approach in Section 3.1. The eighth chapter is a technical appendix with brief and concise overviews of the notation used in the manual, discrepancy functions minimized for the goodness-of-fit indices, measures of fit, and numerical diagnosis of nonidentifiability. Bibliography and a combined author and subject index conclude the volume.

Documentation for EQS

The EQS documentation consists of two volumes. The first, authored by P. M. Bentler (1985), is a 272 page volume entitled *EQS Stuctural Equations Program Manual.* The second, authored by P. M. Bentler & E. J. C. Wu (1995), is a 300-page volume entitled *EQS for Windows User's Guide.* The program manual begins with an overview of the program and an introduction to SEM. The following sections provide information on how to specify program input and information provided in the program output. Chapter 5 introduces models and statistics. Lagrange multipliers and Wald tests are introduced in Chapter 6, followed by multisample covariance structures (Chapter 7), mean and covariance structures (Chapter 8.), and multisample mean and covariance structures (Chapter 9). Chapter 10 describes technical aspects of EQS and the underlying statistical theory. References and a subject index follow. The last part of this manual contains five appendices that describe features incorporated in the more recent program releases.

The users' guide for EQS for Windows introduces users into EQS within a Windows environment. It consists of 11 chapters. The first is an introduction that contains information concerning system requirements, installation, and file conversion. The second chapter provides a tutorial. Chapter 3 covers data preparation and management. It is followed by a chapter on data import and export. Chapters 5 and 6 describe program characteristics that are unique to EQS. These chapters describe program options for a wide variety of plots (Chapter 5) and basic statistics (Chapter 6). What the manual calls "basic statistics" contains more than just means and standard deviations. It contains, among other options, analysis of variance, regression and factor analysis. EQS also contains a module for estimating missing data. Among other methods, this module allows mean substitution and imputation of values estimated by multiple regression methods. Structural modeling with EQS is described in the following chapters. Chapter 7 describes *Build EQS*, i.e., an interactive method to specify a model. This method is unique to EQS. Chapter 8 (and flyers included in the package) describe EQS' method of specifying a model by drawing a program. This method is similar to, yet even more refined than, the comparable option in Amos. LISREL does not provide this option. Chapter 10 describes the Help system, and Chapter 11 provides details concerning the installation and running of EQS under Windows. A subject index concludes the volume.

Documentation for LISREL

The LISREL documentation consists of three volumes, all coauthored by K. Jöreskog and D. Sörbom (1996a–c). The first volume is the 378-page *LISREL 8: User's Reference Guide.* This manual is an only slightly updated version of the 1989 manual for LISREL 7. Thus the new manual leaves plenty of room for revisions. In particular the graphics options now available are not even mentioned in the manuals. The same applies for the robust estimation methods incorporated in release 8.20. The first of the eleven chapters of this manual describes LISREL and its submodels (see Section 1, above). The chapter introduces the Greek name notation for which LISREL is well known; methods for specification; PRELIS (see below); estimation and evaluation methods. As in the other manuals, this chapter has text book character. Chapter 2 describes in detail how to specify a LISREL run. Chapters 3 and 4 present the LISREL Submodels 1 and 2, where Submodel 1 specifies measurement models and confirmatory factor analysis and Submodel 2 specifies models for directly observed variables. Models for latent variables are covered in Chapter 4. Chapter 5 presents the LISREL Submodel 3, a more comprehensive model that, in principle, makes the first two submodels obsolete. Chapter 7 discusses the analysis of ordinal and other nonnormal variables. Chapter 8 presents methods for placing constraints, power calculation, and discusses equivalence of models. Chapter 9 introduces readers to multisample analysis. Chapter 10 discusses analysis of mean structures, and Chapter 11 gives hints on resolving problems with LISREL runs. It explains the meaning of LISREL error messages. Appendix A summarizes program features that are new to LISREL 8, and Appendix B provides a syntax overview. References and the author and subject indices follow.

The PRELIS 2 225-page user's reference guide describes features and use of PRELIS, a program for multivariate data screening and description. PRELIS is an add-on to LISREL. It allows users to "pre-process" data before a LISREL run, which can be started from within PRELIS. The first of four chapters describes the PRELIS procedures. Most important here are, from an SEM user's perspective, the six different types of moment matrix that PRELIS can create. Instructions and commands for PRELIS runs are listed in Chapter 2. Chapter 3 presents PRELIS commands and their keywords. Chapter 4 contains examples and exercises. Four appendices follow. The first, Appendix A, explains error messages and warnings issued by PRELIS. Appendix B explains new features of PRELIS. These include, for instance, a missing-data imputation method that is based on what has become known as the hot deck method. This method imputes missing data using information from the most similar case. Appendix C describes how to perform simulations using PRELIS and LISREL. Appendix D provides a syntax overview. The references and the subject and author indices conclude this volume.

The 226-page manual for the SIMPLIS command language introduces readers to SIMPLIS, a second command language for LISREL. This second language is less cryptic and closer to plain English than is the standard LISREL command language. In terms of performance, the two languages are equivalent. In the first of seven chapters Jöreskog & Sörbom (1996b) present simple examples, detailing how one can use SIMPLIS to convince LISREL to perform such pedestrian tasks as estimating a single manifest variable regression equation. The chapter also contains more complex examples on topics such as recursive path models with latent variables. Chapter 2 covers multisample examples. Path diagrams are the topic of Chapter 3. Chapter 4 covers fitting and testing of models. Chapter 5 explains the LISREL output. Chapter 6 contains general syntax rules for the SIMPLIS command language and all options of the program. Chapter 7 presents computer exercises. The references and the author and subject indices complete this volume.

As was indicated before, the documentation provided for Amos, EQS, and LISREL goes far beyond standard documentation for statistical software packages. Each of the packages comes with instructions on how to install the program, with a reference list that contains commands and options, and examples. In addition, each of the documentations for the three programs explains SEM. This is exemplary.

The manuals differ in the distance in time from the latest releases of their programs. The programs are typically more capable than the program manuals indicate. The Amos manual seems to be the most up to date, as measured by the (subjective) distance between program capabilities and manual description. The manuals for EQS and LISREL provide more room for improvement. Here are examples. The EQS manuals are up-to-date as far as most of the statistical and mathematical options are concerned. The latest developments of the estimation methods (Yuan & Bentler, 1997) are implemented, at least in part, but still need to be described in the manual. The same applies to the options provided by the new and very capable graphics module. These options are described only in a flyer that comes with the program and should be incorporated in the manuals. As far as LISREL is concerned, there has been no comprehensive update of the manual since LISREL 7. The new options provided with LISREL 8 and its updates are only sparsely described and not integrated into the manual. This applies, for instance, to estimation methods (the new 8.20 robust estimation methods are nowhere mentioned in the manuals), to the nonlinear capabilities of the program, and to the options provided by the new graphics module.

In spite of these shortcomings, which, according to our experience, are common place in software packages that keep pace with the state of the art, the manuals are among the highlights of these three packages. Most importantly, the manuals for Amos, EQS, and LISREL not only tell users how to invoke yet another option. In addition, they explain to the user the technical background of methods and options to the extent that understanding is enhanced.

Data management and entry

Data management and, even more so, data entry are among the strongest features of Amos, EQS, and LISREL. Each of these programs offer at least two radically different options for data entry. However, the programs differ in their data management capabilities. With *data management* we mean operations that can be performed on the data in addition to structural modeling. Examples of such operations include transformations, ordering, calculation of moment matrices and standard statistical analysis such as exploratory factor analysis and analysis of variance. With *entry* we mean the procedures provided for input of data and program commands.

Data management in Amos

Amos is the most SEM specific of the present three programs. Therefore, data manipulation options are minimal. The program will calculate such descriptive summary measures as the mean or moment matrices during a modeling run. These measures can be made part of the output by invoking the appropriate commands. For instance, the command $samplemoments causes the program to report the sample covariance matrix for each group of an analysis. If means are modeled, they will be reported also. However, Amos does not provide extensive data manipulation facilities.

Data management in EQS

EQS is at the other end of the spectrum of data management capabilities. It provides a wide spectrum of options, including: the creation of data files; calculation of correlation and covariance matrices; various missing data identification, estimation, and imputation options; variable transformations and recoding; case selection; file merging; and data smoothing. In addition, EQS includes a mini statistical software package that allows the user to run standard statistical analyses such as regression, analysis of variance (ANOVA), frequency tables, exploratory factor analysis, *t*-tests, missing data estimation, and correlations. Another data manipulation feature is an extensive plot module that allows one to create a large number of plots of categorical and continuous data, including a three-dimensional spin plot. In this respect, EQS seems the most complete of the three packages compared in this chapter. EQS also allows users to create artificial random data for simulations.

Data management in LISREL

Just as the other programs do, LISREL allows the user to create a number of descriptive summary measures as part of the modeling process. In addition, PRELIS provides the option to calculate polychoric correlation matrices for ordinal variables, logarithmic transformation, or to estimate asymptotic variances and covariances. In addition, PRELIS allows one to merge files, select cases and delete variables, estimate and impute missing values, perform regression analysis, homogeneity tests, or estimate thresholds. PRELIS also allows users to create artificial random data for simulations.

Data input in Amos

Amos supports the four binary database file input formats dBase III and IV, Foxpro 2.0 and 2.5, MS Access 1 and 2, and SPSS. In addition, ASCII data can be read in either fixed or free format. The data themselves can be raw data or various forms of matrices. Specifically, Amos can read correlation matrices and covariance matrices. Optional input can contain means and standard deviations. For ASCII data, the original Amos format is also supported.

Data input in EQS

EQS supports a large number of input formats for data. These formats include binary files created by EQS, BMDP/PC, PC-90, 386 Dynamic, dBase III Plus (or earlier), Lotus 1-2-3 (release 2.2 or earlier), and SPSS SYSTEM files. The last option is new to EQS 5.7 and mentioned in a flyer that comes with the program. In addition, EQS can read ASCII files in fixed or free format. The files can contain raw data or full symmetric or lower triangular covariance matrices. When reading raw data, EQS creates an appropriate covariance matrix. The program can handle categorical data.

Data input in LISREL

LISREL can read raw data, nine types of matrices, weight matrices, and standard deviations. The matrix types that can be read are moment matrices, augmented moment matrices, covariance matrices, correlation matrices based on raw or normal scores, correlation matrices of optimal scores created by PRELIS, polychoric correlation matrices, polyserial correlation matrices, Spearman rank correlation matrices, and Kendall's τ–**c** correlation matrices. Data are expected in ASCII fixed or free format. There is no provision for data input in other formats. (It should be noted, however, that this comparison uses only the stand alone version of the three programs. Earlier versions of LISREL were made available in conjunction with SPSS. These versions were capable of reading SPSS system data. This is no longer correct for the more recent versions of LISREL.)

Modeling options

Modeling used to be a complicated matter that only the wizards of the alchemic art of computing were able to handle, and that for money. No longer! The programs reviewed here have made it incredibly easy to set up even complex programs with large numbers of variables and complex structures. Each of the three programs provides at least two options for program specification.

Modeling using Amos

Amos allows users to create text files with the commands and data needed for model specification. Within the text module, many commands can be expressed in different, yet equivalent ways. For example, consider the case where a variable *A*, is specified to be a linear combination of variable *B* and an error term. This relationship can be expressed as follows:

```
$structure
  A = B + (1) error
```

Alternatively and equivalently, one can write

```
$structure
  A ← B
  A ← error (1)
```

In addition, Amos was the first of the major SEM programs to provide a very attractive and useful option to start program runs from user-drawn diagrams. When a user sets up a diagram from which to start a program run, Amos creates the text code automatically. Thus corrections can be done to the text code or to the diagram. While extremely convenient for problems with small numbers of variables, the graphical input mode can become tedious to use when the number of manifest variables and latent factors increases. Section 2 of this chapter provides more detail on both input modes.

Modeling using EQS

EQS provides the user with three qualitatively different input modes for model specification. These are: (1) the text mode; (2) the graphical mode; and (3) unique to EQS, a matrix point-and-click mode. The text mode is comparable to the text modes in Amos and LISREL, although it uses a different, simpler grammar. The module CALIS in SAS (SAS Institute, 1990) has adopted a grammar very similar to EQS' grammar. Basically, the text mode of EQS requires users to formulate relationships in the form of regression-type equations. Consider, for example, the variables V1 and V2 that are assumed to load on factor F1. The measurement part of this model can be specified in EQS' text mode as:

```
/EQUATIONS
V1= + 1F1 + 1E1;
V2= + *F1 + 1E2;
```

Here, the '1' indicates parameters fixed to the value of 1 and the asterisk indicates free parameters. The numbering of residuals and variables goes in parallel.

The graphical mode in EQS is comparable to the graphical mode in Amos. In both programs the user draws a diagram using the program graphics tools. In EQS the estimation can be started directly from the screen with the diagram. We perceived both the drawing and the invoking of the estimation as more refined and easier to use in EQS than in Amos.[3]

The matrix-oriented model specification option is unique to EQS. Sooner rather than later, most users familiarize themselves with the matrix notation that makes it possible to present structural models in a very compact form. These matrices, which will be illustrated in Section 2.2, contain the information needed to uniquely specify a structural model. This information consists of (1) relationships between latent and manifest variables, (2) relationships among latent variables, (3) relationships among manifest variables, and (4) relationships that involve residuals and disturbances. The Easy Build module in EQS presents all matrices needed to express these relationships. The users have only to single-click cells in these matrices if a relationship is part of their model and must be estimated, or double-click cells when a parameter is fixed to the value of 1. Selection of variables, estimation method, etc. can all be done from within Easy Build. While the user specifies a model using nothing but mouse-clicks, EQS creates the equivalent text code. This code is then available for modifications. Even the estimation process can be started from within Easy Build. This module provides the fastest and easiest way yet to specify a structural model both with or without latent variables.

Modeling using LISREL

LISREL is currently the only one of the three programs discussed in this chapter that does not provide a graphical model specification option. However, LISREL provides two different text modes, and within these text modes there are several options to express model specifications. The first text mode is the original mode for which LISREL is known. Users express variable relationships using LISREL's Greek-character matrix notation. To be able to specify a model, users have to learn the meaning and characteristics of nine[4] matrices. Within this text mode there are several equivalent options to express model characteristics. Consider again the example used

[3] A new user interface for Amos has been developed since the time of this writing. This interface is more elegant, easier to use, and provides more options.

[4] The manual (Jöreskog & Sörbom, 1996a, p. 11) lists only eight of these matrices. However, as is indicated only in the Preface to the manual of release 8, LISREL 8 now incorporates an extension of the LISREL model that allows users to relate the residuals of manifest variables from the dependent and the independent variables sides to each other. This is done by declaring elements of the ninth matrix, TH ($\Theta_{\delta\varepsilon}$), free. This matrix must not appear on the Model line.

to illustrate the language used in EQS. The example involves two manifest variables, V1 and V2, that load on one factor, F1. To express this loading pattern one can explicitly free the loadings by specifying

```
FR LX 1 1 LX 2 1
```

This expression implies that there is a factor loading matrix, $\mathbf{\Lambda}_x$. Alternatively and equivalently, one can write

```
PA LX
2(1 0)
```

This expression specifies the pattern of matrix $\mathbf{\Lambda}_x$ (PA) by identifying the first two elements in the first column as free parameters. (For didactical reasons we presented in this last expression a matrix $\mathbf{\Lambda}_x$ with two columns.) The second option for specifying loading patterns is most parsimonious when large matrices need to be specified, i.e., matrices with many manifest and many latent variables. The order of variables must match the order of free matrix elements in $\mathbf{\Lambda}_x$. All this applies accordingly when a pattern for $\mathbf{\Lambda}_y$ is specified. As an alternative to these two options, the user can explicitly specify each cell of $\mathbf{\Lambda}_x$.

The second text mode in LISREL is called SIMPLIS. SIMPLIS is a command language that simplifies model specification for LISREL. Within SIMPLIS, variable relationships are expressed in terms of paths or relationships. The language used to express variable relationship is a shorthand version of plain English. To illustrate, consider again the two variables, V1 and V2, and factor F1. Suppose a researcher wishes to let the residuals of these variables correlate with each other. The SIMPLIS expression for this correlation is

```
Let the Errors of V1 and V2 Correlate
```

The expression for the above loading pattern is either

```
V1 V2 = F1
```

or, within a block of `Paths`,

```
F1 ← V1 V2
```

In sum, each of the three programs, Amos, EQS, and LISREL provides various extremely convenient and flexible ways to express measurement models and structural models. The graphical modes provided by Amos and EQS are very useful when models are small. As soon as a regular computer screen can no longer present a diagram in a legible way, the usefulness of such modes can become limited. EQS' matrix approach Build EQS is faster but also limited in the sense that it becomes harder to keep an overview of the model to specify when a matrix does not fit on a computer screen. Therefore, all three packages provide text modes that allow researchers to specify models using very flexible command languages.

One option for model modification in interactive LISREL deserves special mention. When a model converges, the LISREL command PD (or path diagram), inserted on the line before the output line (OU) leads to the presentation of a diagram on the screen. LISREL provides the option that paths can be added or deleted on this diagram on the screen and the model be re-estimated. This can be done by clicking the start and end-points of the paths of interest. The results of the re-estimation runs are then appended to the original result file. More detailed illustrations follow in Section 2.2.

Output options

All three programs offer a large number of useful output options. For instance: residuals can be printed and plotted, estimated and standardized parameters can be printed, along with their standard errors and *t*-values; modification indices or their equivalents can be printed; the iteration process can be reported; the number of decimals can be manipulated, etc. Amos and EQS provide the option to insert parameter estimates or *t*-values in existing or program-created diagrams. LISREL can create diagrams with estimates or *t*-values. LISREL can also create diagrams with modification indices. The diagrams of all three programs are very good quality as long as the number of elements is within tight limits. The LISREL diagram comes in color. There is no additional manipulation needed to create a color output. It should be noted that the graphical displays presented by LISREL 8.20 differ from the displays presented by earlier releases. LISREL 8.20 now includes residual correlations and the structural part of a model in one graph rather than three.

Ease of use

The following brief comments are highly subjective. They reflect solely our own experiences and impressions. Readers are recommended to try out the programs or, even better, take a course in SEM before making a decision as to which program to purchase.

Ease of use can be viewed from many perspectives. In this section we incorporate three aspects of ease of use. The first aspect is speed of input of model specifications. The second aspect is understandability of output. The third is communication with other researchers.

As far as *input of model specifications* in the form of command files that are written by the user is concerned, the programs are very similar. The original LISREL command structure may be the hardest to learn, SIMPLIS, however, has made the life of users easier. The ease of use of text commands for Amos and EQS is comparable to SIMPLIS. The graphical input modes of Amos and EQS are equally easy to use, with the advantage going to EQS. The easiest way to specify small- to medium-size models is provided by EQS' Build EQS module. However, the user of Build EQS needs to master some elementary matrix notation. In each of the programs the text mode seems necessary when the size of problems increases. Therefore, users are well advised to learn more than one mode of program specification for each of the three programs.

Understandability of output is the same for the three programs. Each needs some getting used to. However, the information provided is largely the same. Unique information is explained in the manuals. Therefore, we do not see that any of the three programs provides more understandable output than the others. This applies to both the printed output files and the graphics. It should be noted that none of the three programs uses McArdle's RAM notation. The diagrams are displayed in the standard format introduced with LISREL.

The third criterion, *communication with other researchers*, concerns the discussion of models and their specification with colleagues and other researchers around the world. Here we note that LISREL is the oldest and most widely distributed program, followed by EQS and Amos. Thus, it can be expected that communication using the LISREL terminology will find the largest number of researchers who understand this language. When it comes to solving problems such as when a model does not converge, command files need to be inspected. To be able to discuss command files, one needs to master a command language.

A data example: Iris or the struggle for admissibility

In this section we present a data example. We use the data set that is arguably the best known in biometrical circles, Fisher's Iris data. These data have been analyzed many times and are provided by many statistical software packages. The data describe the four variables, Petal length, Petal width, Sepal length, and Sepal width in three Iris species. The total number of Iris flowers for which these data were gathered was 150, 50 for each species. In the following analyses we apply the three programs, Amos, EQS, and LISREL, to estimate a model with the following features:

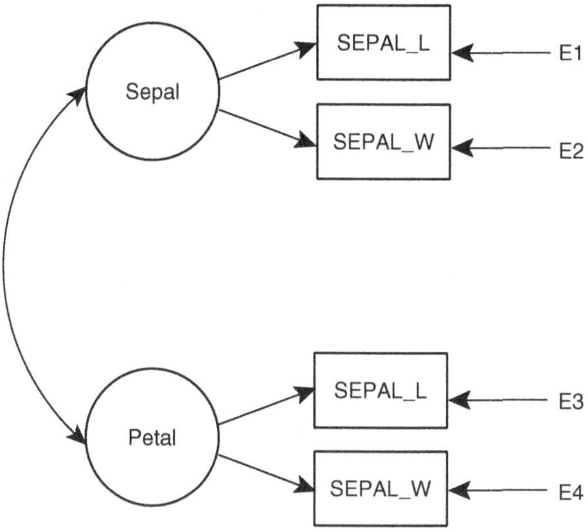

Figure 14.1. Structural path model linking Iris sepal and petal information.

1. There is one factor, Sepal, and a second factor, Petal.
2. The first factor has Sepal length and Sepal width as indicators; the second factor has Petal length and Petal width as indicators.
3. The two factors are correlated.

Figure 14.1 displays this model. It was drawn using LISREL, and could have also been drawn using Amos or EQS (or, for that matter, WordPerfect, Harvard Graphics, Corel Draw, or many other graphics programs). (Please note that this figure resulted from a sample run, different from the run started later for program comparison; Figure 14.2, Section 3.2, displays the model used for program comparison.)

The measurement equations for this model follow in LISREL notation. We specify the model "on the y-side", i.e., all factors have the same status. An equivalent model can be specified "on the x-side".

$$\text{Sepal length} = \text{Sepal} + \epsilon_1,$$
$$\text{Sepal width} = \lambda_{21}\,\text{Sepal} + \epsilon_2,$$

where the ϵ_i are the error terms and λ_{21} is the loading of Sepal width on the factor Sepal. The loading of Sepal length on the factor Sepal is fixed to 1.0 and, therefore, does not appear in the equation. The equations for the second factor are

$$\text{Petal length} = \text{Petal} + \epsilon_3,$$
$$\text{Petal width} = \lambda_{42}\,\text{Petal} + \epsilon_4,$$

where the parameters are analogous to the ones for the Sepal factor. These two pairs of equations can be expressed in the more compact matrix notation as follows. To do this we first denote the observed measures Sepal length, Sepal width, Petal length, and Petal width by y_1, y_2, y_3, and y_4, respectively. We denote the factor Sepal by η_1 and the factor Petal by η_2. Then, the two pairs of equations are in matrix form

$$\begin{bmatrix} y_1 \\ y_2 \\ y_3 \\ y_4 \end{bmatrix} = \begin{bmatrix} 1 & 0 \\ \lambda_{21} & 0 \\ 0 & 1 \\ 0 & \lambda_{42} \end{bmatrix} \begin{bmatrix} \eta_1 \\ \eta_2 \end{bmatrix} + \begin{bmatrix} \epsilon_1 \\ \epsilon_2 \\ \epsilon_3 \\ \epsilon_4 \end{bmatrix}.$$

As was indicated before, the covariance of the two η factors is expressed using the matrix $\mathbf{\Psi}$. More specifically, we write

$$\mathbf{\Psi} = \begin{bmatrix} \psi_{11} & \\ \psi_{21} & \psi_{22} \end{bmatrix},$$

where the diagonal elements are the variances of the factors and element ψ_{21} is the covariance of the two factors. The errors of the observed measures appear in the diagonal elements of the matrix $\mathbf{\Theta}_\epsilon$, or, specifically,

$$\mathbf{\Theta}_\epsilon = \text{diag}(\theta_{11}, \theta_{22}, \theta_{33}, \theta_{44}).$$

In the following sections we present the results of our attempts to fit this model using Amos, EQS, and LISREL. We present text command files and selected output in Section 3 and in the Appendices. In addition, we discuss how the three programs handle the admissibility problem that appears for this data example.

A comparison of features of Amos, EQS, and LISREL

A comparison of the three SEM programs would not be complete without an inspection of the output that each of these programs generate. Table 14.2 displays the various programming and output features of each of the three programs. The reader will see that each of the programs have similar features in some areas, but different options in others. A discussion of this will follow. Three appendices are included at the end of this chapter. Each contains a copy of the program command text file, created to fit the Iris data model. Table 14.3 gives a comparison of the different outputs and Table 14.4 indicates various programming and output options that are available in each program. The programs' strategies to handle admissibility problems are discussed and then a review of the figures generated on these programs completes this section.

Programming features

As was indicated before, one of the main differences in these programs is in the way in which they handle missing data. Amos has the most sophisticated approach for handling missing cases called the full information matrix method. This uses a maximum likelihood solution based on the complete and incomplete cases. The method does not impute missing values as the EQS and LISREL programs do, but rather estimates paths and covariances from the implied moments from the raw data. This method is superior to either listwise deletion or pairwise deletion when running SEM models. Another SEM program, Mx (Neale, 1994), also uses this method to handle missing data.

The other two programs use approaches that impute missing data. EQS offers three methods of computing replacement values for missing data: (1) mean imputation, which calculates the mean of that subject based on the entire sample; (2) group mean imputation based on the other subjects in a group that may be specified as long as that variable is not categorical in nature (i.e., sex) and has no missing values in the set; and (3) regression imputation in which an estimate of a missing value is generated by estimating what this value may be on the basis of several user-selected variables in the data set. If the selected variables are good predictors of the missing value, the replacement value will be of a higher quality.

LISREL's approach to mean imputation uses what is known as the hot deck method. This procedure, which is a part of PRELIS, estimates missing data by examining the patterns of specified matching variables in the matrix. The imputation procedure requires the programmer to specify a set of matching variables (x) and a list of variables that have some missing values on at least one variable (y). These variables are standardized for each case i with the equation

$$z_{ij} = \frac{x_{ij} - \bar{x}_j}{s_j},$$

where i indexes cases (e.g., individuals) and j indexes variables. The program finds all of the matching variables in which no values are missing for each variable in the missing variable list. This solution is minimized with the following equation

$$\sum_{j=1}^{n}(z_{bj} - z_{aj})^2,$$

where a and b indicate cases and n is the sample size. According to the PRELIS 2 manual, two events should occur: one is that a single case a will satisfy the above equation and be replaced by b. Otherwise, if there exist more than one matching case of b with the same minimum value of the above equation, then these missing values $(y_1^{(m)} \ldots)$ will have a mean of

$$\bar{y} = \frac{1}{n}\sum_{i=1}^{n} y_i^{(m)}$$

and a variance of

$$s_m^2 = \frac{1}{n-1}\sum_i \left(y_i^{(m)} - \bar{y}_m\right).$$

Imputation will only be done if $(s_m^2/s_y^2) > v$, where s_y^2 is the total variance of y estimated from all complete data on y. Further, v is the variance ratio below which no imputation will be done. This essentially means that the matching cases will predict the value of missing cases with a reliability of at least $1 - v$. The default specification is 0.50 (Jöreskog & Sörbom, 1996c, pp. 155–157). This procedure has the advantages that (1) it gives identical values even if the variables are transformed in some way, and (2) it will be performed in the same way regardless of the order of cases in the data.

Another difference between these programs is in the graphing procedures that are contained in the programs. Not only can EQS run several additional statistical procedures not found in the other packages (i.e., ANOVA), but it also has several unique plotting and graphing features that make data exploration highly useful (see Section 2.1.3). EQS also has a feature to map missing data so that cases and single observations of missing data can be examined to determine whether or not the data are missing at random. There are also several procedures for mapping missing data and outliers together. Table 14.2 gives a comparison of the features of the three programs and the options available for the user.

Table 14.2. *Overview of program features of Amos, EQS, and LISREL*

Options	Amos 3.6	EQS 5.7	LISREL 8.20
On-line help	Yes	Yes	Yes
Data manager			
Spreadsheet	Yes	Yes	Yes
Data exploration and manipulation			
Plot univariate and bivariate statistics	No	Yes	Yes
Histograms and 3D graphs	No	Yes	No
Advanced statistical procedures			
ANOVA	No	Yes	No
Exploratory factor analysis	No	Yes	No
Regression	No	Yes	Yes
Case selection (analyzing a subset of cases using a logical or arithmetic function)	No	Yes	Yes
Transforming data	No	Yes	Yes
Data smoothing and removing autocorrelations	No	Yes	No
Tests of normality	Yes	Yes	Yes
Data analyzed			
Raw	Yes	Yes	Yes
Covariance matrix	Yes	Yes	Yes
Pearson matrix	Yes	Yes	Yes
Polychoric/polyserial correlation matrix	No	Yes	Yes
Handle missing values	Yes	Yes	Yes
Model specification			
Matrix input	No	No	Yes
Equation input	Yes	Yes	Yes[1]
Speadsheet equation input	Yes	Yes	Yes
Graphical input (GUI)	Yes	Yes	No[2]
Missing data			
Plots outliers	No	Yes	No
Plots missing data	No	Yes	No
Displays map of missing cases	No	Yes	No
Missing data imputation methods			
Mean imputation	No	Yes	No
Group mean imputation	No	Yes	No
Regression imputation	No	Yes	No
Hot deck method[3]	No	No	Yes
Full information matrix[4]	Yes	No	No

Table 14.2. (*cont.*)

Options	Amos 3.6	EQS 5.7	LISREL 8.20
Estimators			
ML	Yes	Yes	Yes
GLS	Yes	Yes	Yes
AGLS	No	Yes	No
SLS[5]	Yes	No	No
LS or ULS	Yes	Yes	Yes
WLS	No	No	Yes
DWLS	No	No	Yes
Elliptical	No	Yes	No
Ridge	No	No	Yes
IV/TSLS[6]	No	No	Yes
Scale free LS	Yes	No	No
Starting values	Yes	Yes	Yes
Robust estimation	No	Yes	Yes
Constraints			
Simple equality	Yes	Yes	Yes
General equality	No	Yes	No
Inequality	No	Yes	No
Equal error variances	Yes	Yes	Yes
Multigroup analysis	Yes	Yes	Yes
Use of mean structures	Yes	Yes	Yes
Use of continuous and categorical data models	No	Yes	Yes
Polyserial and polychoric correlations	Yes	No	Yes
Tests of homogeneity for categorical variables	No	No	Yes
Multivariate multinominal probit regressions	No	No	Yes
Output			
Files with results	Yes	Yes	Yes
Likelihood ratio test	No	Yes	No
Total, direct and indirect effects	Yes	Yes	Yes
Overall fit measures	Yes	Yes	Yes
Residuals	Yes	Yes	Yes
Standardized estimates	Yes	Yes	Yes
Factor score weights	Yes	No	Yes
Retest model automatically with non-significant paths eliminated	No	Yes	No
Correlations of parameter estimates	Yes	Yes	Yes

Table 14.2. (*cont.*)

Options	Amos 3.6	EQS 5.7	LISREL 8.20
Model modification indices	Yes[7]	Yes	Yes
Lagrange multiplier test	No	Yes	No[8]
Control of the LM test (i.e., Set, Block and Lag)	No	Yes	No
Wald test	No	Yes	No
Data presentation			
Output camera-ready figures	Yes	Yes	Yes
Simulation capabilities			
Monte Carlo, jack-knife, bootstrap	3	2 and 3	1 and 3

[1] Available in SIMPLIS language format.
[2] The graphical interface that is available is a way to make modifications to a model after it was run with one of the other language formats. This feature has a "point and click" method of operation similar to that of Amos and EQS.
[3] In the manual for PRELIS 2, the method of data imputation states that the value to be substituted "is obtained from another case that has a similar response pattern over a set of matching variables". (Jöreskog & Sörbom, 1996c, p. 78). The mathematical solution is given on pp. 155–157 of the PRELIS 2 manual (Jöreskog & Sörbom, 1996c).
[4] The Amos manual indicates that even with missing values it does not use any of the conventional methods but can compute full information maximum likelihood estimates as long as the data are missing at random.
[5] Scale-free least squares
[6] Instrumental values and two-stage least squares.
[7] Amos uses the same modification indices that LISREL does, based on Jöreskog & Sörbom (1984).
[8] In the LISREL 8 manual, the "modification indices" are described as being "approximately equal to the difference in χ^2 between two models in which one parameter is fixed or constrained in one model and free in the other... The largest modification index shows the parameter that improved the fit most when set free" (Jöreskog & Sörbom 1993, p. 127; cf. Sörbom, 1989). This essentially accomplishes the same task as a Lagrange multiplier.

Amos 3.6

An Amos program is given in Appendix 14.1. The program is easy to use and is based largely on the construction and manipulation of a path model figure in the graphics editor. Commands to control output and specify data are entered in another file by clicking on a list of commands.

For users interested in using the command language throughout, the $Structure command and the following statements are used to specify the model. The $ commands do not appear to require a particular order save that some commands are nested within others. The manual specifies methods for using the program to read external SPSS files but does not indicate how to input text files. The manual suggests that users enter data in the text editor using the $Raw Data command and use spaces as delimiters in the data file. The authors of this chapter suggest using the cut-and-paste method available in most word-processing programs to transfer an ASCII file with line breaks into the text editor. The command window also accepts covariance matrices and correlation matrices along with standard deviations and means.

The final product produces a listing that appears to be less output than that given by either EQS or LISREL. An interesting observation is that when the program is run the output window appears with all of the relevant statistics yet it opens in the middle of the listing. This feature allows the user to quickly examine the χ^2 fit statistics and parameter estimates easily. The developers obviously constructed the program in this way so that upon repeated runs of a similar model the user does not have to scroll past the initial estimates and minimization history to obtain the more important information. A similar option is available in LISREL. The modification indices that are available through Amos are the same as in LISREL, with the exception that a threshold must be given in order for the modification indices to be printed. These are requested with the $Mods command. A comparison of output appears in Table 14.3.

EQS 5.7

Appendix B contains the command language generated by EQS with most of the available options specified. The parameter estimates are given mainly in the form of structural equations consistent with the Bentler–Weeks model (Bentler & Weeks, 1980, 1985). EQS gives unique output related to multivariate kurtosis and elliptical theory estimates that are not available for LISREL or Amos. Further, a note of clarification for the Wald and LM tests is needed. Even though these tests were specified, the output contains no usable information derived from these commands. Because the model used currently has only one degree of freedom, to free a fixed parameter would lead to problems of identification. Further, to set a nonzero path to a fixed value would, in two out of three cases, lead to factors with no estimated indicators. Therefore, the use of these tests is not appropriate.

Table 14.3. *Output in Amos 3.6, EQS 5.7, and LISREL 8.20*

	Amos	EQS	LISREL
Program command file	√	√	√
Univariate statistics		√	
Multivariate kurtosis estimates		√	
Elliptical theory kurtosis estimates		√	
Case numbers with largest contributions to kurtosis		√	
Matrix to be analyzed		√	√
Determinant of matrix		√	
Residual covariance matrix	√	√	√
Standardized residual matrix	√	√	√
Rank order of largest residuals		√	√
Distribution of standardized residuals		√	√
Fit statistics	√	√	√
Iterative summary	√	√	√
Parameter estimates	√	√	√
Standardized solution	√	√	√
Modification indices	√	√	√
Correlation/covariance matrices of parameter estimates			√
Factor score regression	√	√	
Total effects	√		√
General normality statistics	√		
Outlier assessment	√		
Implied correlation and covariance matrices	√		
Tests of significance for unstandardized paths	*t*-test	*t*-test	CR

CR, Cressie–Reid test.

LISREL 8.20

Appendix 14.3 contains command language for LISREL. The specifications were made using the text language but similar models could have been generated from using the interactive features of LISREL. The output command at the end of the program requested all of the output options available in LISREL including modification indices, standardized solution, completely standardized solution, total direct and indirect effects, standard errors, residual matrix, correlation matrix of parameter estimates and factor-scores regression. This output differs from EQS in its format. The parameter

Table 14.4. *Goodness of fit for Iris models*

Variable	Amos 3.6	EQS 5.7	LISREL 8.20
χ^2	2.306	64.52	2.25
p-value	0.129	0.0001	0.130

Table 14.5. *Path coefficients for Iris models*

	Amos 3.6		EQS 5.7		LISREL 8.20	
Path	Estimate	Standardized	Estimate	Standardized	Estimate	Standardized
Petal length from petal	17.96	1.02	17.65	1.00	18.00	1.02
Petal width from petal	7.167	0.943	7.339	0.973	7.18	0.94
Sepal length from sepal	3.95	0.479	7.23	0.945	4.13	0.50
Sepal width from sepal	−1.067	−2.46	−1.855	−0.611	−1.11	−0.26

estimates are given in matrix form in line with the LISREL format. It should be noted that the problem with the LM and Wald tests in EQS also applies to the modification indices that are reported in the LISREL output.

Admissibility

Looking through the goodness-of-fit results from Amos, EQS, and LISREL in Table 14.4 one notices immediately the differing values that are seen for the χ^2 values. The same applies to the parameter estimates in Table 14.5. The χ^2 values for Amos and LISREL differ only slightly whereas the value for EQS is quite a bit larger.

To go beyond software defaults, the model was run a different way by specifying two paths (petal length and sepal length) to 1.00 and setting the variances of the two latent variables free. This produced the estimates presented in Tables 14.6 and 14.7. Figure 14.2 displays the LISREL solution (cf. Figure 14.1, above). Notice that the legend in Figure 14.2 suggests that $\chi^2 = 2.43$ rather than 2.25, as indicated in Table 14.4. This difference is due to the fact that Table 14.4 reports the minimum fit function $\chi^2 = 2.25$, whereas the figure reports the normal theory weighted least squares χ^2. The minimum fit function χ^2 appears only in the LISREL output file, not in the figure.

Table 14.6. *Goodness-of-fit results for constrained Iris models*

Variable	Amos 3.6	EQS 5.7	LISREL 8.20
χ^2	2.306	19.136	2.25
p-value	0.129	0.0001	0.130

Table 14.7. *Path coefficients for constrained Iris models*

	Amos 3.6		EQS 5.7		LISREL 8.20	
Path	Estimate	Standardized	Estimate	Standardized	Estimate	Standardized
Petal length from petal	−0.270	1.021	−0.259	1.00	−0.27	1.02
Petal width from petal	0.399	0.943	0.419	0.972	0.40	0.94
Sepal length from sepal	1.00	0.479	1.00	0.439	1.00	0.50
Sepal width from sepal	1.00	−0.246	1.00	−0.220	1.00	−0.26

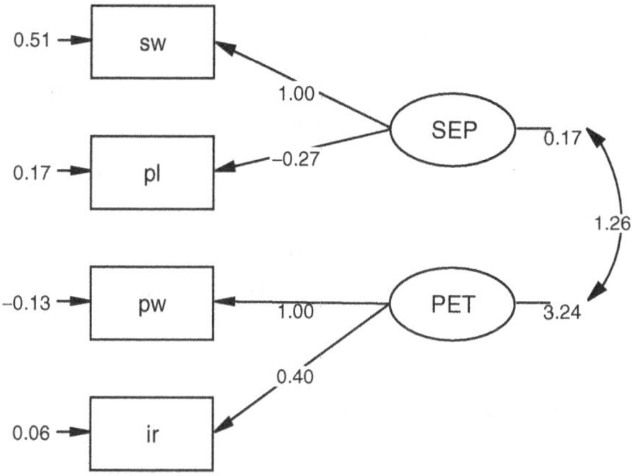

Figure 14.2. LISREL solution for model in Figure 14.1. $\chi^2 = 2.43$, df = 1, p-value = 0.11875, root mean square error of approximation = 0.098.

Again, EQS tends to have χ^2 estimates that are quite different from the other two programs. The χ^2 shows a larger value and smaller p-values than the other two programs. Also, the parameter weights differ slightly in their estimates as well. Further, upon specifying a slight change in the model, the χ^2 estimates remain the same for LISREL and Amos but change dramatically for EQS. The explanation for these differences rests in the handling of *admissibility problems* in the three programs.

SEM solutions are deemed inadmissible if parameter estimates assume values that are logically impossible. As an example of how programs identify admissibility problems, consider LISREL. This program checks whether (1) the loading matrices, $\mathbf{\Lambda}_y$ and $\mathbf{\Lambda}_x$, have full column ranks and no rows of only zeros, and (2) the matrices of interrelationships among latent variables, $\mathbf{\Phi}$ and $\mathbf{\Psi}$, and the residual variance covariance matrices of the manifest variables, $\mathbf{\Theta}_\delta$ and $\mathbf{\Theta}_\epsilon$, are positive definite.

Examples of impossible values include negative variances and correlations greater than $r = 1$. The three programs pursue three different strategies to deal with admissibility problems. Amos prints explicit warnings in the program output, identifies the inadmissible parameters, and presents a solution nevertheless. As a result, a solution may be printed that is not acceptable. Researchers may then modify their model, taking into account the problems with a given model specification. EQS prints explicit warnings in the program output and constrains parameters to their upper or lower boundaries. As a result, EQS will always print solutions that are acceptable. Researchers can then modify their model to obtain estimates rather than constrained values for their parameters. LISREL, as Amos, does not constrain the parameters to be admissible. As a result, negative variances can appear in the diagonals of the four matrices involved in the admissibility check. However, LISREL has a default admissibility check after 10 iterations. If a solution fails the admissibility check, LISREL aborts the run. In the output (OU) line, the user can modify this number or even disable the check entirely. The result will then be a solution with impossible values.

One of the reasons why we selected the Iris data example is that solutions suffer from admissibility problems. This allows us to demonstrate the different admissibility handling strategies in Amos, EQS, and LISREL. Specifically, the solution presented above for Amos was generated using Amos' default parameters. The same applies for the EQS solution. To create the LISREL solution we turned the admissibility check off. Thus the difference between the three solutions can be traced back to the ways these three programs deal with admissibility problems.

Figures

The advances in computer technology and the advent of ever faster desktop computers over the past years, has led to the improvement in the options that are built into SEM and statistical packages. In earlier versions of these programs it was often required to have a line printer or other specialized apparatus in order to print the graphical features of SEM models in a way that could be published in a scientific journal. Also, it used to be that once a computer package provided a solution for a path model, the user would then have to use separate drawing software in order to produce diagrams from the printouts. In a recent article, Miles & Shevlin (1998) reviewed the drawing programs that were available as software independent of SEM packages (e.g., Powerpoint 7, Chartist 1.7, MSPaint, and Visio 1.0) and two statistical programs, Amos 3.51 and EQS 5.3, which had recently added graphical features. In this article the authors suggested that the features of an ideal SEM package would be able to produce shapes with text in them, draw curved and straight arrows with a variety of angles, have control over fonts including subscripts, superscripts and Greek letters, and be able to be object oriented. Their review of Amos and EQS was positive but is currently out of date. The newest versions of both EQS and Amos are much more advanced in their ability to produce this output. Further, LISREL has a graphical feature as well.

The latest versions of all three of these programs include highly advanced and capable software that automatically allows the user to draw figures (except in LISREL, where the graph is generated on the basis of code, and only when the program converges), inserts the parameter estimates, and allows the user to adjust the size, color and captions of the diagram. As icing on the cake, these programs print to the desktop computer's personal printer camera-ready figures of the model in seconds with little more effort than clicking the mouse in select menus. All three of the programs allow Greek notation, bold and italics, and a selection of several different fonts. In fact, the font packages on all three programs are identical (they use the Windows fonts) except for the option of color in LISREL. No subscript or superscript features are available in any of the three programs. All three allow the user to manipulate the diagram by changing the size, location and paths of any variable in the program. When variables are moved, the connections to other variables via paths are maintained. All three use a toolbox and the mouse to draw features of the model, although Amos has the largest toolbox of the three programs.

The drawing features of each have been markedly improved and are similar. All three have features designed to even the size of figures, align figures with each other, all arrows are drawn to the center of the variable boxes, and the curved arrows are fit between two variables without the user having to adjust the curve. As Miles & Shevlin point out, however, if a user wants to specify a path diagram that is not conventional, say an endogenous variable without an error variance, this is a difficult task in these kinds of program because the program specifies much of the model for the user. However, for traditional models, these programs can easily generate all the figures that most users of structural equation modeling would ever need.

In what follows, graphical features of each of the programs will be described. All have different features but produce output comparable to Figures 14.1 and 14.2.

Amos 3.6

Amos uses the path diagram as one of its main input and programming features. The figure can be drawn easily. The program allows the user to select multiple elements of the figure and manipulate the size for a perfect match throughout. The error variances are specified in ovals and are drawn automatically on any endogenous variable. The figure shows only the constrained paths until the model is run. When the button "Groups/Models" is clicked (or selected from the "Model-Fit" menu) the user can select the output format for the estimates. Two default parameter formats can be selected: standardized and unstandardized. Customized formats may also be developed.

The figure caption is specified by the user. If certain values such as χ^2 are desired, there are a series of macros that the user can choose to specify these values in the caption automatically. These are similar to the options for EQS. So, for example, the command to obtain the χ^2 estimate is Chi Square = \cmin. This exact caption will appear until the program is run again, the Groups/Models option is selected and the output is specified. Then, the value for the χ^2 will appear in the caption, replacing everything after and including the backslash. The list of values that can be requested is extensive, including many of the popular fit functions.

EQS 5.7

When programming in the command language, the commands for diagraming (i.e., /Path Diagram) are meaningless on the desktop computer version of this software. These commands are still specified in the manual but a footnote designates them as line printer commands and states that a desktop computer user will not need them. However, if the user is interested in having a path diagram printed from a desktop computer, the he or she will have to draw the diagram in the Diagrammer and run the program from there. Unlike LISREL, the path diagram will not be drawn automatically. One may still make changes in the text such as to add commands such as requesting modification indices; however, each time the program is re-run, the additional commands will be replaced with the command syntax from the Easy Build module. Thus the user can specify a program and an output file that has the modifications in it and any technical additional command such as increasing iterations in a separate file. When the figure is requested, paste this command language into the text file that is generated by the figure so that the parameter estimates will be consistent with the output from the original file. Even though the titles of the variables are specified in the diagram, the command lines only reflect these values as V1, F1, etc. The output, however, uses the labels that were designated in the diagram.

The figures have a variety of options for printing. The default values for the paths are unstandardized regression weights and variances and covariances for the other parameters. The other two options for printing are standardized estimates and start values. Even though the figure is in color, the printed output appears only in black and white, even on a high-quality color printer. One can, however, cut and paste or save the figure and print it from a program that allows color printing, e.g., WordPerfect or Corel. The caption below it is default and contains the title, χ^2 estimates, p-value, Bentler's comparative fit index and the root mean square error of approximation. By choosing the "Diagram Title" option under the "View" menu, the user can customize the caption for any specifications, including changing the position, content, and font of title. Other variables can be requested by a series of macros in the title statement that are similar to those in Amos. This way the caption may correspond to the publication manuals of many disciplines.

LISREL 8.20

When the program in LISREL is run, a path diagram is generated if the keyword PD is specified in the program. The figure appears with a variety of tools for modifying the figure. The final product can be manipulated, respecified, rotated, and flipped. The error variances and variances of exogenous variables are not displayed in boxes or ovals but are simply numbers at the base of an arrow. The figure is displayed and printed in color when sent to a color printer. There are default color schemes that can be altered (e.g., converting the format to black and white for journal submissions) by choosing "Options" under the "View" menu. This menu presents options for changing the shape, color, and variable positions in the path diagram. The "View" menu also has options for the parameters that are displayed on the path diagram. The default displays the unstandardized parameter estimates; however, the standardized estimates, t-values, modification indices and expected changes can be selected as well. A caption appears under the figure when the program is run that contains the normal theory weighted least squares χ^2 estimate, degrees of freedom, p-values and the root mean square error of approximation. This caption disappears as soon as the diagram window and the output window are closed. These figures are for the researcher to evaluate model fit after making modifications to the model.

Time flies

As was indicated repeatedly in this chapter, since the time of writing, new releases of the programs have been made available. Thus this chapter could be rewritten and a good number of modifications could be made. For instance, Amos now has an even more elegant and powerful graphical input mode, EQS contains fewer bugs and improved estimation routines, and LISREL now provides a new volume focusing on multilevel modeling. This volume complements the other, unchanged manuals. The general tenor of the present chapter, however, remains. All three packages for structural modeling are highly capable and can be recommended. Preferences in favor of one package or the other can be explained by prior user experiences and the unique features of each package that may differ in importance to the individual user.

Appendix 14.1. EQS program file for Fisher's Iris data

```
/TITLE
Model created by EQS 5.7 — C:\EQS\FISMODFP.EDS
/SPECIFICATIONS
  DATA='C:\EQS\FISHER.ESS';
  VARIABLES= 5; CASES= 150;
  METHODS=GLS;
  MATRIX=RAW;
/EQUATIONS
V1= + *F1 + 1E1;
V2= + *F1 + 1E2;
V3= + *F2 + 1E3;
V4= + *F2 + 1E4;
/VARIANCES
F1= 1.00;
F2= 1.00;
E1= *;
```

```
E2= *;
E3= *;
E4= *;
/COVARIANCES
F2,F1= *;
/tec
  ITR=100;
/lm test
  set=gvf, pff, pee;
process=simultaneous;
/wtest
/OUTPUT
 parameters;
 standard errors;
 listing;
/OUTPUT
 parameters;
 standard errors;
 listing;
 data='EQSOUT& .ETS';
/END
```

Appendix 14.2. LISREL command code for Fisher's Iris data

```
DA NI=4 NO=150
ra fi=c:\fisher.dat
LA
sl sw pl pw ir
se
1, 2, 3, 4/
MO Nx=4 Nk=2 td=fu, fi lx=fu, fi ph=fu, fi
Lk
SEP PET
FR lx 2 1 lx 4 2 lx 1 1 lx 3 2
FR td 1 1 td 2 2 td 3 3 td 4 4
va 1 ph 1 1 ph 2 2
fr ph 2 1
PD
OU ss sc ef se va mr fs pc pt mi rs tv ME=gl ad=100 IT=250
```

Appendix 14.3. Amos command code for Fisher's Iris data

```
$Smc
$Standardized
$Sample size = 150
$Gls
$Ml
$Normality check
$Residual moments
$Total effects
$Iterations = 200
$Extended

$Input variables
  SEPAL_L
  SEPAL_W
  PETAL_L
  PETAL_W
  IRIS
$Raw data
  50.000   33.000   14.000    2.000   1.000
  64.000   28.000   56.000   22.000   3.000   ...
```

References

Arbuckle, J. L. (1996). *Amos Users' Guide: Version 3.6.* Chicago, IL: SmallWaters Corporation.

Arminger, G., Clogg, C. C. & Sobel, M. E. (eds.) (1995). *Handbook of Statistical Modeling for the Social and Behavioral Sciences*. New York: Plenum Press.

Bentler, P. M. (1985). *EQS: Structural Equations Program Manual.* Encino, CA: Multivariate Software.

Bentler, P. M. & Weeks, D. G. (1980). Linear structural equations with latent variables. *Psychometrika*, **45**, 289–308.

Bentler, P. M. & Weeks, D. G. (1985). Some comments on structural equation models. *British Journal of Mathematical and Statistical Psychology*, **38**, 120-121.

Bentler, P. M. & Wu, E. J. C. (1995). *EQS for Widows: Users Guide.* Encino, CA: Multivariate Software.

Bollen, K. A. (1989). *Structural Equations with Latent Variables.* New York: Wiley.

Bollen, K. A. & Ting, K. (1991). EQS 3.16 and LISREL 7.16. *American Statistician*, **45**, 68–73.

Fraser, C. & McDonald, R. P. (1988). COSAN: covariance structure analysis. *Multivariate Behavioral Research*, **23**, 263–265.

Jöreskog, K. G. & Sörbom, D. (1984). *LISREL VI User's Guide*, 3rd edition. Mooresville, IN: Scientific Software, Inc.

Jöreskog, K. G. & Sörbom, D. (1993). *LISREL 8: Structural Equation Modeling with the SIMPLIS Command Language*. Hillsdale, NJ: Lawrence Erlbaum Associates.

Jöreskog, K. & Sörbom, D. (1996a). *LISREL 8: User's Reference Guide*. Chicago, IL: Scientific Software International.

Jöreskog, K. & Sörbom, D. (1996b). *LISREL 8: Structural Equation Modeling with the SIMPLIS Command Language*. Chicago: Scientific Software International.

Jöreskog, K. & Sörbom, D. (1996c). *PRELIS 2: User's Reference Guide*. Chicago: Scientific Software International.

Marcoulides, G. A. & Schumacker, R. E. (1996). *Advanced Structural Equation Modeling: Issues and Techniques.* Mahwah, NJ: Lawrence Erlbaum Associates.

McArdle, J. J. & McDonald, R. P. (1984). Some algebraic properties of the reticular action model for moment structures. *British Journal of Mathematical and Statistical Psychology*, **33**, 161–183.

Miles, J. N. V. & Shevlin, M. E. (1998). Multiple software review: drawing path diagrams. *Structural Equation Modeling*, **5**, 95–103.

Neale, M. C. (1994). *Mx: Statistical Modeling*, 2nd edition. Richmond, VA: M. C. Neale.

Neale, M. C., Boker, S. M., Xie, G. & Maes, H. H. (1999). *Mx: Statistical Modeling*, 5th edition. Box 126 MCV, Richmond, VA 23298: Department of Psychiatry, Virginia Commonwealth University.

SAS Institute (1990). *SAS/STAT User's Guide, Version 6*, 4th edn. Cary, NC: SAS Institute.

Schumacker, R. E. & Lomax, R. G. (1996). *A Beginner's Guide to Structural Equation Modeling* Mahwah, NJ: Lawrence Erlbaum Associates.

Sörbom, D. (1989). Model modification. *Psychometrika*, **54**, 371–384.

von Eye, A. & Clogg, C. C. (1994). *Latent Variable Analysis: Application for Developmental Research.* Thousand Oaks, CA: Sage.

Wilkinson, L. (1996). *SYSTAT 6.0.* Evanston, IL: SYSTAT.

Yuan, K. & Bentler, P. M. (1997). Mean and covariance structure analysis: theoretical and practical improvements. *Journal of the American Statistical Association*, **92**, 767–774.

Index

Bold numbers indicate tables and *italic* numbers indicate figures